Essentials of Chemical Education

Hans-Dieter Barke • Günther Harsch
Siegbert Schmid

Essentials of Chemical Education

Translated by Hannah Gerdau

 Springer

Prof.Dr. Hans-Dieter Barke
Westf. Wilhelms-Universität
Münster
Institut für Didaktik der Chemie
Fliednerstr. 21
48149 Münster
Germany
barke@uni-muenster.de

Prof.Dr. Günther Harsch
Westf. Wilhelms-Universität
Münster
Inst. für Didaktik der Chemie
Fliednerstr. 21
48149 Münster
Germany
harsch@uni-muenster.de

Dr. Siegbert Schmid
University of Sydney
School of Chemistry
Sydney New South Wales
Bldg. F11
Australia
s.schmid@chem.usyd.edu.au

Translated by
Hannah Gerdau
214 Rue de Charenton
75012 Paris
France

Completely revised and updated English edition of the German title: Chemiedidaktik Heute by
H.-D. Barke und G. Harsch, published by Springer-Verlag Heidelberg 2001

ISBN 978-3-642-21755-5   e-ISBN 978-3-642-21756-2
DOI 10.1007/978-3-642-21756-2
Springer Heidelberg Dordrecht London New York

Library of Congress Control Number: 2011939999

Printed on acid-free paper

Springer is part of Springer Science+Business Media (www.springer.com)

Prologue

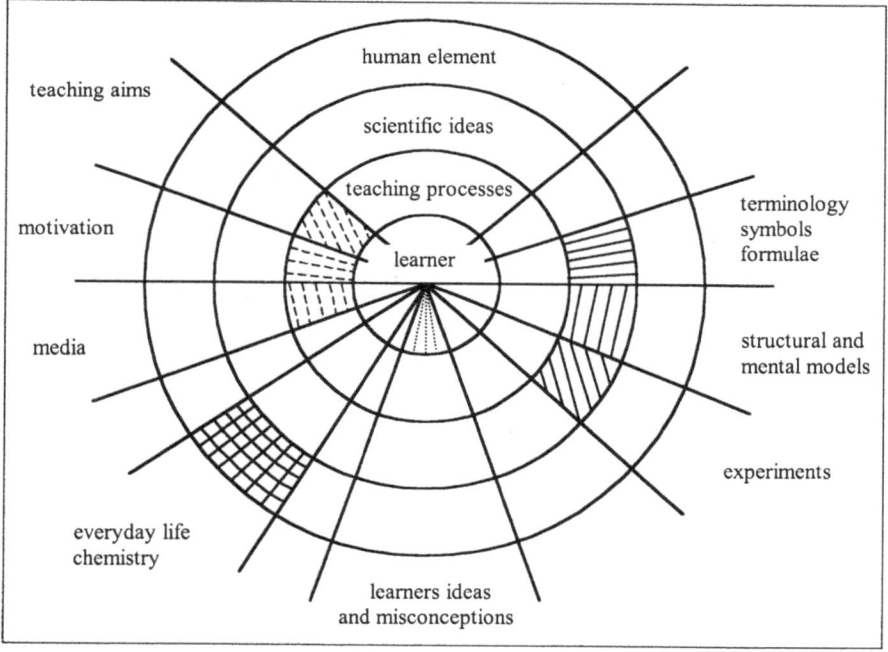

"That was my first lecture for which I did not mind to get up early in the morning, . . . we saw meaningful presentations with many experiments and structural models for school chemistry, . . . a seminar where I really learnt a lot for my later profession, . . . it was good for our orientation to have the pie chart in the first lecture . . ."

These and other statements of students in a one-semester unit towards a qualification as a chemistry teacher at the University of Münster in Germany encouraged us to write this book and to translate it into English. The content covers a broad range of chemical education knowledge and many applications concerning chemistry teaching at secondary level. We hope that apart from lecturers and students at universities and colleges chemistry teachers in secondary schools will also benefit from this book.

There is no single best order of contents in chemical education. Important chapters like "students' misconceptions, teaching aims, motivation, media, experiments, models, chemical symbols and chemistry in everyday life" are linked in many aspects: a linear order does not do justice to all these connections. Therefore a group of scientists in the Society of German Chemists GDCh used a *pie chart* to organise the chapters mentioned before (see first picture): these various chapters can be taught in nearly every order the lecturer likes. The pie chart is explained in detail in the following introduction chapter.

All reflections referring to these levels contain many examples of practical teaching of chemistry at school. Especially a lot of experiments and many models or model drawings showing aspects of the structure of matter should help developing mental models in the cognitive structure of the chemistry-teacher students or later of the learners at school.

Because of the pie-chart metaphor the lecturers can create their own ideas of teaching the "slices of the pie": one may start with "students' misconceptions", the other may start with "teaching aims" – it is up to them, in which order the "slices" are taught. It is also our intention to keep some sectors that are not occupied by a special topic: the lecturer may fill these gaps with their own ideas or contents, for example with the topic "Assessment in chemical education" or "Curriculum development".

At the end of each of the first eight chapters we offer short "problems and exercises", which the students of the seminar may solve and discuss for a repitition or for the final exam. At the end of most chapters there are also short procedures for the experiments, which may be shown during the lecture or seminar. For inexperienced students the descriptions of these experiments are so short that they will need to consult additional lab manuals for a successful performance and for appropriate waste treatment.

Chapters 9 and 10 are in addition to the pie-chart seminar – these chapters show special views of the authors on two important issues of chemical education. In Chap. 9 the method of *Inquiry learning* is linked to the introduction of Organic chemistry in schools or at university level. In Chap. 10 the *Structure-oriented approach* of chemistry education is introduced, especially sphere packings of infinite structures and ions as smallest particles of salt crystals are described and the corresponding structural models are discussed. These structural models may help to train the spatial ability of students so that they are able to develop three-dimensional mental models for a better understanding of formulae and equations in chemistry.

We thank Mrs. Hannah Gerdau very much for doing most of the translation from the original German book "Chemiedidaktik Heute" into the English language. She took a lot of care to preserve the intentions of the original; she also adapted specific German aspects of chemical education to those which are more common and known in the in England, UK and US.

We hope that lecturers of many universities in many countries – where English is spoken in lectures and seminars – will use ideas from our book for their chemical education. Please tell us if any important aspects are missing: we would be very happy to add them in the next edition!

Muenster and Sydney   Hans-Dieter Barke
                                                                                          Günther Harsch
                                                                                          Siegbert Schmid

Contents

Introduction

The Russian pedagogue Itelson [1] once stated: "If engineers building bridges, doctors treating patients and judges deciding verdicts would show such a tendency to superficial argumentation, as it sometimes appears in pedagogy, then all bridges would have collapsed, all patients would have died and many innocents would have been hung alread.".

This quotation might elicit a chuckle, but one has to admit that pedagogues' thoroughness in preparation and argumentation in their field is not developed as much as it is in other fields. The negative effects of mistakes in pedagogy are not as obvious as they are in engineering, medicine or law. The professionalism of prospective teachers should be increased and to form a basis for good lessons, the basics of pedagogy, education and chemistry didactics have to be taught to an adequate extent. Definitions regarding these matters should be imparted before specific aspects of chemistry education are covered. It has to be clear, however, that there cannot be just *the* one didactics! Different opinions and positions exist on this matter: every teacher has to reflect on this in their own time and their environment.

Pedagogy. The term derives from the word "pedagogue," which literally means child or boy leader in Greek. Pedagogy is a collective term for different philosophical or psychological disciplines. Their shared subject is the social action. Roth [2] distinguishes some fields in pedagogy: "They touch the field

1. of educational research (to investigate the methodology and its practices),
2. of lessons at school (and their historical and current aspects),
3. of lessons (in its diverse determinedness as a complex interdependency)" [2].

Didactics. This term stems from the Greek philosophers' schools of ancient times: it derives from the word "didaskein" (Greek: to teach, to prove), or from "didaktos" (Greek: instructive). These terms affected a lot more than just teaching and school. In the seventeenth century Comenius was the one to connect the term "didactics" more and more with teaching. In his "Didactica Magna" (1657) he defined connections to current issues, everyday life and clarity as didactical principles. For Comenius, didactics is the reasoned selection of contents for the art of teaching (docendi artificium). Today, several didactical models are being discussed [3–6]. Blankertz [7] and Ruprecht [8] give a summary of these models.

Aschersleben [9] defines in general: "Didactics is the universe of learning aids, which are given to the student by the teacher or which the student uses independently. The learning aids do not only refer to the selection of topics in education, but also to learning methods. The lesson is the didactical situation where learning takes place and the school is the corresponding institution. The lesson as a didactical situation includes everything that matters to it: participants, teachers and students, learning topics, media, learning material, and so on." One might want to separate didactics and methodology by having didactics answer the question of "what" and methodology the question of "how." The interdependency of these questions [4] as well as the role of didactics to answer questions of "why" (question of reason) and "what for" (question of intention), leads us to combine all these questions in the term "didactics." Two equivalent definitions shall be quoted:

"Didactics deals with the question of	"The following questions arise:
– who	– with which aims should
– what	– which contents under
– when	– which requirements and
– with whom	– which conditions on
– where	– which level with
– how	– which methods in
– with what	– which time with
– why and	– how much success
– for what one has to learn" [10]	– be taught by whom respectively
	– be learned by whom?" [11]

The term "general didactics" becomes clearer, when particular forms of didactics are listed and connections are being made:

– Didactics concerning type of school: elementary-, middle- and high school didactics
– Didactics concerning groups of subjects: science, languages, social studies, and others
– Didactics concerning specific school subjects: mathematics, biology, chemistry, physics

Chemistry didactics deals with the aims of teaching and learning chemistry at all different levels of schools.

Chemistry didactics. At first sight this expression might indicate that there is an additive combination of the corresponding subject and didactical elements: chemistry "plus" pedagogy "plus" general didactics. At second sight it becomes clear that a simple combination is not applicable and that it is hard to grasp all corresponding contents. Chemistry didactics indeed concerns contents of science on one side and contents of general didactics on the other side, but these are only reference disciplines. Chemistry didactics as the real science for teachers is an independent and interdisciplinary science with its own aims, tasks and methodology.

It reflects the contents of its reference disciplines and applies them to questions concerning teaching and learning. Figure 1 illustrates Chemistry didactics as an independent bridging discipline between pedagogy and general didactics on one side and the sciences on the other side [12]. Figure 2 shows the interplay of educational, scientific, and didactical parts in the first phase of teacher training and the independent reflection in the second phase and in class, respectively [13].

To demonstrate chemistry didactics as an interdependent reflection on educational, scientific, and didactical aspects and to highlight the reasoning in chemistry education exemplarily, an interesting experiment will be discussed as a basis (see Fig. 3). It will be conducted in line with different didactical approaches and analyzed accordingly.

"Iron wool on the balance." A ball of gray shiny iron wool, which is fixed and balanced on one side of a beam balance, is to be ignited with the Bunsen burner. A front of red ember that moves inside of the iron wool can be observed. This side of the balance starts to sink and finally a black product remains: iron and oxygen have reacted to iron oxide; the additional mass of oxygen causes the observed change. This school experiment can be carried out and analyzed in different ways – depending on the teacher's educational aim and on the students' skills and abilities. A few different possibilities are summarized in Fig. 3.

In *method 1* the experiment is being demonstrated as aforementioned and observed by the learners. The teacher has to make a decision for their explanation: Do they offer the reaction equation in words, or do they introduce the chemical equation with formulae, or even the redox reaction? Do they choose "exothermal" or "$\Delta H < 0$" for the energy turnover? Do they show or draw models of the chemical structure of iron, oxygen and iron oxide? Or do they even let the students

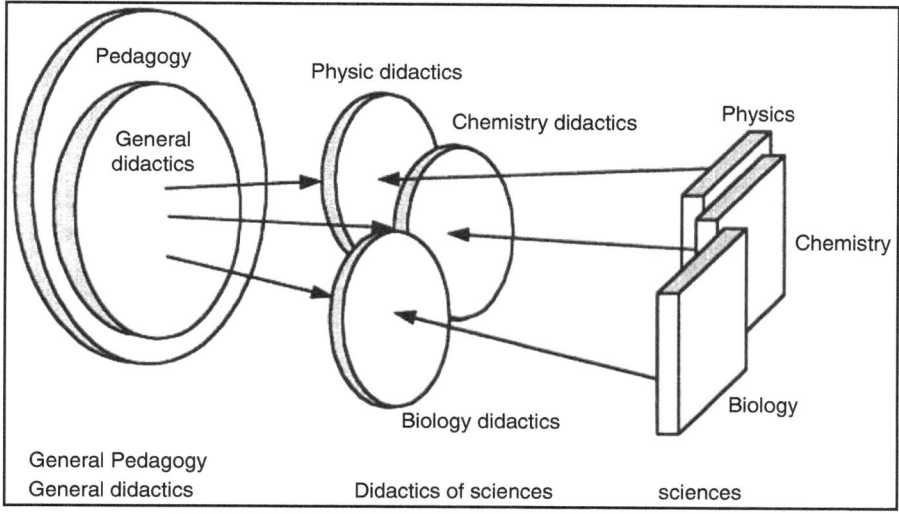

Fig. 1 Chemistry didactics as a bridge between pedagogy/general didactics and chemistry

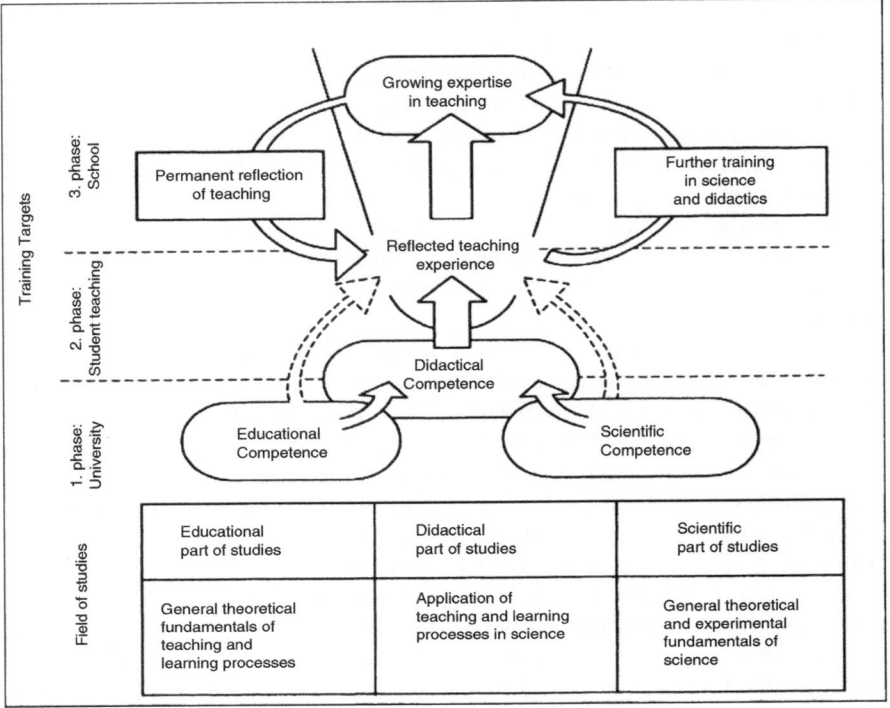

Fig. 2 Connections of studies in pedagogy, science and didactics on the way to the teacher

build these models? Do they formulate an equation as a shortened model on basis of these structural models? These questions of interpretation also arise for methods 2, 3, 4 and 5.

In *method 2* the ball of iron wool is weighed – either with a digital scale or with a beam scale. After igniting the wool the teacher asks the students to predict whether the black substance will be lighter, heavier or remain at equal weight.

The teacher allows the students to hypothesize and develop solutions independently. Because many students already have preconceptions for combustion from their everyday life, they might predict that the product will be lighter and they will be surprised that instead it gets heavier. A problem-oriented approach will be possible and has the pedagogical advantage that the hypothesis "decreasing mass" and the observation "increasing mass" cause a cognitive conflict and motivate students to solve this problem. Curiosity and motivation to develop a solution emerge; students are taking this problem as their own problem and want to solve it!

In *method 3* the problem is stated more openly. The teacher allows the students to discuss their experience with combustion and demonstrates the combustion of iron wool without the scale first. The students express their concepts on this matter – depending on ability and common practice in their class. They might

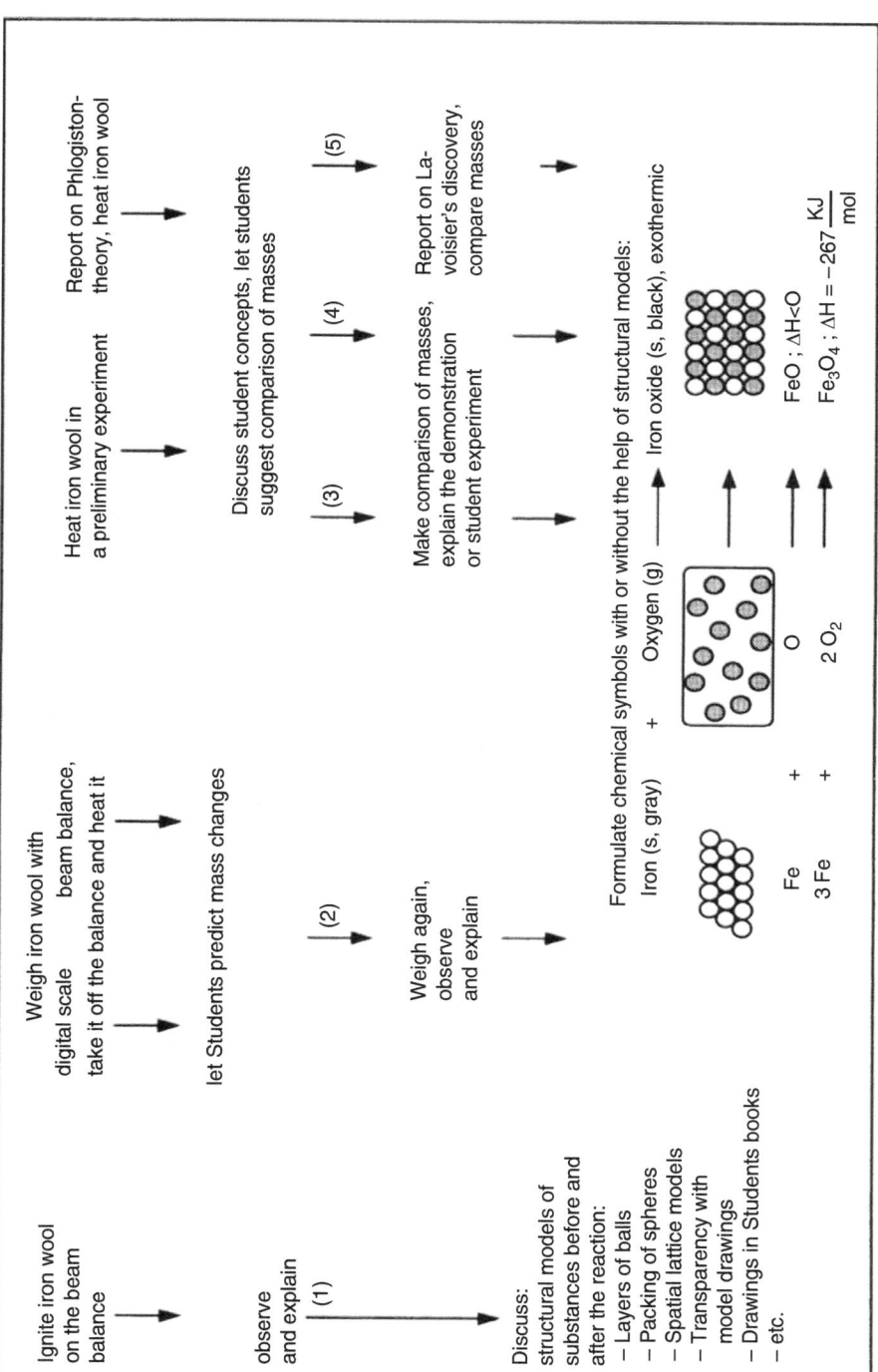

Fig. 3 Five different ways to introduce the "Ironwool at the balance" experiment

spontaneously suggest comparing masses. The teacher compares the masses as a demonstration experiment and analyzes them as described.

They might also invite the students in *method 4* to plan and run such an experiment as a student experiment. The groups of students get different equipment and devise various solutions for the problem – depending on ideas and experimental skills. The advantages of activity-orientation add to the advantages of problem-orientation – another important didactical criterion.

The teachers can also introduce the topic historically-oriented in a different *method 5*. They show the glowing iron wool and question the students about their concept of the combustion process and expect the students to give the following explanation: "something from the combustible passes over to the air". The teacher tells that a few centuries ago scientists thought the same and gave this "something" the name "Phlogiston". They can talk about Stahl's Phlogiston theory and the disproving of this theory by Lavoisier using scales to investigate the behavior of mercury and mercury oxide by heating it on a stove. It is also possible that the teacher lets students give talks on these historical developments.

Conclusion. Reflections concerning this large number of possible experiments and interpretation alternatives of this single issue "iron wool on the balance" show that chemistry teachers have to decide many aspects and ways to plan and to realize lectures in chemistry. These decisions can be made and explained easier by teacher students, when these or similar situations are familiar from the teacher training at university. Didactical decisions have to be made constantly to initiate learning processes and to achieve learning success especially in the beginning of chemistry lessons[14]. Therefore the following thoughts on chemistry education apply to the basics of chemical education at university or college.

A profound knowledge is required for the following fundamental reflections on chemistry didactics. The publication of the German Society of Chemists (GDCh) contains one possible listing of such scientific contents: "Memorandum on the teacher training for chemistry lessons in classes of 10–15-year-old students" [15].

The same memorandum [15] also suggests a way of chemistry educational training: "The aims of the didactical training of chemistry teachers are very diverse. Not all of them can be studied in a fair amount of time and context, not even if the part of chemistry didactics of the total volume of studies is reasonable. For this reason, a choice of important subject areas has to be made. These subject areas can be assigned to one of the four main areas: the learner, teaching processes, scientific ideas and the human element." These four levels of reflection can be found in Fig. 4 in form of concentric circles.

Every didactical subject can be divided into the four aforementioned sections: "If the *learner* is in the focus of every didactical consideration, a bow can be drawn to the *human element*, which influences the learner as an individual and as a member of our society. The *scientific ideas* are made accessible to the learner through *teaching processes*. Considering the interdependence of didactical questions and content-related aspects, learner, teaching processes, scientific ideas and the human element are not assigned firmly to every subject area – in fact the sections within the concentric circles are mostly interchangeable. Therefore

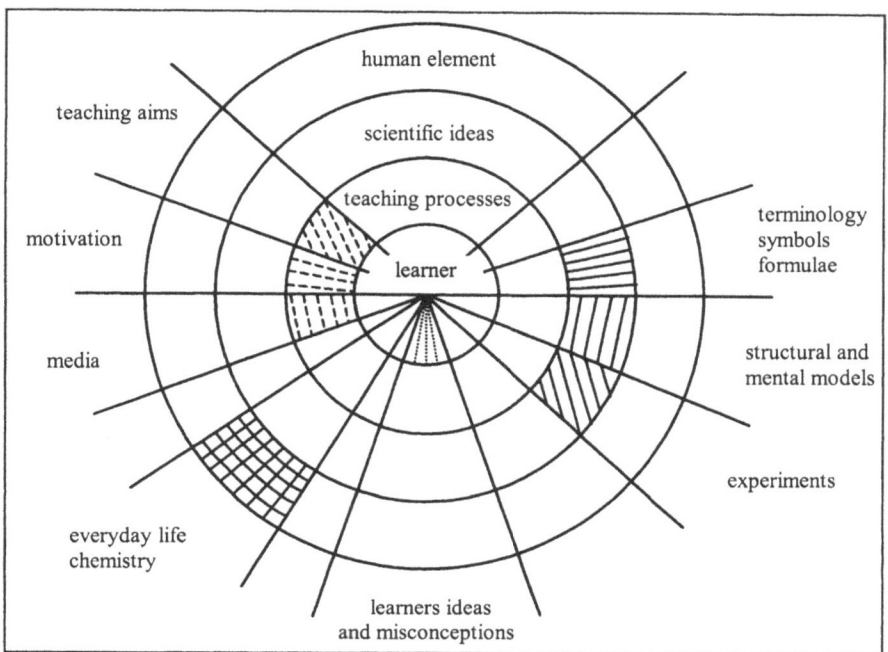

Fig. 4 Main considerations for lectures in chemistry didactics arranged in a "pie chart" [15]

the concentric rings can also be understood as 'rotating rings' with alternate assignments of contents" [15].

The pie chart in Fig. 4 illustrates eight topics exemplarily, that are particularly suitable for a chemistry educational reflection: the classical topics like *experiments, structural and mental models, terminology, symbols and formulae* are hatched to show that they are assigned to the area of "scientific ideas." Modern topics of chemistry didactics are *"learners ideas and misconceptions"* (assigned to the area "Learner") and *everyday-life chemistry* (assigned to the area "human element"). The topics that are assigned to the area "teaching processes" are *teaching aims, motivation and media*, which are relevant for every school subject.

The pie chart includes empty segments without topics: these reflect the fact that there are a lot more topics for a chemistry educational consideration that could be added to the chart. To name a few: performance assessment, role of the teacher, curriculum development, classroom teaching research, history of chemical education. The combination with other standard references of chemistry education [16–24] is possible and highly recommended, due to the conceptual openness of this book. It begins with "Learners ideas and misconceptions" because students have preconceptions of substances and their chemical changes from everyday life that already exist long before chemistry lessons start.

References

[1] Itelson, L.: Mathematische und kybernetische Methoden in der Pädagogik. Berlin 1967

[2] Roth, L.: Handlexikon zur Erziehungswissenschaft. München 1976

[3] Klafki, W.: Didaktische Analyse als Kern der Unterrichtsvorbereitung. Hannover 1969

[4] Heimann, P., Otto, G., Schulz, W.: Unterricht – Analyse und Planung. Hannover 1970

[5] Cube, F.: Kybernetische Grundlagen des Lernens und Lehrens. Stuttgart 1968

[6] Winkel, R.: Die kritisch-kommunikative Didaktik. West. Päd. Beitr. 1 (1980), 202

[7] Blankertz, B.: Theorien und Modelle der Didaktik. München 1973

[8] Ruprecht, H.: Modelle grundlegender didaktischer Theorien. Hannover 1972

[9] Aschersleben, K.: Didaktik. Stuttgart 1983

[10] Jank, W., Meyer, H.: Didaktische Modelle. Frankfurt 1991

[11] Vossen, H.: Kompendium Didaktik der Chemie. München 1979

[12] Riedel, W., Trommer, G.: Didaktik der Ökologie. Köln 1981

[13] Hammer, H.O.: Fachdidaktik der Chemie an der Hochschule. CU 12 (1981), 5

[14] Scheible, A.: Gedanken zum Einführungsunterricht in die Chemie. MNU 19 (1966), 1

[15] Barke, H.-D., Bitterling, D., Gramm, A., Hammer, H.O., Hermanns, R., Leibold, R., Lindemann, H., Wambach, H.: Denkschrift zur Lehrerbildung für den Chemieunterricht in den Altersstufen der Zehn- bis Fünfzehnjährigen. GDCh, Frankfurt 1983

[16] Schmidt, H.J.: Fachdidaktische Grundlagen des Chemieunterrichts. Braunschweig 1981

[17] Christen, H.R.: Chemieunterricht. Eine praxisorientierte Didaktik. Basel, Boston, Berlin 1990

[18] Becker, H.J., Glöckner, W., Hoffmann, F., Jüngel, G.: Fachdidaktik Chemie. Köln 1992

[19] Pfeifer, P., Häusler, K., Lutz, B.: Konkrete Fachdidaktik Chemie. München 1992

[20] Harsch, G., Heimann, R.: Didaktik der Organischen Chemie. Vom Ordnen der Phänomene zum vernetzten Denken. Braunschweig 1998

[21] Lindemann, H.: Einführung in die Didaktik der Chemie. Düsseldorf 1999

[22] Barke, H.-D., Harsch, G.: Chemiedidaktik Heute. Lernprozesse in Theorie und Praxis. Heidelberg 2001

[23] Rossa, E.: Chemie-Didaktik. Praxishandbuch für die Sekundarstufe I und II. Berlin 2005

[24] Anton, M.A.: Kompendium Chemiedidaktik. Bad Heilbrunn 2008

Chapter 1
Learners Ideas and Misconceptions

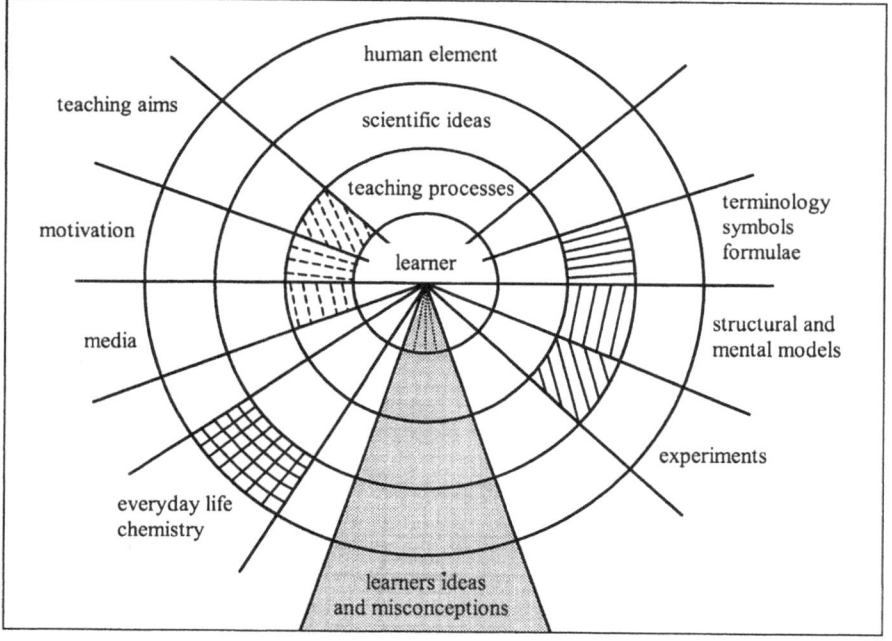

"Acids eat magnesium ribbon, rust eats up the iron nail" – *spontaneous comments about phenomena of metal–acid or metal–oxygen reactions often sound like this. This is acceptable as long as students make these comments in everyday language. Teachers should be familiar with such preconceptions to successfully change them into scientific mental models.*

H.-D. Barke et al., *Essentials of Chemical Education*,
DOI 10.1007/978-3-642-21756-2_1, © Springer-Verlag Berlin Heidelberg 2012

Some decades ago it was assumed that students do not have any preconceptions or knowledge in chemistry: good preparation for a chemistry lesson only had to decide, which new terms were to be introduced in which order and with which experiments or models. Empirical studies, however, showed that learners have preconceptions for many topics and that these preconceptions do not match today's scientific concepts. For that reason a first basic question of chemistry education is: which preconceptions exist for which topics and how can we effect conceptual change? Often the preconceptions are simply called "false" – without considering that students make correct observations and create individual mental models on the base of their observations. Therefore, these conceptions should better be called:

– Conceptions of everyday life
– Primary or prescientific conceptions
– Student precomprehension or preconception
– Misconcepts or misconceptions

Iron wool example. One example might back up this opinion. A ball of light gray shiny iron wool is to be weighed and heated in a roaring Bunsen flame (see "Introduction"). It can be observed that the iron wool glows and turns black. If students are asked whether this black substance is of equal weight or lighter or heavier than before, then the majority of students reply that it should be lighter. The reason for this assumption is that wood or charcoal disappear while burning and only a small amount of ash remains; alcohol even burns completely and without any residue left behind.

This experience has been gained for 12–15 years and is now transferred to every combustion process: therefore it should not be called false, but primary or prescientific. It is advantageous to talk about preconceptions with the students before introducing the scientific concept, or to introduce the scientific idea first and to come afterwards to the images of the involved students. In every case apart from talking it is necessary to show experiments – they may convince students more than talking.

For the iron–wool experiment the mass of a portion of metal is to be weighed before and after combustion, the higher mass is to be discussed on the background of the mass of oxygen which reacts with iron and forms solid black iron oxide. Increasing mass can also be observed in the experiment "candle on the balance," when the invisible gaseous combustion products water vapor and carbon dioxide are chemically bound. A discussion on the basis of this experimental experience may reduce primary conceptions for the benefit of lasting scientific concepts. But these learning processes cannot be made in one lesson – they have to be developed in continuously problem-oriented lessons on the combustion process during a number of weeks.

Student conceptions on combustion astonishingly show elements of the historic Phlogiston theory: apparently there exist parallels between student conceptions and the development of historic perception processes in science. Therefore, it makes sense to study the development of historical theories which made radical changes through the centuries, and to analyze which parallels exist in the student's thinking. Such theories, for example, refer to

- Primary matter theories of the Greek philosophy schools in ancient times
- Transformation concepts of alchemists
- The Phlogiston theory
- The "horror vacui" and the development of knowledge of air and air pressure
- Theories of atomism, the structure of matter, and others

Since reflections on this matter are of a scientific-historical kind, they will be introduced first in the section "scientific ideas" (see pie chart metaphor). After that, empirical results on student conceptions will be discussed and compared to historical theories in the section "learner." Recommendations for lessons will be given in "teaching processes." The influence of everyday language and commercials in media on chemistry lessons will be taken up in "human element."

1.1   Scientific Ideas: Theories of Science History

The Greek philosophers of ancient times deeply considered many issues of human life and established acknowledged theories in many fields, also in today's cultures and human values.

Primary theories of matter. The question of a primary matter arose for Greek philosophers: What does the world consist of? What does primary matter, primary substance or element mean? Equally important was the second fundamental idea that this primary substance should be of eternal existence: "nothing can emerge from nothing nor can anything vanish to nothing, it is only the appearance that changes." Thereby attention was drawn to the following problems:

1. The materiality of earth
2. The fact that matter cannot be created nor destroyed
3. The transformability of matter while keeping the primary substance [1]

Aristotle was the first to teach the strict separation of a substance and its properties. The Greek philosophers before Aristotle were not aware of this differentiation; they did not discuss the substance as the "bearer" of characteristics on one side and the characteristics themselves on the other side. Based on his understanding, Aristotle proposed: "development and change, formation and transience are nothing else but a transition from one form of being to another form of being" [2].

Transformation concept of the alchemists. The age of alchemy stretched from the fourth to the sixteenth century, with the development of alchemy particularly strong in the Arabic regions. Alchemy in Arabic was just another term for chemistry, derived from the Arabian "al" and the Greek "chyma": cast metal. This term reflects the great importance of metals and their production for humans including their wish to turn base metals into gold.

Many Arabian writings already include instructions for artificial gold production with the help of the ferment of ferments, the elixir of elixirs. The correct mixture of the four elements is essential, the "spirit" (mercury vapor) has to enter the "body" (lead or copper). The mysterious elixir can only be created with the right combination of the four elements, the body (metal) and the spirit (mercury), the masculine and the feminine. It assimilates the bodies and colors them (therefore it is called "tincture") by turning them into an up to thousand fold amount of gold. Even the scientist Albertus Magnus believed that artificial gold can be produced, but he did not find an alchemist who succeeded in transforming any substance into gold [3].

Alchemy did not lack in practical and apparent proves up to the eighteenth century. Coins were shown that were supposed to be of alchemical gold or nails that consisted of iron in one half and gold made from iron in the other half. Even court decisions were handed down that benefited alchemical operations: clever swindlers made sure that the belief in the possibility of metal transformation was revived again and again through their successful transmutations [1].

In 1923 the scientific world was excited once again, "when a professor from Berlin conveyed that he had turned mercury into gold by treating it with electric current. The correctness of this discovery was not only confirmed by several sides, but different scientists (even from Japan) also reported that they found the same earlier. Through a thorough experimental retesting 2 years later, it came to light that the small traces of gold originated from the electrodes. The gold dream that cast a spell over so many bright minds for more than a 1,000 years is finally over" [3].

The Phlogiston Theory. People have always observed the disappearance of fuel during a combustion process, for example through burning of coal, alcohol, or sulfur. The German Georg Ernst Stahl published his interpretation of these observations in 1697 and introduced the term "Phlogiston" (Greek: phlox, the flame): "He started with the combustion of sulfur and believed that sulfurous acid, which was produced following the combustion process, was in fact sulfur robbed of its combustible character, i.e., robbed of Phlogiston" [4].

Stahl assumed that every combustible and lime forming substance contained phlogiston, therefore combustion was a process during which phlogiston left the body. Air was required to absorb the phlogiston, from the air it could get into leaves and wood and could be given back by burning wood. Also by heating metal lime (today: metal oxide) on a piece of charcoal Stahl observed the reduction to metals. Subsequently, reactions of metals in air and reductions of metal lime to metals were recognized as mutually dependent processes [1]. The supporters of the Phlogiston theory formulated similar equations:

$$metal \rightarrow metal\ lime + Phlogiston$$

$$metal\ lime + coal(Phlogiston) \rightarrow metal$$

Stahl had to consider the pure metal as a compound of metal and phlogiston, and metal lime as an element which will not decompose. Furthermore experiments using a balance did not support his theory: "He did not ascribe importance to the weight increase of the formed metal lime, instead he tried to dismiss this problem with the assumption of a 'negative weight' of the phlogiston and chemists were mostly focused on the qualitative phenomena, not on the weight of substances. The Phlogiston theory turned out to be an excellent model for explanation and systematization of qualitative conversions of matter" [1].

The view of the supporters of the Phlogiston theory becomes clearer, when it is seen from the point of energies rather than substances. Not only the flames and substances can be observed during combustion, but also thermal energy. So if one replaces the "substance Phlogiston" by heat or energy, many observations of the combustion process can be explained better [3]. In this context people often thought of a special "heat substance" that is leaving the burning fuel; energy was thought of as matter, as substantial.

Lavoisier threw light on this issue with measurements of masses observed for the synthesis and decomposition of mercury oxide. For this decomposition he took even a glass bowl with water, fixed a special volume of air under a glass cover and observed that the volume decreased by 20% (see Fig. 1.1). He did not only find oxygen as a new element and the oxidation theory, but also the law of conservation of mass. Prior to these results of Lavoisier oxygen had also been called "fire substance" (Empedocles), "Phlogiston" (Stahl), and "fire air" (Scheele). "A remnant of these terms was still in use as 'heat substance' for several decades, until the reason for heat was found in the movement of smallest particles" [1].

Horror vacui and air pressure. In ancient times experiments with pipettes and wine cradles drew the natural philosophers' attention to the nonexistence of an absolute vacuum on earth. As soon as a substance leaves a space, another substance, mostly air, takes its place. "In this context a statement from Canonicus came to be known, which says that nature has a horror vacui, an aversion against the vacuum" [5].

Even Galilee was aware of this phenomenon and knew from well builders that it is not possible to pump up water from a depth of more than 10 m. He assumed this measure to be the uttermost power of nature to prevent a vacuum. He invented a

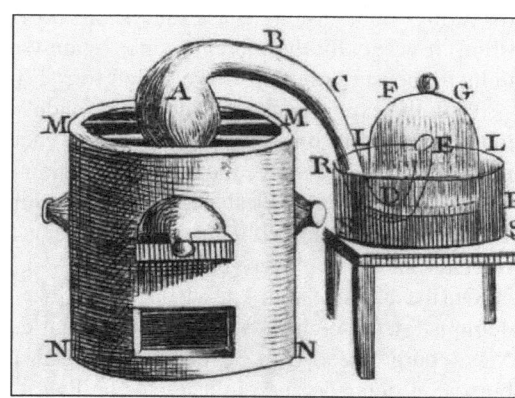

Fig. 1.1 Apparatus of Lavoisier to form mercury oxide and to determine the amount of used air

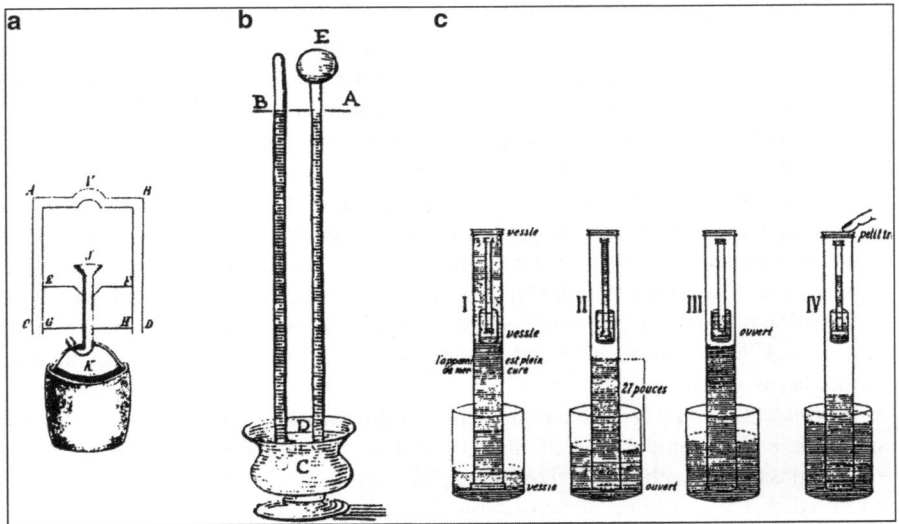

Fig. 1.2 Apparatuses of Galilee (**a**), Torricelli (**b**), and Pascal (**c**) to discover air pressure

thought experiment to measure this power, the "resistenza del vacuo" [5] in 1643 (see Fig. 1.2a): a cylinder was supposed to contain water, a big weight of heavy stones was to be attached to the movable piston until – besides the water – an empty space formed in the cylinder. No one knows, whether this experiment was only described on paper or if it was really performed [5].

This apparatus inspired his student Torricelli to use fluid mercury with high density instead of constructing a cylinder with a movable piston. He filled a 1-m-long glass tube with mercury, opened the tube under mercury in a glass bowl, and observed mercury gliding down about 24 cm like a piston in a cylinder (see Fig. 1.2b). With this experiment, first described in 1643 [5], he could prove the existence of air pressure: a 760-mm column of mercury pushes like about 20,000 m of air. He could also prove that above this column an empty space exists, a vacuum: by tilting the 1-m-long glass tube more horizontally, the glass tube fills totally, tilting it vertically the mercury sinks again (see E.1.1). In honor of Torricelli, the units to measure pressure were called torr: 1 Torr = 1 mm mercury column.

With the experiment "du vide dans le vide" (see Fig. 1.2c), Pascal finally proved that the apparatus does show the Torricelli vacuum and that there is no special "air" in the empty space above the mercury column [5]. Various experiments in different heights above sea level show that air pressure decreases with the altitude: on a 1,000 m high mountain the mercury column is 690 mm high. Subsequently, the first mercury barometers were built.

On the basis of this knowledge, Guericke developed powerful air pumps and demonstrated the air pressure in spectacular experiments with the "half-spheres of Magdeburg" (see Fig. 1.3): air was pumped out of one pair of half-spheres, eight horses on one side and eight horses on the other side sometimes managed to pull

Fig. 1.3 Half-spheres of Magdeburg: Guericke's demonstration of air pressure

those half-spheres apart by using all their power. This power was necessary to overcome the pressure of air on both empty half-spheres!

Theories of atomism and the structure of matter. These theories have their origin in interpretations of the Greek philosophy schools. Two schools of thought existed: one school around Democritus and Leucippus believed that repeated partition of a piece of matter had an end and that matter is built of undividable particles, called atoms (Greek: atomos, undividable). This conception assumed discrete particles and empty space around them – therefore today it is called the *discontinuity* hypothesis.

Aristotle and other philosophers thought that a repeated partition of matter does not lead to an end. Especially the mental impossibility of an empty space, which separated one particle from another – the "horror vacui" – convinced them of a continuously structured matter: the *continuity* hypothesis was born. Due to the wide influence of the Aristotelian school the continuity idea was taught everywhere, Democritus's atoms disappeared for almost 2,000 years.

After Torricelli disproved the idea of the "horror vacui" in a macroscopic way and the vacuum became imaginable, Gassendi transferred this knowledge on the existence of a vacuum to the submicroscopic level. He broke away from the Aristotelian conception and took up Democritus's idea: the atoms and the empty space between the atoms are the big principles of nature [6]. After a disruption of 2,000 years scientists could now take the discontinuity hypothesis as a basis to think about smallest particles and the structure of matter. In 1808, Dalton formulated his atomic hypothesis that there are as many kinds of atoms as kinds of elements and found the first table of atomic masses: The scientific way of chemistry started with these ideas!

This fascinating story of the discontinuity hypothesis and the disprovement of the "horror vacui" theory should be transferred to our students and to the planning of chemistry lessons regarding the structure of matter and the involved particles.

1.2 Learner: Empirical References of Student Conceptions

According to Piaget students starting with chemistry lessons can mostly be assigned to the period of concrete operations in their intellectual development – their thinking is tied to concrete reality. Therefore, they explain phenomena in a concrete-pictorial and in a magical-animistic way of speaking. This becomes apparent in the following examples:

- Pieces of wood *don't want* to burn, the flame *wants* to go out
- The flame *eats* the candle, substances *like to* react, hydrogen *likes* to react with oxygen
- Acids *attack* other substances, acids *eat* base metals, rust *eats up* iron, etc.

Student interpretations are often basic analogies; reasons are being personalized and equalized to purposes:

- Sodium reacts with water "like a fizzy tablet"
- When copper sulfate dissolves it looks "like distributing red-cabbage juice in water"
- Crops grow on fields *so that* humans can eat it
- Wood burns *so that* one can warm themselves, etc.

Parallels between the historical development of chemistry and students' way of thinking can be found. For this reason, the concepts from science history stated in Sect. 1.1 will be taken up and compared to empirically observed examples of today's student comments.

Substances as a bearer of properties. Chemical reactions do not necessarily produce new substances, but new properties are being added according to student conceptions [7]:

- Copper roofs turn green, silver turns black, a solution turns deep blue, etc.

As for philosophers of ancient times, a "primary matter or a bearer of properties" seems to exist that survives and only changes the outward appearance again and again. Therefore, it should be made clear to the students that:

- The green layer on a copper roof can be removed: it is copper carbonate
- The black layer on a silver spoon is not silver: it is silver sulfide
- A solution changes color because of a new dissolved substance

It can be shown in an experiment (E1.2) that the metal can be recovered from black silver sulfide or black copper oxide. Additionally, it can be demonstrated that the oxides or sulfides can be formed again in a reaction of metal with oxygen or sulfur: chemical reactions are reversible. But this leads to the question of what survives in the compound and what can be formed again. The answer is difficult and can only be formulated on the level of atomic models: the nuclei of metal atoms and nonmetal atoms do not alter in the reaction. Also the question of the difference between red copper oxide (Cu_2O) and black copper oxide (CuO) can only be answered on the basis of Dalton's atomic model: with the ratio of atoms or ions in different compounds and with different chemical structures of these substances (see also Chap. 6).

Mixing and unmixing. These images play an important role in students' interpretation of the conversion of materials [7]:

– Silver sulfide is *mixed* from silver and sulfur
– Copper oxide *contains* copper and oxygen
– Water *consists* of hydrogen and oxygen
– A hydrocarbon *is built of* carbon and hydrogen

These expressions suggest a concept of mixture, similar to what has been discussed by the Greek philosophers. A heterogeneous mixture of two substances, like one of copper crystals and sulfur crystals, has to be shown to the learner and is to be compared to homogeneous copper sulfide. The formulation "copper sulfide is a compound of the elements" or "hydrogen and oxygen exist chemically bound in water" is to be preferred to the ones stated above. Later, on the level of Dalton's atomic model one can talk of H_2O molecules "consisting of" H atoms and O atoms or "containing" those atoms.

Conservation of mass. Most students do not know of the alchemists and their wish to turn mercury or lead into gold. But they see the "green copper roof," and are thinking that red copper transformed over a long period of time into "green copper," that silver turns into "black silver." These arguments show that students have the idea of the change of substances, but no idea of a chemical reaction conserving the elements in their compounds, or gaining back the elements from compounds by chemical reactions.

Similar preconcepts exist concerning the evaporation of water or ethanol to the disappearance of metals in the reaction with acids, or to the removal of a fat stains from clothes: "the water, the metal, the fat has gone away" [7]. Young students do not think of water or ethanol vapor going into the air, of metals reacting with acids forming a salt solution and hydrogen gas, or of fat that dissolves in petrol: the substances have simply "gone away." So the teacher has to demonstrate that, for example, acetone is evaporating at room temperature forming a big volume of acetone vapor in a syringe (E1.3), that magnesium is dissolving in hydrochloric acid forming a magnesium chloride solution and the gas hydrogen (E1.4), that the disappearance of fat is interpreted by dissolving of the fat in petrol (E1.5). Through these experiments the "destruction" idea of young students can be turned into a "conservation" idea.

The same holds for the concept of *energy*. Not only students but also everyday people are thinking that energy is "away," that a battery is "empty," that the fuel container of the car has to be filled up because "the fuel is gone." The teacher has to demonstrate that energy is transformed into another kind of energy all the time: electric energy coming into our homes can be converted into thermal energy (coffee machine), into light energy (light bulb), into mechanical energy (lawn-mower). *Chemical energy* is the most sophisticated idea: it is hard to understand that sugar and starch are compounds with high chemical energy, converting to substances with lower energy like water vapor and carbon dioxide. Also fuel is a high energy compound which turns into carbon dioxide and water vapor through the reaction with oxygen. The teacher has to point out with exothermic reactions that chemical energy of the educts is converted into lower chemical energy of the products, that the differences in those energies are turned into heat and sometimes also into light.

The biggest problems result from *combustion processes*. Throughout their lives students have observed that wood or paper burn with hot flames, and that wood and paper are "away" after the combustion [7]. Driver [8] showed students a balance and asked about the mass of the black substance after heating iron wool in the Bunsen flame. Most of the students answered: "the mass is less than before" (see Fig. 1.4). The arguments of some pupils are the substance is "lighter because by heating it certain things were being burnt away and so making it lighter" [8].

Later on Driver put a stopper on a flask, ignited a piece of white phosphorus and dissolved formed white smoke in water (see Fig. 1.5). Although the flask was closed all the time students supposed that the flask is lighter after burning and dissolving:

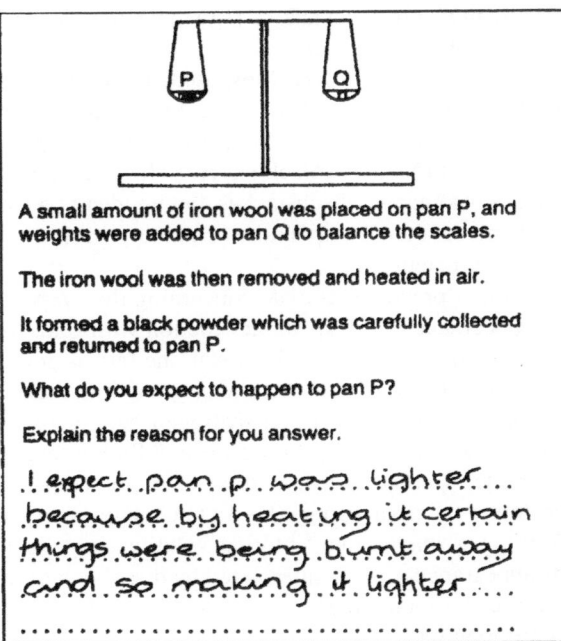

Fig. 1.4 Masses before and after combustion of iron wool [7]

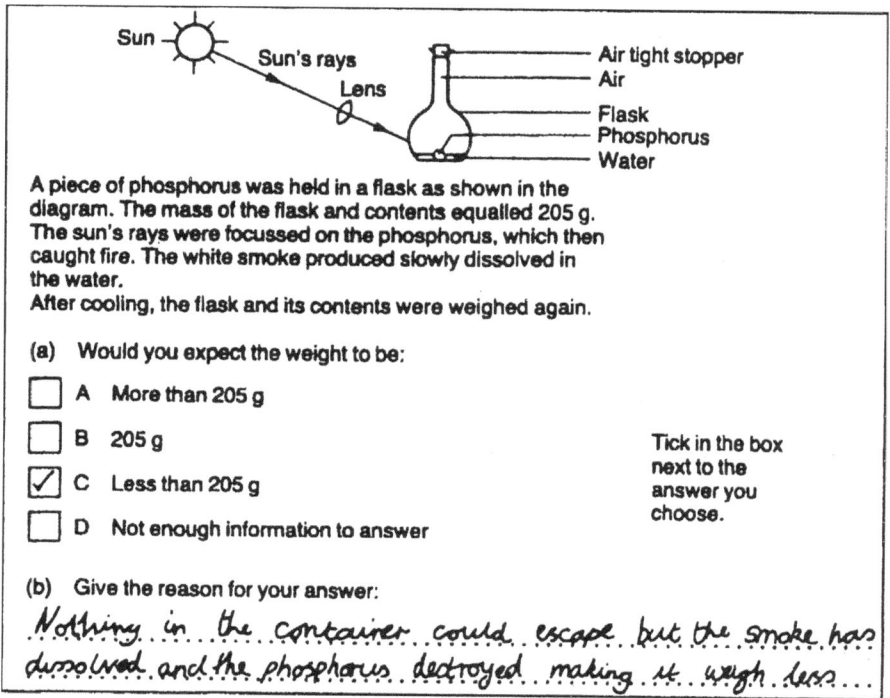

A piece of phosphorus was held in a flask as shown in the
diagram. The mass of the flask and contents equalled 205 g.
The sun's rays were focussed on the phosphorus, which then
caught fire. The white smoke produced slowly dissolved in
the water.
After cooling, the flask and its contents were weighed again.

(a)   Would you expect the weight to be:

☐   A   More than 205 g

☐   B   205 g   Tick in the box
                                                             next to the
☑   C   Less than 205 g   answer you
                                                             choose.
☐   D   Not enough information to answer

(b)   Give the reason for your answer:

*Nothing in the container could escape but the smoke has
dissolved and the phosphorus destroyed making it weigh less*

Fig. 1.5   Masses before and after combustion of phosphorus [8]

"Nothing in the container could escape, but the phosphorus has been destroyed and
the smoke dissolved making it weigh less" [8].

Barke [7] asked students about the equation representing the combustion of
magnesium and most of the grade 10 students answered correctly (see Fig. 1.6).
Because they knew the atomic model and the idea about atoms, ions, and molecules
students were asked to describe "what happens to the small particles of magnesium
during combustion" (see Fig. 1.6).

Some students answered the question in this way: "magnesium contains two
kinds of particles, one evaporates by combustion, the other one remains as magnesium oxide." They even drew their imagination in the same way (see Fig. 1.6). This
example shows that students learn the equation only formally, but keep their mental
model from everyday life that they developed in 10–15 years of observations.
Teachers should know that they cannot effect complete conceptual change from
students' preconcepts in 1 or 2 h of instruction: for several weeks one has to show
new experiments concerning combustion. The observations have to be explained
with oxygen reactions to metal or nonmetal oxides, and one has to visualize the
structure of substances before and after oxidation with sphere packing or model
drawing [8].

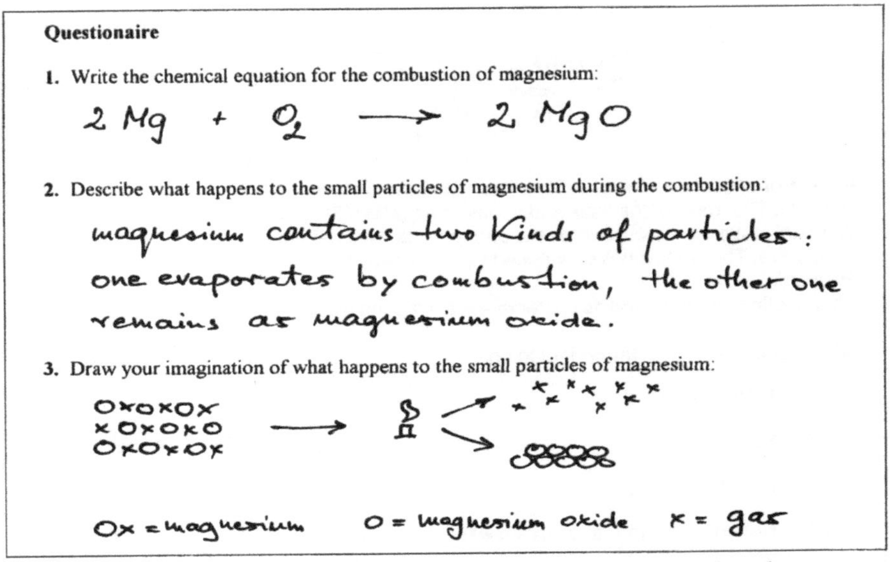

Questionaire

1. Write the chemical equation for the combustion of magnesium:

$$2\,Mg + O_2 \longrightarrow 2\,MgO$$

2. Describe what happens to the small particles of magnesium during the combustion:

magnesium contains two Kinds of particles:
one evaporates by combustion, the other one
remains as magnesium oxide.

3. Draw your imagination of what happens to the small particles of magnesium:

Ox = magnesium   O = magnesium oxide   x = gas

Figure 1. Questionnaire about the combustion of magnesium (teacher questions in block letters, student responses hand-written)

Fig. 1.6 Mental model of a student regarding the combustion of magnesium [8]

Even advanced students with 3–4 years of chemistry instruction sometimes cannot give up their destruction theory. Pfundt [9] reported about a student who stated: "concerning the formula CO_2, the black substance carbon should be produced from carbon dioxide, but it is impossible to gain a black solid from a colorless gas" [9].

In every case the preconcepts concerning combustion processes have to be discussed in lectures, especially regarding these two facts: the reaction with oxygen as a part of air, and the formation of colorless and invisible gases through the combustion of candles, wood, or paper. It is possible to show with experiments the mass of oxygen that combines with portions of metals: iron wool reacts to solid iron oxide and gets heavier by the mass of oxygen, and magnesium reacts to solid magnesium oxide and the mass increases by reacting with oxygen (E1.6).

To show the same effect for candles or tea lights the masses of the gaseous products water vapor and carbon dioxide are to be measured with an apparatus absorbing both gases with the help of calcium oxide and sodium hydroxide (E1.7). Both effects with metals and candles are shown in an *open apparatus*: the mass of oxygen adds to the mass of substances before.

The most important step is to show the conservation of mass in a *closed apparatus*. If the combustion of iron wool is carried out in a test tube and the opening is closed with an air balloon, the educts iron and air can be weighed. After heating the test tube and observing the formation of black iron oxide the mass is determined a second time: the same mass can be observed (E1.8). In the case of burning wood these measurements are also possible: some matches are ignited in the same way as the iron wool, and the combustion products weigh the same as the educts (E1.8). The most convincing experiment is burning a few small pieces of charcoal in a closed flask containing oxygen and observing that the pieces disappear totally: although the coal "is gone" and nothing is to see – the flask weighs the same before and after combustion (E1.9). Students will explain this with the formation of the colorless gas carbon dioxide and the mass of this gas portion.

Air and other gases. Even scientists from previous centuries did neither perceive air as a substance nor did they differentiate the colorless gases mixed in the air. It is similarly difficult for children today. Since air is always weightlessly around us and warm air rises up, air is not perceived to have a mass and is not being seen as a substance in children's eyes.

Münch [10] could show with an empirical inquiry that half of the students between 10 and 16 years of age believe that a soccer ball, which they pump up hard with a regular air pump, is lighter than the one that is only pumped up to a smaller extent. The mass of a certain amount of air and thus air density can be demonstrated to the students quickly and convincingly: a glass ball is being evacuated with a water jet pump and weighed with an analytical balance. After 100 mL of air are transferred to the glass ball from a gas syringe, it is weighed again: a mass of 0.13 g is measured (E1.10). So air is identified as a substance with the density of 1.3 g/L.

If densities of other gases are determined in the same way, air and other gases can be differentiated by their densities. Oxygen, nitrogen, hydrogen, carbon dioxide, or butane can also be introduced experimentally with the help of a burning or glowing wooden splint (E1.11). Additionally, the oxygen content of air of about 20% by volume can be shown by demonstrating the reaction of iron wool or phosphorus to form solid oxides in a closed apparatus (E1.12). One can even show that from gaseous, colorless carbon dioxide the black solid carbon can be obtained: magnesium reacts with carbon dioxide by forming white magnesium oxide and black solid carbon spots (E1.13).

Many false conceptions regarding gases are derived from everyday language. Weerda [11] listed the following student comments:

- Fresh air is "good" air; air without oxygen is "bad"
- Chimneys need "fresh air" and give "used air"; cars give off "exhaust gas"
- Colorless gases are "air" or "like air"; water evaporates "to form air"
- Gases are combustible, are there to cook and to heat
- Gases are dangerous, explosive, and toxic
- Gases "can be liquid"; lighters contain "liquefied gas"

A simple experiment can be run to clarify the term "liquefied gas." By air displacement butane is to be filled into a gas liquefaction pump (E1.14), strong pressure has to be placed on butane gas with the help of the piston: a drop of liquid butane forms in the presence of gaseous butane. It needs to be clarified that lighters and camping gas cartridges contain liquid *and* gaseous butane. It is also helpful to show the big volume of gaseous butane compared with the volume of the liquid: the drop of butane fills the whole pump with gaseous butane.

Structure of matter. It is possible to take up the natural philosophers' discussion from 2,000 years ago: do you get to an end when you divide a sample of a substance again and again? At the end of all divisions, are there smallest particles of the substance? Since students do not have experience with particles of matter, the continuity hypothesis is more obvious to them. After teaching the particle model students were asked, if they could imagine water particles. They spontaneously said: "no, because you can spread a drop of water as wide as you want."

Even when young pupils accept the term "particle" in a discussion, difficulties in their imagination still arise: they do not use the particle concept consequently. Pfundt [12] demonstrated the dissolution of a blue copper sulfate crystal in water and asked students about their mental models. The answers were not only variable regarding a continuity or discontinuity concept, but also the possibility that particles may appear during the dissolution process (see Fig. 1.7) or that existing particles disappear: they "melt together," and the new substance forms crystals without any particles – students prefer the continuity theory if they are not sure.

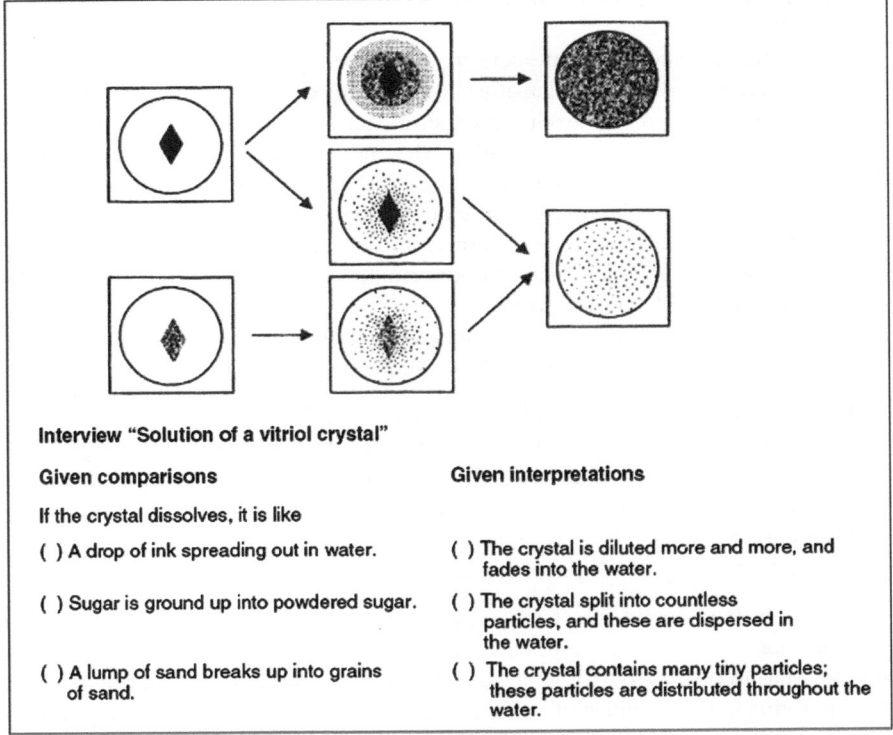

Interview "Solution of a vitriol crystal"

Given comparisons

If the crystal dissolves, it is like

( ) A drop of ink spreading out in water.

( ) Sugar is ground up into powdered sugar.

( ) A lump of sand breaks up into grains of sand.

Given interpretations

( ) The crystal is diluted more and more, and fades into the water.

( ) The crystal split into countless particles, and these are dispersed in the water.

( ) The crystal contains many tiny particles; these particles are distributed throughout the water.

Fig. 1.7 Part of a questionnaire concerning the particle concept of the solution process [12]

In this context, Pfundt created the term "not prebuilt particles" [12]: they can appear or disappear. In the other case, there are "prebuilt particles" which exist in any matter, in crystals, liquids, gases, solutions, etc.

The result of the inquiry was that pupils from grade 7 to 9 mainly chose the continuity hypothesis and considered "not prebuilt particles." Only few students ticked mental models of prebuilt particles and reasoned consequently with the particle concept for questions of state changes, dissolution and crystallization processes. For successful teaching the particle concept has to be used in lectures consequently, it has to be deepened with as many examples as possible.

With model drawings of the structure of matter, Pfundt found out also that students prefer squares to circles as models for particles. When she asked for the reason, students answered that models "have to be compatible and it has to be possible to put them together without gaps" [12]. Gaps, which do not exist in students' conceptions, occur when circles are being used: "the horror vacui in their mental models made them prefer squares to circles as models of particles."

Novick and Nussbaum [13] made inquiries on the particle model of gases and asked: "What is there between the particles"? They offered answers like "no material, air, vapor, oxygen or pollutant." They found that the correct answer "no material" was given by 20–40% only (see Fig. 1.8), the majority of high school and university students in the USA has the conception of air or other matter being between the gas particles.

Subsequently, other inquiries have been made to find out to what extent the "horror vacui" exists, concerning the space between the particles in gases [14]. An experiment with butane was demonstrated (E1.14) and students were asked for model drawings and their thoughts about gaps between butane particles in butane gas (see Fig. 1.9). The results of the inquiry showed that students of grade 9–11 could all reproduce the model drawing correctly, but only about 50% of students ticked "nothing" or "empty." That means that the other 50% have the conception of butane, air, or other matter filling the gaps between the butane particles – these students believe in the "horror vacui"!

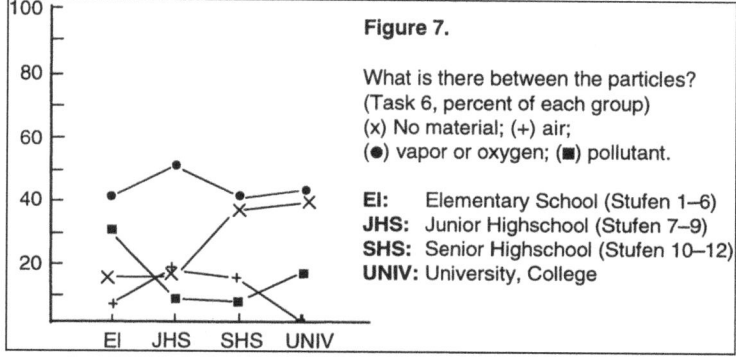

Fig. 1.8 Novick and Nussbaum's results of asking about space between particles in a gas [13]

Models for the evaporation of liquid butane (camping gas)

The cylinder of a gas pump or syringe is filled with colorless butane gas (see picture). The piston is pressed by hand and liquid butane appears. The piston can be released and liquid butane turns into gaseous butane.

Draw your mental model of the formation of butane particles:

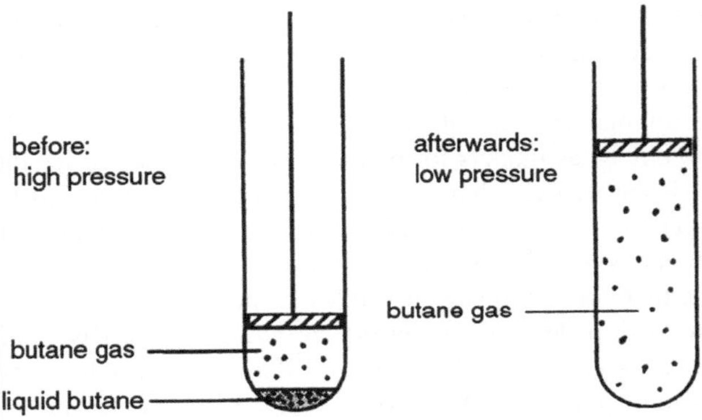

before:
high pressure

afterwards:
low pressure

butane gas

butane gas

liquid butane

How do you imagine the space between the particles of the butane gas? I think.....

( ) there is also butane gas between the particles,

( ) the space between the particles is empty,

( ) there is nothing at all between the particles,

( ) there is air between the particles,

( ) there is a special invisible substance between the particles.

Fig. 1.9 Part of a questionnaire concerning the "horror vacui" of students [14]

Especially the explanations for their answers show this:

- "Space between the particles cannot be empty, there cannot be nothing
- I cannot imagine there to be nothing
- If there is no air, there has to be a vacuum and this I cannot imagine
- Something has to be there, there is no place, where there is nothing
- The space cannot just contain nothing
- Something has got to be there" [14]

Apart from the introduction of smallest particles themselves it is important to discuss the empty space between the particles when teaching the particle model. A convincing experiment may be the evaporation of ethanol (E1.15). An empty balloon is filled with 1–2 mL of ethanol and closed with a knot. Water is boiled in a big beaker; the balloon is thrown into the boiling water: the balloon gets bigger and bigger, 200–300 mL of ethanol vapor are formed. After taking the balloon out of the water, the volume decreases to the former small amount of ethanol. Explaining the big gas volume one has to point out that the movement of ethanol particles is increasing and distances between the ethanol particles are growing. There is no material between the particles: the fast movement of particles produces the big volume.

1.3 Teaching Processes: Considering Students' Conceptions

"All teaching should begin with children's experiences – each new experience children have in a classroom is organized with the aid of existing concepts" [15]. "Without explicitly abolishing misconceptions it is not possible to come up with sustainable scientific concepts" [16]. "Lessons should not merely proceed from ignorance to knowledge but should rather have one set of knowledge replace another. Chemical education should offer a bridge between students' preconcepts and today's scientific concepts" [17].

These statements make it quite obvious that teachers should not assume students to enter their classroom with no knowledge or ideas whatsoever. Lessons, which take not into account that students have existing concepts, usually push them to barely following the lecture until the next quiz or exam. After that, newly acquired information will gradually be forgotten: students tend to return to their old and trusted concepts.

Nowadays, teachers and pedagogy experts agree that one should be aware of students' ideas before the "bridge can be successfully built between the preconcepts and the scientific ones" [17]. Therefore, an important goal is to allow students to express their own preconcepts during a lesson or, in the attempt to introduce new subject matter in a lesson, to let them be aware of inconsistencies regarding their ideas and the up-to-date scientific explanation. In this way, they can be motivated to overcome these discrepancies. Only when students feel uncomfortable with their ideas, and realize that they are not making any progress with their own knowledge they will accept the teacher's information and thereby build up new cognitive structures.

For the teaching process, it is therefore important to take developmental stages of the students into account according to:

- Existing discrepancies within students' own explanations
- Inconsistencies between preconcepts and scientific concepts
- Discrepancies between preliminary and correct explanations of experimental phenomena
- Possibilities of removing misconceptions
- Possibilities of constructing acceptable explanations [18]

One should take into consideration especially that, regarding constructivist theories, it is only possible to change from preconcepts to scientific concepts if

- Individuals are given the chance to construct their own learning structures
- Each student gets the chance to actively learn by themselves
- "Conceptual growth" can occur congruent to Piaget's assimilation principle, or even
- "Conceptual change" can occur congruent to Piaget's accommodation principle [18]

Nearly every student has to undergo a big conceptual change concerning the interpretation of burning coal, candles, or alcohol. The teacher can talk and talk – they will still not convince young students to overcome their destruction theory. The only way to achieve this is by interpreting convincing experiments and structural models (see Sect. 1.2). For example, the conservation of mass should be shown by burning pieces of coal in a closed flask and weighing before and afterwards (see E1.9). Structural models of carbon and oxygen before the reaction and carbon dioxide afterwards show that atoms are only rearranged, no atom is missing, no atom is added – so the mass has to be the same. Over weeks and months the students may start to change their destruction concept in favor of the conservation of mass, however, some students will not change their old conviction at all.

Concept Cartoons. For diagnosis and challenge of misconceptions, Temechegn and Sileshi [19] developed concept cartoons (see an example in Fig. 1.10). Concerning the conservation of mass they asked: "What is the mass of the solution when 1 kg of salt is dissolved in 20 kg of water?" [19]. The correct answer "21 kg" is given by one of the students in the cartoon (see Fig. 1.10), but three other answers are also given, especially the answer "20 kg" for those who are thinking: "the salt is gone and the water tastes salty – but weighs 20 kg as before." Students can discuss all four answers and the teacher may diagnose the different kinds of misconceptions in their class. Since the question is "what do you think" even more than four answers can appear. After some hours of instruction with experiments and structural models, students may have a second look at that concept cartoon to compare the just learnt scientific answer with the other misconceptions – they can even tell what is wrong with those alternatives. Through discussions of concept cartoons both teaching processes are possible: diagnosis *and* challenge of misconceptions. For other topics, some more cartoons are shown at the end of this chapter.

Learning Doctor. Taber developed the picture of a "Learning Doctor" as a means of discovering individual misconceptions and a suitably related science lesson regarding conceptual growth or conceptual change [20]. "A useful metaphor here might be to see part of the role of a teacher as being a learning doctor (a) diagnose the particular cause of the failure-to-learn; and (b) use this information to prescribe appropriate action, designed to bring about the desired learning. . . Two aspects of the teacher-as-learning-doctor comparison may be useful. Firstly, just like a medical doctor, the learning doctor should use diagnostic tests as tools to guide action. Secondly, just like medical doctors, teachers are 'professionals' in the genuine sense of the term. Like medical doctors, learning doctors are in practice (the 'clinic' is the classroom or teaching laboratory). Just as medical doctors find that many patients are not textbook cases and do not respond to treatment in the way the books suggest, many learners have idiosyncrasies that require individual treatment" [20].

Including misconceptions into lectures. In a project in progress, Barke and Oetken intend to diagnose preconcepts and school-made misconceptions [21], but in addition they will integrate them into lectures to develop sustainable understanding of chemistry. For 20–30 years educators have been observing nearly the same misconceptions of students, and that corresponding lectures at school are not changing much. Hence, being convinced that preconcepts and school-made misconceptions

Fig. 1.10 Concept cartoon concerning the conservation of mass [19]

have to be discussed in chemistry lectures, there are two hypotheses to influence instruction: (1) one can discuss the misconceptions first and come up with the scientific explanation afterwards, (2) one teaches the scientific concept first and afterwards students compare it with their own or other misconceptions from the literature.

Petermann, et al. [22] use the first hypothesis for their study concerning the well-known preconcept of combustion: "Something is going into the air,. . . some things are going away." In their lectures they showed the burning of charcoal and discussed alternative conceptions like: "charcoal is gone, some ashes remain." Afterwards they used the idea of a cognitive conflict: little pieces of charcoal are deposited in a big round flask, the air is substituted by oxygen, the flask is tightly closed and the whole assembly is weighed using an analytical balance. Pressing the stopper on the flask and heating the area of the charcoal, the pieces ignite and burn until no charcoal remains. The whole contents are weighed again, and the scales present the same mass as before.

Working with this cognitive conflict the students discover that there must be a reaction of carbon with oxygen to form another invisible gas. After testing this gas using the well-known lime water test one can conclude: the gas is carbon dioxide.

Presenting misconceptions first and teaching the scientific concept afterwards can enable students to compare and investigate by themselves what is wrong with statements like "some things are going away" or "combustion destroys matter, mass is going to be less than before." Integrating preconcepts in lessons in this way will improve sustainable understanding of chemistry, and by comparing misconceptions with the scientific concept students will internalize the concept of combustion. More results in line with these hypotheses will be published in the future.

Barke and Doerfler [23] planned lectures concerning the subject acids, bases, and neutralization. Because of the known misconceptions (see concept cartoons at the end of this chapter) experiments and structural models are involved to avoid most misconceptions. Instead of taking the usual equation "$HCl + NaOH \rightarrow NaCl + H_2O$" for the reaction, $H^+(aq)$ ions for acidic solutions and $OH^-(aq)$ ions for basic solutions were introduced, the ionic equation for the *formation of water molecules* was explained: "$H^+(aq)$ ions + $OH^-(aq)$ ions $\rightarrow H_2O(l)$ molecules." Later students were told that some students are thinking of "HCl molecules" in hydrochloric acid – the students rejected this statement and corrected it by pointing out the ions. With regard to the neutralization other students are thinking of a "*formation of salt*" because "NaCl is a product of this neutralization." Students discussed this idea with the result that no solid salt is formed in the neutralization. $Na^+(aq)$ ions and $Cl^-(aq)$ ions do not react, but are left behind by the neutralization reaction. Ions like these are usually called "spectator ions."

So students were taught the scientific idea of the new topic first, and afterwards confronted with well-known misconceptions. By comparing the scientific idea and the presented misconceptions the students could intensify the recently gained scientific concept. Preliminary data show that the consideration of misconceptions in class is successful in understanding chemistry and preventing misconceptions. More empirical research investigating whether this method is the most sustainable strategy for teaching and learning will be forthcoming.

Teaching structure of matter. With regard to teaching ions and ionic bonding, Barke, Strehle, and Roelleke [24] evaluated lectures for the introduction of "atoms and ions as basic particles of matter" on the basis of Dalton's atomic model (see Fig. 10.4): scientific ideas concerning chemical structures of metal and salt crystals are discussed. Following this way of instruction all questions regarding chemical bonding are reduced to nondirectional electrical forces surrounding every atom or ion – no electrons or electron clouds are involved at this time. However, the structure of elements and compounds can be discussed because 3D models or model drawings are possible on the base of Dalton's atomic model (see Fig. 1.11).

In the first 2 years of teaching chemistry only the structure of matter should be regarded (see Chap. 10) – the detailed questions according to chemical bonding should be answered later after the introduction of the nucleus-shell model of the atom or the ion. By combining ions to salts students learn the scientific idea about their composition: cations and anions, their electrical attraction or repulsion, their arrangement in ionic structures. Through this strategy of combining ions and using ion symbols most of the related misconceptions, which can be found worldwide, can be prevented [21]!

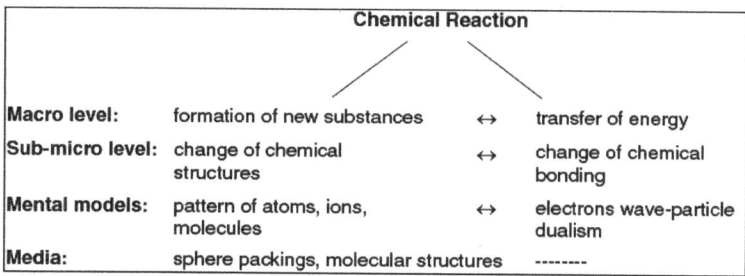

Fig. 1.11 Chemical structure and bonding with regard to the chemical reaction [21]

Tetrahedral-ZPD metaphor. Last but not least Sileshi [25] formulated a "Tetrahedral-ZPD" chemistry education metaphor as another framework to prevent students' misconceptions (see Fig. 1.12). This metaphor rehybridizes the very powerful 3D-tetrahedral chemistry education concept proposed by Johnstone [26] and Mahaffy [27]: macroscopic, molecular, representational level, and human element. With the idea of the "Zone of Proximal Development (ZPD)" of social constructivist Vygotsky [28], ZPD should describe "the distance between the actual development level as determined by independent problem solving and the level of potential development as determined through problem solving under adult guidance or in collaboration with more capable peers" [28].

The basic elements of this metaphor are what Shulman [29] has labeled "Pedagogical Content Knowledge (PCK)" integrated with contextual and research knowledge: "Pedagogy-Content-Context-Research Knowledge (PCCRK). Content knowledge refers to one's understanding of the subject matter, at macro-micro-representational levels; and pedagogical knowledge refers to one's understanding of teaching-learning processes; contextual knowledge refers to establishing the subject matter within significant social-technological-political issues; and research knowledge refers to knowledge of 'what is learned by student?', that is, findings and recommendations of the alternative conceptions research of particular topics in chemistry" [28].

Sileshi further conducted an empirical study to evaluate the effects of the Tetrahedral-ZPD metaphor on students' conceptual change (see Fig. 1.12). Knowing by own experiences that high school students in Ethiopia are mostly memorizing chemical equations without sufficient understanding, that they are not used to think in models or to develop mental models according to the structure of matter, new teaching material and worksheets for the application of the particle model of matter and Dalton's atomic model were prepared [25].

In pilot studies lasting for 6 weeks, the research was carried out with an experimental-control group design: pretests and posttests were used to collect data before and after the intervention. First results from the posttests indicated that the students in the experimental group, taught with the new teaching material concerning the structure of matter, show significantly higher achievements compared with the students in the control group: students' misconceptions in the experimental group after they were taught using the new teaching material are less than in the control group [25].

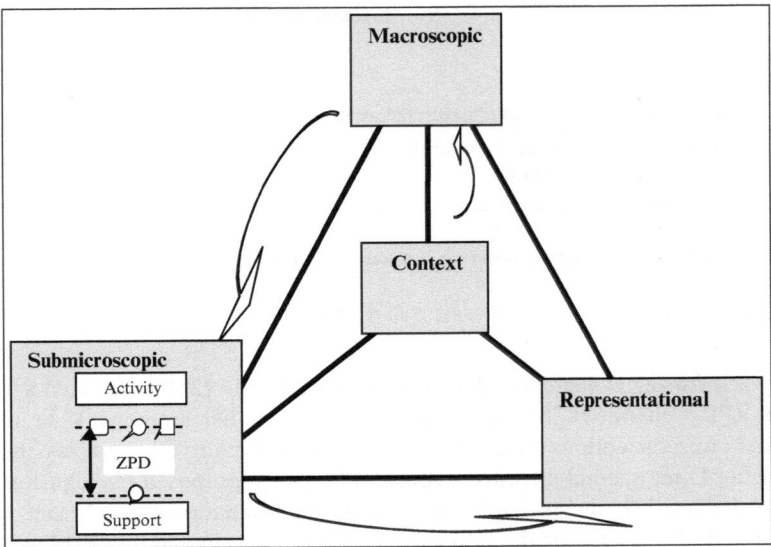

Fig. 1.12 Tetrahedral-ZPD chemistry education metaphor by Sileshi [25]

1.4 Human Element: Students' Conceptions and Everyday Language

New ideas do not survive for all time, but might get altered soon. Theories, which children build up in their cognitive structure over a couple of years are more deeply held than new theories, which were only taught during some weeks or even in a few lessons at school. Therefore, new ideas have to be used again and again during teaching situations to deeply embed them in the students' thinking.

It should be clear that discussions with friends and family can irritate the student with their new concepts: everyday language remains opposite to these new concepts. Students still have to deal with everyday comments like "candles are away after burning," "stains are being removed," or "electrical current is being consumed." Students should be motivated to start reflecting on these comments and offer their ideas in discussions with friends and family: the candle reacts with oxygen of air and forms invisible carbon dioxide gas, stains are dissolved in petrol and the fat exists still in the wrap of paper after petrol is evaporated, electric energy is transformed into thermal energy (electric heater), or into light energy (light bulb), or into other kinds of energy.

These students would acquire a competence, which also supports much-desired critical thinking skills. Such expertise could have a positive influence on our society and more scientific issues would be described in a scientifically adequate way and passed on by the students.

Also the influence of media on new conceptions exists. There are radio and TV commercials which sometimes convey diffuse conceptions for scientific phenomena and take students away from their new acquired scientific ideas. Objective consumer guidance can take up positive aspects and support the new conceptions. In this sense, Becker subjects consumer questions in a TV show to a scientific reflection [30].

Problems and Exercises

P1.1. Some students' misconceptions are also conceptions from scientists of past centuries. Give three examples and show parallels between historic conceptions and preconceptions of students today. Explain the scientific bases of the given examples.

P1.2. In explanations of natural processes or laboratory phenomena, children often reason in a "magical-animistic" way. Name three examples for such reasoning. Explain the scientific bases and suggest corrected explanations that are suitable for children.

P1.3. Chemical compounds are often described by sentences like "they contain the elements" or they "consist of elements" ("water consists of hydrogen and oxygen"). Which problems hide behind such comments? Suggest expressions that are correct from your point of view. Consider Dalton's atomic model and propose correct expressions from that point of view as well.

P1.4. Concerning the structure of matter terms like "continuum" and "discontinuum" on one hand, "preformed and nonpreformed particles" or the "horror vacui" on the other hand are used in chemical education. Explain these discussions. Which teaching ideas, which experiments and models do you suggest for teaching these issues?

P1.5. Experiments are convincing instruments to make students realize their misconceptions and to motivate them to reduce or to change their misconceptions in favor of scientific ideas. Describe three examples of misconceptions and the experimental procedure to outline the reduction of these mistakes.

Experiments

E1.1. Torricelli's Experiment on the Existence of the Vacuum

Problem: Air pressure is neither palpable nor concrete for scientists of past centuries or today's students, even when the weather forecast gives the current air pressure in millibar or hectopascal every day. Torricelli's historic experiment is a good opportunity to illustrate the balance of a 20-km-air column and a 760-mm-mercury column. On the basis of this experiment students are able to understand a mercury barometer, and so the air pressure.

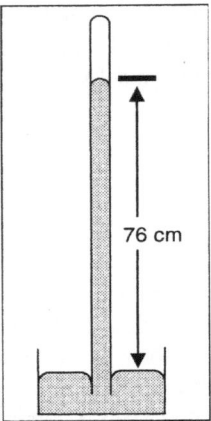

Material: Safety bowl, small crystallizing dish, glass tube (90 cm, sealed on one end), pipette, small funnel (to fit into the opening of the glass tube), folding ruler; mercury (T).

Procedure: The crystallizing dish in the safety bowl is to be filled with mercury up to 3 cm. The glass tube – over the safety bowl – is to be totally filled with mercury with the help of the funnel and the pipette, sealed with the index finger and opened under mercury in the crystallizing dish. The height of the metal column is to be measured with the ruler. The glass tube is to be put into an inclined position and back into an upright position.

Observation: The mercury column sinks – depending on current air pressure – to a height of 750–770 mm. In the inclined position the glass tube is totally filled with mercury, the mercury column sinks to the metered value again in the upright position.

Disposal: The mercury is to be put back into the container. If part of the metal remains in the safety bowl it is to be picked up with the mercury tongs. The glass tube is to be sealed with a stopper and stored in the cabinets for further experiments.

E1.2. Reaction of Copper Oxide and Hydrogen

Problem: This experiment exemplarily shows that elements can be obtained from a compound: they are not "gone away" after their reaction. Redish-brown, shiny copper can be obtained from the reaction of black copper oxide with hydrogen – although copper cannot be seen in the black substance. It is possible to discuss "bearer of properties" versus "new substances." Simultaneous oxidation and reduction can be illustrated as well as redox reactions in terms of oxygen transfer.

Material: Combustion tube with plug and outlet tube, porcelain ship, test tube; black copper oxide (Xn), hydrogen (F^+), cobalt chloride paper.

Procedure: Hydrogen is to be passed through the combustion tube, which contains a porcelain ship filled with copper oxide. The hydrogen has to be ignited on the outlet tube after a negative test of hydrogen–oxygen mixtures. The copper oxide is to be heated with a burner. As soon as the reaction sets in, recognizable by the formation of metallic copper, the burner can be switched off. After the reaction the tube has to be cooled off in the hydrogen stream – otherwise the copper reacts back to copper oxide. Then the hydrogen stream can be stopped.

Observation: By the glowing substance red brown copper appears, drops of a liquid can be observed in the outlet tube; these drops can be identified as water with cobalt chloride paper.

Disposal: The copper can be used for further experiments or it is being oxidized in air to reuse copper oxide for the next experiment.

E1.3. **Mass Change During Acetone Evaporation**

Problem: Students often understand the evaporation as a disappearance of the substance. The evaporation and the well-known mass decrease can be observed on a balance in a first step, in a second step the vapor can be collected in a gas syringe to show that a gaseous substance remains, that acetone only changes from the liquid state of matter to the gaseous state.

Material: Analytical balance with display, watch glass, Erlenmeyer flask with glass beads and drilled stopper, gas syringe; acetone (F) or ether (F⁺).

Procedure: Put some drops of acetone or ether on the watch glass and weigh it on the balance. Wait 1 min. Put some drops into the Erlenmeyer flask with glass beads, attach the gas syringe, shake the Erlenmeyer flask, and observe the moving piston.

Observation: The balance shows a decreasing mass until the liquid has evaporated. The gas syringe fills bit by bit until the gas volume is constant.

E1.4. **Dissolution of Alkaline Metals in Water**

Problem: The popular reaction of lithium or sodium and water leads students to the assumption that the metals "disappear irretrievably." One has to show to students that the metals and water react to new substances: on the one hand a gas is formed (hydrogen) and acid–base indicators switch their colors indicating an alkaline reaction. On the other hand white sodium hydroxide or lithium hydroxide can be obtained after the evaporation of water from the solution. From these white solids one could obtain the pure metals by electrolysis of the melts.

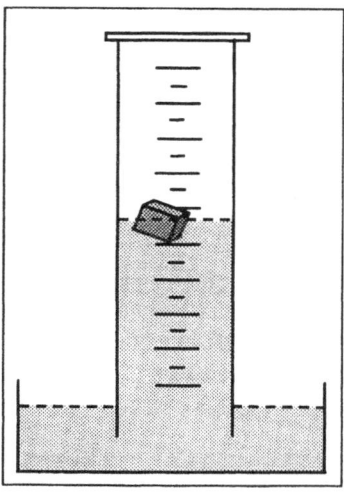

Material: Big glass bowl, plain cylinder with cover glass, test tubes, beaker, tweezers, knife, burner, filter paper; sodium (F/C), lithium (F/C), phenolphthalein solution in ethanol (F), universal indicator solution (F).

Procedure: (a) Fill the glass bowl half-full with water and place it on the overhead projector. Put a piece of sodium on the water surface and observe the sodium ball. Repeat the experiment a couple of times.

(b) Fill the cylinder with water and open it under water with help of the cover glass. Taking tweezers push a piece of lithium (not sodium!) into the water under the cylinder and observe it. After the reaction invert the plain cylinder and ignite the formed gas.

(c) Test both solutions in a test tube with each indicator solution. Let the water from a part of the solution evaporate in the beaker.

Observation: In the bowl, the moving sodium piece, a fizzing sound, and gas formation can be observed during the reaction. In the cylinder the colorless gas is collected and ignited: a red flame can be observed for a short time. The indicators change their colors: phenolphthalein from colorless to red, universal indicator from red to blue. After evaporating the water solid white substance remains in the beaker.

Disposal: Put remains of sodium or lithium metal in ethanol and wait for their dissolution. Dilute all solutions so that they can be disposed off down the drain.

E1.5. Comparison of Petrol and a Solution of Fat in Petrol

Problem: Students think that the fat disappears when grease spots are being removed from clothes by petrol: "the spots are gone irretrievable." Students have to understand that spot removing is nothing else but the dissolution of fat in petrol and the evaporation afterwards, after the evaporation the fat remains in the cloth.

Material: Petrol ether (F/XN/N), solution of olive oil in petrol ether, filter paper.

Procedure: Put some drops of petrol ether on one filter paper and some drops of the fat solution on another filter paper at the same time. Observe both papers for a minute.

Observation: The spot of the pure dissolvent shrinks and evaporates residue free, the spot of the fat solution also shrinks, but the fat remains.

E1.6. Burning Metals on a Balance

Problem: Due to everyday experience students believe that substances loose mass and become lighter during combustion – like in the familiar combustion of alcohol, wood, paper, or candles. Metal combustion shows that due to the bound part of oxygen in the solid metal oxide instead of a mass decrease there actually happens a mass increase. This can be demonstrated in experiments on a balance. The masses of candles during combustion will be observed later (E1.7).

Material: Beam balance, digital balance, porcelain crucible with cover, tripod and clay triangle, crucible tongs; iron wool, magnesium ribbon (F).

Procedure: (a) Iron wool that hangs on one side of the balanced beam is to be ignited; blow gently if necessary to accelerate the reaction and to see the metal glowing.

(b) A strip of 10 cm magnesium ribbon is to be weighed in the porcelain crucible. The crucible is to be heated with a roaring flame until the magnesium ignites. To catch the white smoke of magnesium oxide the crucible is to be covered and opened and covered again several times during the reaction. The cooled crucible is to be weighed again.

Observation: The balance arm with the red-glowing iron wool sinks: formation of black iron oxide. The magnesium forms a white combustion product, while it glows brightly: magnesium oxide. The balance shows a bigger mass after the combustion. A green substance appears, when the white combustion product is being broken up: magnesium nitride.

E1.7. Burning Candle on a Balance

Problem: Students are able to understand the mass increase during the combustion of metals and the formation of solid metal oxides. But they will argue that this does not apply to spirits, paper, or candles. To convince students that the mass increase also applies to these substances, a candle can be burnt on a balance and the invisible gaseous combustion products caught with special chemicals. Carbon dioxide and water vapor can be bound with soda lime, a mixture of sodium hydroxide and calcium oxide. The combustion gases are being absorbed in the device for this experiment (see picture). Due to the bound oxygen in the invisible gases, their mass is bigger than the burnt part of the candle before.

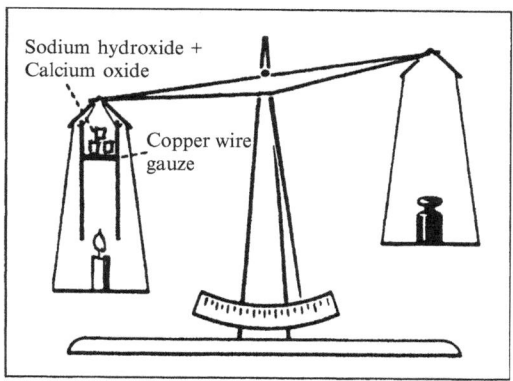

Material: Candle or tea light, beam balance, jar with soda lime (C) or sodium hydroxide (C).

Procedure: The upper part of the jar is to be filled loosely with soda lime so that the formed gas can flow through the pieces. The pans are to be balanced and the candle is to be lit. If smoke development occurs in the jar (the absorbent is packed too tightly!), the experiment has to be started again.

Observation: The balance pan with the burning candle sinks.

E1.8. Conservation of Matter by Burning Iron Wool and Matches

Problem: The previous experiments have shown the increasing of mass due to bonding of oxygen in combustion products – for these experiments an *open apparatus* was chosen. If one takes a *closed apparatus*, weighs all educts first and later all reaction products, one will see that masses are not changing, but are conserved. Lomonossov and Lavoisier were the first to have found this law of conservation of mass in the eighteenth century. Dalton explained it in 1808 by combining the idea of elements with the idea of atoms, and stating that no atom can be created nor destroyed, instead atoms are rearranged in every chemical reaction.

Material: Digital scales, test tubes, balloons, burner; iron wool, matches.

Procedure: (a) A test tube is filled with iron wool, closed with a balloon and weighed on the balance. After heating the test tube with the burner and observing the iron wool, the test tube is weighed again once it cooled down to room temperature.

(b) A test tube is filled with five to eight matches, closed with a balloon and weighed on the balance. After heating the test tube close to the match heads with the burner and observing the matches, the test tube is weighed again after cooling it down to room temperature.

Observation: The balloon gets bigger and bigger; the iron wool glows red and black iron oxide is formed. The match heads are burning with a bright flame. In both cases the mass is the same before and after the reaction.

E1.9. Conservation of Matter by Carbon Combustion

Problem: The experiments with iron wool and matches show that combustion products like solid iron oxide or the wooden remnants of the matches remain. In the case of carbon or charcoal nothing will remain and young students may think: "if the carbon is totally gone there is nothing to be weighed, the law of conservation of mass cannot be valid." So it seems very important that the gaseous combustion product carbon dioxide must be shown and weighed to convince students that the law is valid even in this case. The next experiment (E1.10) will also show that gases have specific densities, that the density of carbon dioxide is higher than that of air.

Material: Digital scales, 2 L round bottom flask, balloon, burner, gas syringe with 20 cm glass tube, test tube; small pieces of charcoal, oxygen, lime water.

Procedure: The flask is filled with oxygen and 3–5 small pieces of charcoal, then closed with a balloon and weighed exactly. With the burner the pieces of carbon are heated from outside until they ignite, then the flask is turned around and around to show the glowing carbon pieces. After complete combustion the flask has to be cooled down and weighed again. The flask can be opened, with the syringe 100 mL of the gas is taken out of the flask and bubbled through 1–2 mL of lime water in the test tube.

Observation: The balloon gets bigger and bigger, and the carbon pieces glow very bright and disappear totally. The gas produces a milky precipitation in lime water: carbon dioxide.

E1.10. Density of Air and Carbon Dioxide

Problem: Students are well aware of the existence of air around them, but not aware of air as a space-filling substance with special properties and measurable density. This density should be measured and compared to the density of carbon dioxide. The density of air can also be discussed in the context of air pressure (see E1.1).

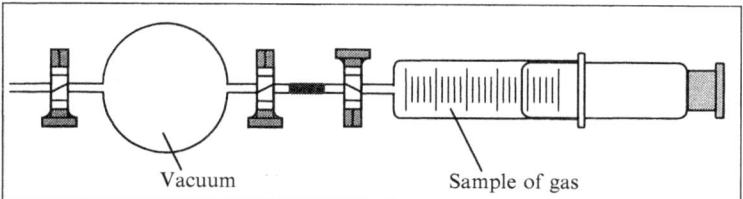

Vacuum   Sample of gas

Material: Digital balance, 250-mL glass ball with valve, gas syringe, water jet pump, hose; air, carbon dioxide.

Procedure: The gas syringe is to be filled with 100 ml of air exactly and closed. The glass ball with valve is to be evacuated by the water jet pump and weighed exactly. The gas syringe is to be attached to the glass ball, and the air inside the gas syringe is transferred into it by opening the valve. The glass ball is to be weighed again. The density of air can be derived from the mass difference and the given volume. The experiment is to be repeated with carbon dioxide, both measured densities are compared.

Observation: 100 mL of air weigh 0.13 g; 100 mL of carbon dioxide weigh 0.20 g. The densities are 1.3 g/L for air (tabular value: 1.29) and 2.0 g/L for carbon dioxide (1.97).

Note: The experiment can also be run with the help of an empty plastic bottle with plug and valve. The bottle is to be weighed exactly with plug and valve. 100 mL of gas are to be pumped into the bottle with the help of the gas syringe. The bottle is to be weighed again.

E1.11. **Properties of Hydrogen and Other Colorless Gases**

Problem: Students usually call colorless gases "air," in everyday language they do not differentiate between air, oxygen, nitrogen, or carbon dioxide. Some colorless gases should therefore be introduced with some easy detection reactions, which show the different properties of these gases. Emphasis should be laid on hydrogen reactions, since the properties of hydrogen and the safety rules are especially new to students.

Material: Five gas cylinders with cover glass, wooden splint, balloon, combustion spoon, glass tube, beaker, empty tin can with a concentric hole (diameter: 1 mm); air, oxygen, nitrogen, carbon dioxide, butane (F^+), lime water (Xi), hydrogen (F^+).

Procedure: The first mentioned five gases are being filled into the cylinders by air displacement, and the gas jars are to be covered and labeled. A burning splint is to be held in every cylinder first, later only a glowing splint. For the distinction of nitrogen and carbon dioxide, a small amount of lime water is added to both jars and shaken. The jar with butane can be taken and poured out in the presence of a burning splint.

Observation: In air the burning splint is not changing its flame. In oxygen the splint burns very bright and even a glowing splint ignites (test for oxygen). The flame and the glowing splint go out in nitrogen and also in carbon dioxide, but in carbon dioxide a white milky substance appears from the colorless lime water (test for carbon dioxide). Butane ignites and burns with a calmly yellow flame. When pouring butane out of the cylinder it ignites and a big flame appears.

Procedure of Hydrogen Experiments

(a) A balloon is to be filled with hydrogen, sealed and released
(b) A candle is fixed on a 1-m-long glass tube and ignited, and the burning candle is to be put near the balloon until the reaction sets in (caution: bang)
(c) Hydrogen streaming out of the steel container is to be ignited at a glass tube: a nearly invisible flame appears
(d) A dry beaker with the opening down is to be held above the flame, the formed water vapor condenses and a layer of water can be seen inside the beaker

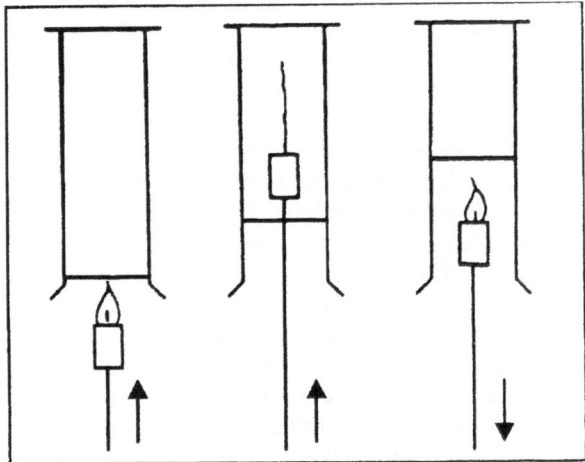

(e) Hydrogen is to be filled in an upside-down gas jar by air displacement. A burning candle, which is fixed on a glass tube, is to be held inside the gas jar, the candle is to be taken out and held inside again (see picture)
(f) A gas jar is to be filled with hydrogen upside-down and put on another gas jar of the same size which contains air. Both gases are being mixed by turning the gas jars. The gas jars are to be separated with cover glasses, the gas mixtures to be ignited with a burning splint (caution: bang)
(g) The empty tin can with a concentric hole is to be put on a flat surface with the opening on the bottom and to be filled with hydrogen by air displacement. The gas is to be ignited at the hole and observed exactly (caution: very loud bang after about 20 s)

Observation: (a) The balloon rises, (b) the balloon burns with a loud bang, (c) the pure hydrogen burns calmly, (d) water vapor condenses in the beaker to form liquid water, (e) the candle goes out in the jar, but ignites again, when it is taken out of the jar, (f) the mixture of hydrogen and air (oxyhydrogen gas) burns very fast and forms water steam with a loud bang, (g) first hydrogen burns calmly (a sheet of paper can be put above the hole: it ignites), a buzzing sound can be heard after about 20 s and shortly after a loud bang (prepare students for that bang!).

E1.12. Composition of Air

Problem: The students use the terms "good air" and "stale air" in their everyday language, but they do not think about the oxygen content. For this reason it is important to run experiments demonstrating the composition of air. It has to be explained to the students why a pure metal or phosphorous is being used: both are reacting with oxygen to form solid oxides, they take the oxygen out of the air and release other gases of the air mixture.

Material: Two 100 mL gas syringes, combustion tube, glass bowl, small gas cylinder with cover glass, splint, bell jar, combustion spoon with plug, ruler; iron wool, red phosphorus (F).

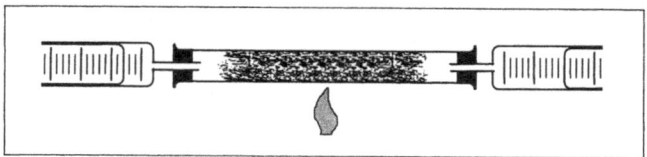

Procedure: (a) A combustion apparatus is to be built (see picture). The trapped air with a volume of 100 mL is to be lead over the heated iron wool a couple of times. The volume of the remaining gas is to be measured. The remaining gas is to be collected in the small gas jar pneumatically and tested with the burning splint.

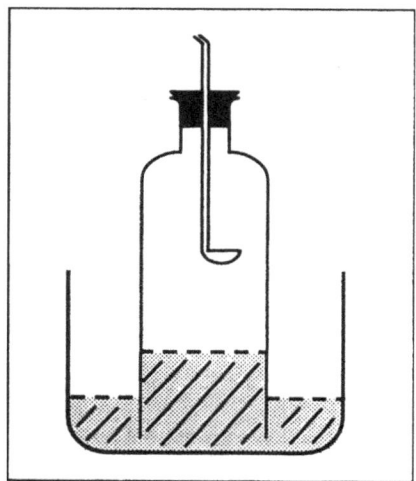

(b) A bell jar stands in the water of the glass bowl (picture). A spatula tip of phosphorus is to be put into the combustion spoon and ignited; the spoon is to be fixed inside the bell jar. The rising water level can be observed in the gas jar, and the remaining part of gas can be estimated.

Observation: (a) Iron wool glows and forms a black product, the gas volume decreases to 80 mL, and the remaining gas smothers a burning splint.

(b) Phosphorus burns for some time, a white smoke forms, the flame goes out, the water level in the bell jar rises, and the volume of the remaining gas is about 80% from before.

E1.13. **Carbon from Carbon Dioxide**

Problem: Students may accept that metals can be obtained from the corresponding metal oxides. But they cannot imagine that carbon in form of a black solid can be obtained from the colorless gas carbon dioxide. The reaction of carbon dioxide with burning magnesium should be shown to convince students.

Material: Glass cylinder with cover, crucible tongs, burner; magnesium ribbon (F), carbon dioxide, sand.

Procedure: Some sand is to be put on the bottom of the gas jar for protection, and it is filled with carbon dioxide by air displacement and to be covered. With the crucible tongs a 10-cm-long ribbon of magnesium is to be ignited in the flame of the burner and put deep into the gas jar.

Observation: The flame does not go out and it burns with a crackling sound. Black spots can be observed on the inside of the gas jar and they prove to be soot when wiping them off with a finger.

E1.14. **Condensation of Butane under Pressure**

Problem: Students know butane lighters and the term "liquid gas," maybe they already observed the fluid and the gaseous butane phases in a transparent lighter. Despite such observations the term "liquid gas" remains as a misconception and one has to point out that an equilibrium exists between liquid and gaseous butane in a butane lighter or in a cartridge of camping gas. This can be demonstrated with the following experiment. The experiment also shows – besides some properties of butane gas – that the movement of butane particles must take a bigger volume than in the liquid, that there is no "butane," "air," or "other material" between the particles in a portion of gas (see also the next experiment E1.15).

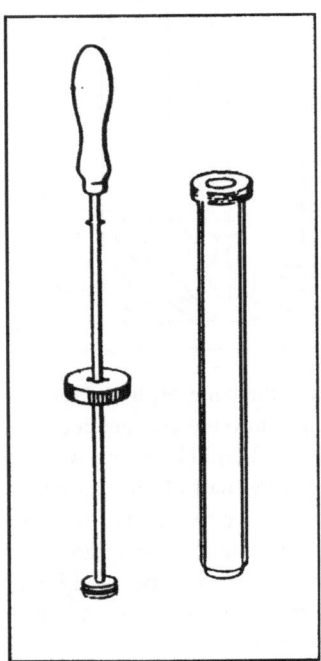

Material: Gas liquefaction pump, camping gas cartridge, hose; butane (F^+) from the camping gas cartridge.

Procedure: The pump is to be opened and filled totally with butane by air displacement (get the hose deep into the cylinder and take it out slowly). The piston is being attached, pushed firmly into the cylinder of the pump and locked. The locking mechanism is being loosened and the liquid butane is being observed. This process can be repeated many times.

Observation: A big drop of liquid butane forms, when the gas is being compressed, the gas volume decreases to one-tenth. After the locking mechanism is being loosened, the piston moves out of the cylinder to the same extent as before, the drop disappears and cools down the bottom of the cylinder: the gas volume is the same as at the beginning.

E1.15. **Evaporation and Condensation of Ethanol**

Problem: Young students accept that matter consists of smallest particles: ethanol particles exist in ethanol, water particles exist in water. When they have to interpret small particles in a gas and are asked about the space between the particles, they mostly think of "air" or "other material." To demonstrate that there is nothing between those particles, ethanol is evaporated first to show the big volume compared with the liquid volume and then it is cooled down and condensed to the same very little volume as before: no air or other material is observed. The bigger volume of ethanol vapor should be interpreted by the higher movement of the ethanol particles and bigger distances between the particles.

Material: Beaker (2 L), electric heater, balloon, pipette, tweezers; ethanol.

Procedure: 1 L of water is heated in the beaker until it keeps boiling (at 100°C at normal pressure). 1–2 mL of ethanol are given into the balloon, closed by a knot and thrown into the boiling water (ethanol boils at 79°C). After some time the balloon is taken out of the water and thrown again into the boiling water.

Observation: The balloon increases to a volume of about 200–300 mL, after cooling down the same small volume of 1–2 mL as before is to be observed. Throwing the balloon again into the boiling water, the same big volume appears as before.

Some More Concept Cartoons for Diagnosis and Challenge of Misconceptions

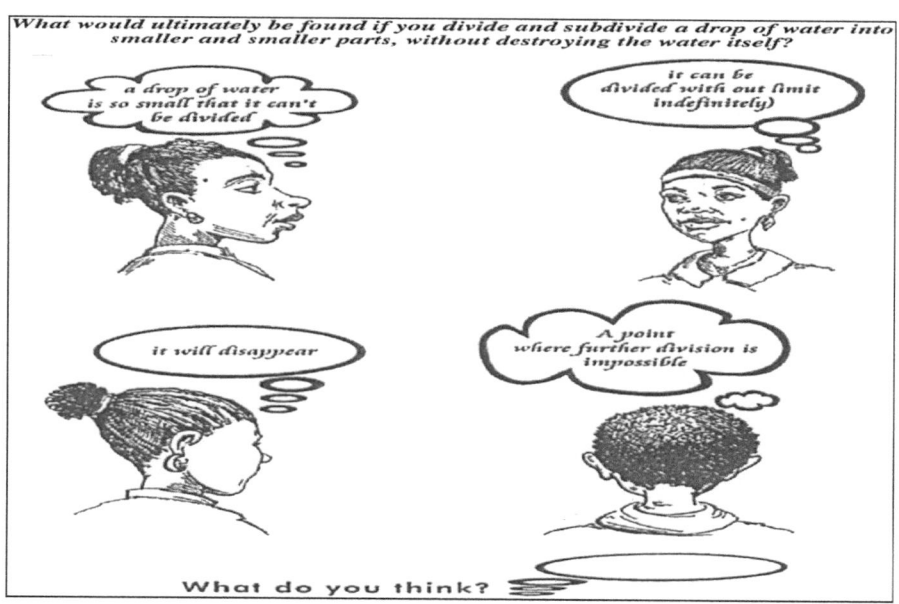

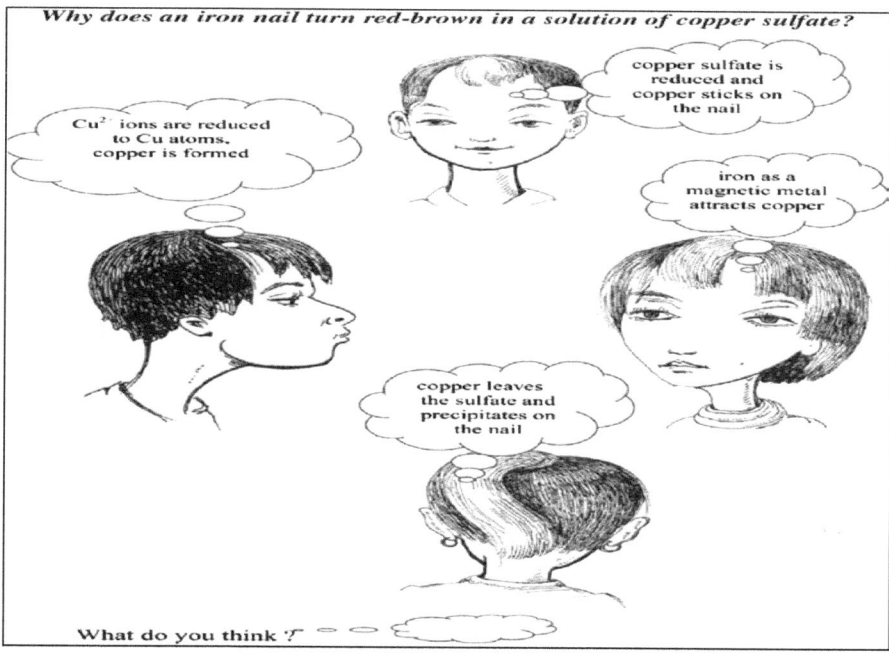

References

1. Strube W (1976) Der historische Weg der Chemie. Grundstoffindustrie, Leipzig
2. Reuber R, Wellens H, Gruss K (1972) Chemikon – Chemie in Übersichten. Umschau, Frankfurt
3. Lockemann G (1950) Geschichte der Chemie. de Gruyter, Berlin
4. Bugge G (1955) Das Buch der Grossen Chemiker. Band 1. Chemie, Weinheim
5. Dijksterhuis FJ (1956) Die Mechanisierung des Weltbildes. Springer, Berlin
6. Lasswitz K (1890) Geschichte der Atomistik. Bände 1 und 2. Voss, Hamburg
7. Barke H-D (1995) Strukturorientierter Chemieunterricht und Teilchenverknüpfungsregeln. Chemie in der Schule 42:49
8. Driver R (1985) Children's ideas in science. University Press, Philadelphia
9. Pfundt H (1975) Ursprüngliche Vorstellungen der Schüler für chemische Vorgänge. MNU 28:157
10. Münch R et al (1982) Luft und Gewicht. NiU-P/C 30:429
11. Weerda J (1981) Zur Entwicklung des Gasbegriffs beim Kinde. NiU-P/C 29:90
12. Pfundt H (1981) Das atom – Letztes Teilungsstück oder Erster Aufbaustein. Chimdid 7:75
13. Novick S, Nussbaum J (1981) Pupils' understanding of the particulate nature of matter. Sci Ed 65:187
14. Barke H-D (1987) Irgendwas muß doch da sein – der Horror vacui in den Schülervorstellungen vom Aufbau der Gase. GDCh-Mitteilungsblatt Nr. 10, Frankfurt
15. Ausubel DP (1974) Psychologie des Unterrichts. Weinheim, Beltz

16. Piaget J, Inhelder B (1971) Die Entwicklung des raeumlichen Denkens beim Kinde. Klett, Stuttgart
17. Pfundt H (1975) Urspruengliche Erklaerungen der Schueler für chemische Vorgaenge. MNU 28:157
18. Duit R (1996) Lernen als Konzeptwechsel im naturwissenschaftlichen Unterricht. In: Lernen in den Naturwissenschaften. Kiel, IPN
19. Temechegn E, Sileshi Y (2004) Concept cartoons as a strategy in learning, teaching and assessment in chemistry. Addis Ababa, Ethiopia, University Print
20. Taber K (2002) Chemical misconceptions – prevention, diagnosis and cure, vol 1, 2. Royal Society of Chemistry, London
21. Barke H-D (2009) Misconceptions in chemistry – addressing perceptions in chemical education. Springer, Heidelberg, New York
22. Petermann K, Friedrich J, Oetken M (2008) Das an Schuelervorstellungen orientierte Unterrichts-verfahren. Chemkon 15:110
23. Barke H-D, Doerfler T (2009) Das an Schuelervorstellungen orientierte Unterrichtsverfahren: beispiel neutralisation. Chemkon 16:141
24. Barke H-D, Strehle N, Roelleke R (2007) Das Ion im Chemieunterricht – noch Vorstellungen von gestern ? MNU 60:366
25. Sileshi Y (2007) The particulate nature of matter. Diagnosis of misconceptions and their remedy in chemical education. Schueling, Muenster
26. Johnstone AH (1997) Chemistry teaching – science or alchemy? J Chem Educ 74:268
27. Mahaffy P (2004) The future shape of chemistry education. Chem Educ: Res Pract 5:229
28. Vygotsky LS (1978) Mind and society: the development of higher mental processes. Harvard University Press, Cambridge
29. Shulman LS (1986) Those who understand: knowledge growth in teaching. Educ Res 15:4
30. Becker H-J (1988) Verbraucherfragen im RIAS-Telefonstudio: Gegenstand fachdidaktischer Forschung? Chim did 14:69

Further Reading

Andersen B (1990) Pupils conceptions of matter and its transformation (Age 12–16). Stud Sci Educ 18:53–85
Benson DL, Wittrock MC, Baur ME (1993) Students' preconceptions of the nature of gases. J Res Sci Teach 30:587–597
Ben-Zvi R, Eylon BR, Silberstein J (1986) Is an atom of copper malleable? J Chem Educ 63:64–66
Boo HK (1998) Students' understanding of chemical bonds and the energetics of chemical reactions. J Res Sci Teach 35:569–581
Cakmakci G, Leach J, Donnelly J (2006) Students' ideas about reaction rate and its relationship with concentration or pressure. Int J Sci Educ 28:1795–1815
Carson EM, Watson JR (1999) Undergraduate students' understanding of enthalpy change. Univ Chem Educ 3:46–51, Available from http://www.rsc.org/images/Vol_3_No2_tcm18-7037.pdf
Chandrasegaran AL, Treagust DF, Mocerino M (2007) The development of a two-tier multiple-choice diagnostic instrument for evaluating secondary school students' ability to describe and explain chemical reactions using multiple levels of representation. Chem Educ Res Pract 8:293–307, http://www.rsc.org/images/Chandrasegaran%20final_tcm18-94351.pdf
Driver R, Leach J, Scott P, Wood-Robinson C (1994) Young people's understanding of science concepts: implications of cross-age studies for curriculum planning. Stud Sci Educ 24:75–100
Horton C (2001) Student preconceptions and misconceptions in chemistry. [Online] Accessed May 27, 2004 from http://daisley.net/hellevator/misconceptions/misconceptions.pdf

Ivarsson J, Schoultz J, Säljö R (2002) Map reading versus mind reading: revisiting children's understanding of the shape of the earth. In: Limón M, Mason L (eds) Reconsidering conceptual change. Issues in theory and practice. Kluwer, Dordrecht, pp 77–99

Johnson P (1998) Children's understanding of changes of state involving the gas state, Part 1: Boiling water and the particle theory. Int J Sci Educ 20(5):567–583

Johnson P (1998) Children's understanding of changes of state involving the gas state, Part 2: Evaporation and condensation below boiling point. Int J Sci Educ 20(6):695–709

Kind V (2004) Beyond appearance: students' misconceptions about chemical ideas (2nd Edn). School of Education Durham University, Durham

Nakhleh MB (1993) Are our students conceptual thinkers or algorithmic problem solvers? Identifying conceptual thinkers in chemistry. J Chem Educ 70:52–55

Nakhleh MB, Mitchell RC (1993) Conceptual learning versus problem solving: there is a difference. J Chem Educ 70:190–192

Othman J, Treagust DF, Chandrasegaran AL (2007) An investigation of the relationship between students' conceptions of the particulate nature of matter and their understanding of chemical bonding. Int J Sci Educ 30:1531–1550

Özmen H, Ayas A (2003) Students' difficulties in understanding conservation of matter in open and closed-system chemical reactions. Chem Educ Res Pract 4:279–290

Read JR (2004) Children's misconceptions and conceptual change in science education. Available from http://acell.chem.usyd.edu.au/Conceptual-Change.cfm

Sewell A (2002) Constructivism and student misconceptions: why every teacher needs to know about them. Aust Sci Teach J 48(4):24–28

Singer JE, Tal R, Wu H (2003) Students' understanding of the particulate nature of matter. School Sci Math 103(1):28–44

Sneider CI, Ohadi MM (1998) Unraveling students' misconceptions about the Earth's shape and gravity. Sci Educ 82:265–284

Stavy R (1988) Children's conception of gas. Int J Sci Educ 10(5):553–560

Stavy R (1990) Children's conception of changes in the state of matter: from liquid (or solid) to gas. J Res Sci Teach 27(3):247–266

Taber K (1997) Students understanding of ionic bonding: molecular versus electrostatic framework. School Sci Rev 78(285):85–95

Taber K (2002) Chemical misconceptions-prevention, diagnosis and eure Volume I: theoretical background. Royal Society of Chemistry, London

Taber KS (1996) Chlorine is an oxide, heat causes molecules to melt, and sodium reacts badly in chlorine: a survey of the background knowledge of one A-level chemistry class. Sch Sci Rev 78:39–48

Treagust DF (1988) Development and use of diagnostic tests to evaluate students' misconceptions in science. Int J Sci Educ 10:159–169

Zoller U (1990) Students understandings and misconceptions in College Freshman chemistry (general and organic). J Res Sci Teach 27(10):1053–1065

Chapter 2
Motivation

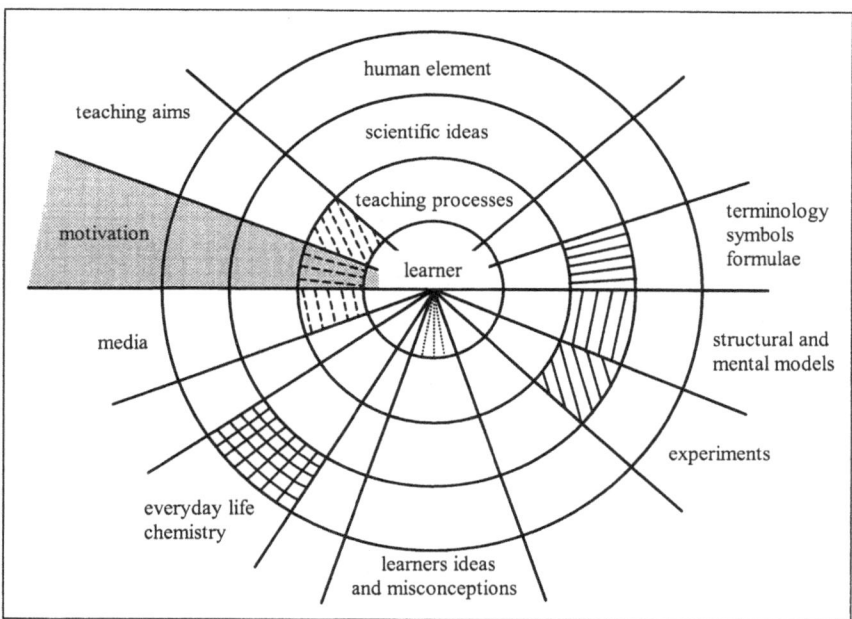

These and similar remarks are quite often heard by young people in school. In particular, due to the importance of school grades for student achievement in society, teachers are often led to threaten with the award of poorer school grades to discipline the students.

H.-D. Barke et al., *Essentials of Chemical Education*,
DOI 10.1007/978-3-642-21756-2_2, © Springer-Verlag Berlin Heidelberg 2012

Some teachers and parents think they can motivate young students with the pressures of school grades, as they see it as the only way to get them to learn. They do not see that students, based on this short lasting extrinsic motivation, only learn until the next test, for the reward from teachers or parents, to reach the target class. After that they forget everything: a desired, long-term learning does not take place!

An important task of the teacher is to consider other measures which encourage students to learn without forcing them. The creativity of teachers is not limited to prepare lessons, to motivate students over a longer period, so that intrinsic motivation will come up. The basic question of any teaching methodology in this regard is what options are possible to motivate intrinsically and inspire sustained interest in the subject.

In the natural sciences it is relatively easy to motivate through natural and laboratory phenomena and to kindle curiosity and interest in students. If the short-term interest in certain contexts comes to a longer period of interest, you lead the students "from situational to individual interest" [1]. In this way you can achieve substantive motivation for longer periods and do not need to motivate extrinsically with praise and blame, good or bad grades.

Interest will unfold easier in favor of a cognitive conflict, if it is accompanied by positive emotions: one seeks to establish consistency between affection and cognition [2]. So positive emotions are necessary to develop long-term motivation and will cause positive attitudes towards the involved subject. In this context, the chemistry teacher does not have difficulties: demonstration experiments and those done by the students themselves, structural models on display and those built by the students themselves evoke positive emotions in most cases and are well suited to produce the desired long-term intrinsic motivation. The following discussion will explain this in detail with many examples.

2.1 Learner: Development, Attitudes, and Preconcepts

The basic question in this regard of teaching chemistry is firstly to reflect on the conditions, which must be observed to build up the desired intrinsic motivation:

- State of the intellectual development of learners
- Present attitudes to chemistry or chemistry lessons
- Preconcepts on natural or laboratory phenomena

Development. According to the theory of Piaget [3], secondary school students are at the stage of concrete or formal operations of thought (see also Sect. 3.2) in terms of their cognitive skills. The adopted age limits, however, may vary significantly in this regard: researchers found that only 25% of the 16-year-old students of grade 10 achieve the stage of formal operations [3].

Measures to motivate must address such development stages. Accordingly, for young students at the stage of concrete operations special phenomena should be preferred. Some examples: students can examine the constant melting temperature of an ice–water mixture (see E2.1) or the lowering of the boiling temperature of water with decreasing pressure (see E2.2). After demonstrations or even experiments carried out by the students they are likely to be more motivated to discuss the phenomena regarding the dependence of pressure and boiling temperatures of water, about the vapor pressure curve of water. This well-known curve could be motivating also for older students at the stage of formal operations of thought, but the discussed phenomena are also important for motivating even this group of older students.

Attitudes. Motivation can only be developed if a neutral or even positive attitude exists to the appropriate school subject or the facts to be learned [4]. With a negative attitude students would not be ready to turn to the desired situation or to think about the offered facts.

As part of an empirical survey of attitudes in the early eighties, Heilbronner and Wyss [5] asked Swiss students aged 11–15 years to paint their image of chemistry. The pictures showed smoking industrial chimneys, contaminated rivers, tanks of toxic chemicals with skull symbols, and animals dying in experiments. The authors reported a very negative attitude to chemistry, given the dominance of these negative paintings and asked: "What teacher will teach chemistry successfully if students start with such a negative image of this subject?" [5] (see Chap. 8).

At the end of the nineties, Barke and Hilbing [6] repeated this study with German students in the same age group. These students were also asked to express their attitudes through painting their images of chemistry and scrutinizing them through a questionnaire. Compared with the earlier study the attitudes had improved much: beside some critical paintings of environmental problems students painted chemistry teachers doing experiments, painted explosions of hydrogen–air mixtures, painted the name chemistry in nice colors. It is possible that a number of big accidents in chemistry factories in the eighties caused the negative image of Swiss students.

The attitudes of young people are also influenced positively through nice and exciting experiments, although these may not always be used for learning chemistry. For this reason it is recommended to surprise the students from time to time with a show experiment or to let students perform it themselves. There is no long-lasting intrinsic motivation from those show experiments, but students might like their teachers more if they perform such experiments and students may be motivated by those persons.

Spectacular experiments alone are not sufficient for motivation – the most beautiful effects fizzle out quickly and are devalued in our media-saturated world through sensory overload and the effects habit quickly. Therefore, intrinsic motivation is developed more successfully if students can discover chemistry relationships through experiments by themselves. An example for inquiry learning in organic chemistry through networked experiments is shown with the phenomenological-integrative network approach (PIN concept, see Chap. 9).

Student preconcepts. As noted in Chap. 1, young people prefer their own explanations that do not necessarily coincide with scientific concepts of today: same substance and different properties ("red copper change to green copper"), weightless gases, no image concerning conservation of mass or energy, combustion and the destruction concept (substances are "unretrievable destroyed"). It was pointed out that it is favorable to involve students' original ideas and preconcepts for the first steps into chemistry instruction, and realizing the conceptual change to the scientific knowledge with convincing experiments or models (see Chap. 1).

This approach is advantageous especially for intrinsic motivation: due to the cognitive conflict between existing preconcepts and the scientific explanation the experimentally demonstrated phenomena may cause curiosity and interest. The students recognize the problem as significant for themselves and are motivated to find the solution to this problem. A number of experimental examples will be used to demonstrate this.

2.2   Teaching Processes: Opportunities to Create Intrinsic Motivation

Motivation is not to be understood as a single act that takes place at the beginning of a lesson. The teacher often tries to make a problem palatable for their students by starting with their experiences, but he uses this exclusively as a vehicle to convey content that is not insightful to the students. The teacher may ask, for example, what the students know about a rusting iron nail, but immediately turns to corrosion of metals, with standard electrochemical potentials, etc. Students will see through this procedure very quickly and will reject it – this attempt of motivation is directly counterproductive.

Successful motivation is based on other measures:

1. To design comprehensible chemistry lessons for students
2. To realize genetic learning (Wagenschein [7])
3. To create continuous references to everyday life
4. To produce cognitive conflicts in the mind of students
5. To show striking experimental effects
6. To enable students to deal with experimental or model building materials

Comprehensible lessons. Regarding chemistry classes at school you often hear the argument that students did not understand the chemistry because of working mostly with incomprehensible formulae and reaction equations, because of many calculations of special amounts of substances and the permanent use of the mole.

As long as students do not understand the actions of the teacher and cannot understand the lessons, they do not develop motivation for further cooperation: so students just work to get through the next test or they memorize the key phrases to obtain sufficient chemistry grades.

To achieve long-term motivation, it is therefore most important to go out of one's way to make chemical education in the classroom understandable: students should have the feeling of understanding the chemical facts, of learning chemistry successfully – that may motivate them to continue the work for chemistry happily. Chapters 5–7 on experiments, models, and chemical symbols will offer concrete help to design an understandable introduction to chemistry.

Genetic learning. Reading the books of Wagenschein [7] you can sense what special questions are used to enter into scientific problems, to motivate long-lasting thinking, and to stimulate successful learning. An entry into a problem should not be "too complex nor too easy," the intellectual work should be more exemplary, and the intellectual work should lead "down to elementary questions and up to complex answers" [7].

In this context, we can use an example from school physics explained in Wagenschein's own words: "A possible start into mechanics is the harmless question: Where does a stone land that is held out of the window of a high rising tower and then released? It seems trivial at first. But then the question confuses in a most captivating way, if one thinks about the curvature of the Earth and the Earth's rotation. First you hold a lag to the west for granted, then because of the earth's rotation you doubt: the rotating air is responsible for co-rotating of the stone. But why is that the fact? Why is there no constant east wind? Analogous experiences in the railway carriage with open and closed windows come up. . . These issues can cause hours of heated debate. They end with the discovery of the law of inertia and, finally, and this is a sensation, with a deviation to the east. Finally, the students really believe that the earth revolves" [7].

This *exemplaric* (starting with concrete experience), *socratic* (establishing a question–answer game), or *genetic* (developing the progress psychologically step by step) way to doubt or question the alleged self-evident, to favor the confusion of students and thereby create a productive tension characterizes the approach of Wagenschein. This approach will not only give motivating insights but also a way of gaining knowledge: "There are two very different teaching styles: whether the teacher wants to show the student only that it is what it is, or if they also want to teach how mankind could achieve all knowledge. Only the second way, the genetic way, has to do with successful exemplary learning" [7]. It should not be underestimated, that this teaching method puts high demands on teachers and students.

References to everyday life. Typical chemistry classes – especially in high schools – mostly implement a scientific structure: teachers bring this orientation from their courses at universities and in accordance with these experiences they also base their chemistry lessons on this way. Students who come to school compulsorily, and may expect to learn for their everyday life, realize that there is no relation to their environment. They are forced to learn quite formally to get the desired good grades: this approach will lead almost exclusively to extrinsic motivation.

To provide the necessary scientific structure for chemistry while also motivating intrinsically, it is useful for the teacher to integrate everyday life into lessons. Additional to the laboratory chemicals, substances and reactions from the kitchen, bathroom, garden, or garage should be taken into consideration. Chapter 8 "Everyday life and chemistry" offers many exemplary references to everyday life: they may motivate at the beginning of a lesson, or at the end of the lesson, or as repetition. Many textbooks [8] provide excursions on matters of everyday life and of environmental problems that can be studied concomitantly with the classical topics of instruction in the classroom or by students studying at home.

For some issues there is the possibility to leave the scientific way of instruction and to consistently use material from everyday life – in such cases the conditions for motivational chemistry lessons are optimal. Wanjek [9] showed with a scientific evaluation of the teaching unit "acids and bases" that food and cleaning substances as examples for acids and bases increase motivation and interest in chemistry. Particularly girls, who reported much less interest in chemistry than boys in a questionnaire prior to these lessons, increased their interest to the level of the boys following this instruction [9].

It will not always be possible to move everyday materials and references to life into the foreground, so that the structure of discipline can be developed itself. Harsch and Heimann [10] show many examples in their PIN concept how to integrate everyday life topics into a genetically growing subject classification: detection of basic organic substances in food and well-known products, relationships between test tube synthesis and biochemical metabolic processes, model experiments to understand chemical recycling including energy balances [10].

Generation of cognitive conflict. The classic way to develop intrinsic motivation according to Piaget is through "generation of a cognitive conflict and equilibration." Lind [11] formulated the same connection with the incongruity theory: the difference between a perceived stimulus and the stimulus expected by an individual is called incongruity. This mismatch may be the reason for motivation because students are determined to eliminate those anomalies, to close the gap between the expectation and actual observation.

In the preparation of lessons, the teacher has to

– Know or estimate the prior knowledge of students
– Select a presenting event, which is incongruent with student expectations
– Determine the presentation mode of the anomaly
– Set type and strength of the incongruence or anomaly [11]

In chemistry class, it is then possible to create mismatches by observation of some natural phenomena or laboratory experiments and to motivate students to solve the problem. It is left to the creativity of each teacher to select a topic appropriate for phenomena or experiments – some examples are outlined below.

Melting temperature of ice. Young students think that heating substances always leads to an increase of their temperature – that is not true, however, for substances at their melting or boiling temperature. So a mixture of ice and water is heated, stirred and the thermometer observed: the temperature remains at a constant value of $0°C$ (see E2.1). The observation that the temperature remains at an unexpected $0°C$ motivates students to think about it. With the help of the teacher they learn that the added energy is needed to first melt the ice, and only after all ice is melted the temperature will increase with the amount of added energy. This experiment is very suitable to differentiate between the idea of energy and measured temperatures.

Boiling temperature of water. Students usually know 100°C as the boiling temperature of water, but they are not aware of the fact that this value is only valid for normal air pressure. A number of impressive mismatches can be generated to introduce this relationship. You can tell the story of high altitude climbers, observing in the altitude of 5,000 m the boiling temperature of water at 92°C instead at 100°C. In addition you might demonstrate a corresponding experiment, which shows the dependence of the boiling temperature on pressure: you decide the presentation mode of the anomaly, and type or strength of the incongruence. The relationship between pressure and boiling point can be demonstrated by the direct connection of the water jet pump to an apparatus for boiling water. You can also take a flask half filled with water, fill the flask totally with water vapor by boiling the water and expelling all air, closing the flask and cooling it with a wet cloth: the water begins to boil at temperatures below 100°C. This "boiling by cooling" is an interesting and very extensive cognitive conflict for students (see E2.2).

Solubility of carbon dioxide. Students know the sparkling appearance of the gas carbon dioxide when dissolving a mineral tablet in water and they have the idea that each tablet will generate the same specific volume of gas. Fill a measuring cylinder with water and open it under water in a half-filled water bowl. Dissolve a mineral tablet under the cylinder and observe a gas volume of 70 mL (see E2.3). At this point, the students may predict the volume of gas, if a second tablet is put into the cylinder: they will say: "the same volume." The second tablet develops almost 200 mL of carbon dioxide – the students will be astonished by the different volume and begin to think, to remove the occurring anomaly.

Combustion. The experiment "burning steel wool on the balance" (E1.6) described in Chap. 1 exemplifies a classic cognitive conflict for young people who throughout their life have always observed the decrease of mass during combustion: a big bunch of wood becomes a little amount of ash, alcohol disappears completely during combustion. You have to weigh a portion of steel wool, ignite it with the burner, and then you ask the students to predict whether the portion of burnt iron wool is lighter, heavier, or shows no change in mass. The likely expectation "lighter" turns out to be wrong after the sample is weighed again: the portion of burnt iron wool is heavier than before. The students will be highly motivated to eliminate this incongruity by thinking with the help of the teacher of solid iron oxide, a compound formed by oxygen from the air.

Extinguishing fires. Students know very well that all fires in everyday life are extinguished by water. When asked, how burning fat of a deep fryer or burning metal shavings should be treated, the response of most students will be given on the basis of their experience: "with water, of course." If you do this, explosions of violent flames occur in both cases (see E2.4), and the pupils are very surprised that the fires are not extinguished. They are very motivated to consider appropriate explanations for these observations.

Striking experimental effects. "No motivation without emotion" is a shortened statement of the previously described fact that interests and attitudes develop optimal with positive emotions. Especially in the sciences, experimental effects may trigger positive emotions for persons of all ages. Therefore, many demonstrations were shown in the past and are shown today to the public – mostly in the area of physics and chemistry.

Liebig held evening lectures at the University of Munich/Bavaria in the years around 1853 with spectacular demonstrations, so that even the king and the queen of Bavaria came as guests. One evening the queen was so surprised by the beautiful blue flash of the nitric oxide–carbon disulfide reaction (E2.5) that she wanted to see it again. She asked Liebig to repeat this experiment, but the glass flask exploded due to the mistake of his assistant (he filled the round flask with oxygen and carbon disulfide – see E2.5) and all three, the king, the queen, and Liebig were wounded [12]. Today we use upright cylinders and not round flasks for those reactions: the explosion will never destruct a cylinder!

Our students may also want to see beautiful or spectacular experiments a number of times, so we should plan for their safe repetition in the classroom. Concerning these show experiments there are two different concepts. First, like the above-mentioned nitric oxide–carbon disulfide reaction, these experiments are an event and supposed to entertain people – they are not explained or interpreted scientifi-cally. These experiments will mainly provide extrinsic motivation, they are suitable for "Christmas Lectures" or other experimental shows – many books are available for those shows (see [12] or [13]).

On the other hand, show experiments can also be used scientifically. If facts about "density of different substances" are introduced, then the usual mass and volume measurements on metal samples can be performed, and the metals can be identified from the table of densities. For a better motivation, the density issue can also be introduced by an effect that students probably do not know. A tin of "Coca Cola" and one of "Diet Coke" of the same size (330 mL) are placed in ice-cold water: the first can sinks to the bottom of the glass bowl, the second one floats on top (see E2.6). If a little story would be told to this effect that – at the last party – you have to grasp very deep into the cold water when someone wants "Coca Cola," while "Diet Coke" simply is to be taken from the water surface, then the students are even more motivated to think about this effect: it causes motivation through reference to everyday life. The debate about the different densities will eventually discuss the sugar contents: one can of "Coca Cola" contains about 20 g of sugar, "Diet coke" only a small amount of sweetener.

Three other effects, which can trigger a substantive motivation, might be men-tioned: the experiment "Ice breaks a bottle" (see E2.7) is an experiment that can show the anomaly of water and may lead to the discussion of the structure of ice. "Black carbon from white sugar" (see E2.8) may take the issue of the composition of sugar with carbon as one of the elements. The amazing effect "Electricity from a lemon" (see E2.9) sparks the discussion of the electrochemical potential series of metals. Similar motivating experiments can be found for nearly any issue – the creativity of the teachers is not limited in any way!

Hands-on with experimental or model material. Especially for children, but also for young students it is always important that they do not have to sit down quietly on their chairs in school, but be allowed to move, to run around, or do something manually: this is motivation in the *psychomotorical area*. For this reason, student experiments are so important. The students are motivated not only to learn through hands-on experiments, but to understand it and keep the chemistry on action-oriented ways far better than through a demonstration conducted by the teacher or even without any experiment. The discussion of student experiments in planning and implementation takes place in Chap. 5 "experiments." If student experiments are linked together, they support each other regarding validity and motivation. This will be discussed in more detail in Chap. 9 "Inquiry learning of Organic Chemistry."

Also the construction of structural models, such as sphere packing or space lattices for crystal structures and molecular models for the structure of molecules is perceived by students as beneficial: on action-oriented ways of motivation students build their own structural models and illustrate themselves the structure of different substances. If on the basis of such structural models the formulae of corresponding substances are understandable, then in this case, the psychomotoric motivation promotes even the good understanding of the chemical symbols. Relevant examples are presented in Chap. 6 "Structural and mental models" or Chap. 10, "Structure-oriented approach in chemical education."

If students prepare certain products in the classroom that they can even take home, then the motivation effect is particularly high. For example, if students are asked to make a name plate of brass for their front door at home, they are very strongly motivated to do this. They coat the plate with wax, write their names carefully into the wax surface, and etch the bare metal areas with nitric acid (E2.10). They will show such a name plate to family and friends and can explain exactly how they made it: the motivation goes far beyond the classroom.

The same applies to the construction of structural models. If they take the sphere packing model home they may even explain and they take it home for their desk, so they build the sphere packing not only very carefully, but they explain the sodium chloride structure to their friends and family – in such cases, the motivation goes into the private domain of students' families. Even teachers who attend our training courses "structural models and chemical understanding" tell us that the driving motivation for the visit of the course is the offer to take the constructed models home for their lectures!

2.3 Scientific Ideas: Experimental Skills

The reflection on the level of "teaching processes" has shown the potential to motivate students through experiments – according to both the intrinsic and extrinsic aspects as well. In both categories, experiments with special effects are proposed and thus often dangerous experiments. Even the great experimenter Liebig could not avoid such mistakes in his evening lectures for the Bavarian royal couple and himself – that could have led to a disaster [12].

Experimenters must therefore have good experimental skills and the ability to perform fast combustion reactions and deal with big flames (see E2.4) or blast effects (see E2.5) safely. In an experimental training for school experiments, students have to prove these skills and abilities to demonstrate those experiments and to perform experimental lectures at school safely. Already working teachers should also test new spectacular experiments for themselves, before going into the classroom with those experiments. Only after elimination of potential hazards one can go into school classes and perform those experiments, albeit with all required safety equipment: goggles, safety screen, splinter basket, etc. To be able to survive a whole "Christmas lecture" unharmed, the teacher will need long experience with experimentation. Finally, they must follow the relevant regulations with regard to hazardous materials (see Chap. 5).

To exploit the many opportunities for motivation, a well-founded scientific education in chemistry is required besides the experimental skills. You need not only to assess spontaneous students' remarks scientifically correct – you will also try to produce a suitable cognitive conflict for motivation. You also want to be flexible and create suitable experiments for motivation in many teaching situations. The beginning teacher will first demonstrate only scheduled or carefully planned experiments – the experienced teacher should be capable to demonstrate spontaneously created experiments or to start students' own experiments for a special question, and to contrast students' misconceptions with suitable experiments.

2.4   Human Element: Motivation Through Everyday Language and Media

Expressions of the everyday language often obscure the scientific facts on the one hand – on the other hand there are occasions to arise motivation to think about these issues.

Statements such as "the copper roof is green" are made in everyday language and lead to the idea that copper could be red-brown on one side and appear green on the other side. If we start from the knowledge of specific properties that distinguish certain substances, it is clear that only one of the colors should be specific for copper. In this way, the cognitive conflict or inconsistency is produced and students may be motivated to reflect on the "change of copper color." The result of those reflections should be the finding that the green substance is a compound of copper – in this case a basic copper carbonate, which forms by the reaction of carbon dioxide containing rain water a green layer on the red-brown metal. An experimental verification of this assumption can follow and eliminate the incongruence.

The media often provide statements that may unintentionally lead to motivating discussions. If a journalist commented about a factory fire in the words: "there are no chemicals involved in that fire" students may use their knowledge and conclude that all fuels are chemicals, and that even the oxygen of the air is regarded as a chemical, which is involved in every fire. The discrepancy between the statement of the journalist and their own knowledge may cause a strong motivation for the students to reveal the error.

Haupt [14] has compiled a collection of newspaper clippings on many subjects of chemistry. Many articles have spectacular headlines that make "chemistry" appear mysterious or dangerous. Teachers and students may be motivated to read such articles accurately and to interpret and correct these headlines and articles with their knowledge. Those articles are also a big opportunity to design problems and tasks for the students and motivate them to find the errors and propose better expressions [15].

Appendix A. Problems and Exercises

P2.1. Indicate examples of extrinsic and intrinsic motivation and discuss the differences between the two types of motivation. In what ways is intrinsic motivation possible in chemistry lessons? Describe three examples of teaching situations and appropriate motivation.

P2.2. Students' misconceptions are particularly useful to create incongruities and anomalies and thus to offer motivating discussions in the classroom. Explain three self-selected examples in this context and illustrate the intended incongruity.

P2.3. Spectacular experimental effects easily motivate students to observe carefully. Choose two experiments, which are likely to create extrinsic motivation only, and find two other experiments which help to develop an intrinsic or substantive motivation.

P2.4. The start into a new topic of chemistry should be motivating for the students. Choose a topic and describe an introduction that (a) follows up on the previous knowledge of the students, (b) is characterized by an incongruity, (c) takes an everyday life reference, (d) is particularly motivating through self-activity of the students.

P2.5. The everyday language contains idioms that are not always scientifically correct, but provide motivation to think about and to correct them. Discuss this with three examples of your choice and suggest proper expressions for these issues.

Appendix B. Experiments

E2.1. Constant Melting Temperature

Problem: Students observe in their daily live that heating a substance increases the temperature. If a pure substance melts during heating, the temperature remains constant until the substance is completely melted: during the melting process the supplied energy is needed to breakup the crystal lattice of the solid substance. Students should recognize this in the following experiments and are motivated to think about energy and temperature.

Material: Thermometer (200°C, with digital display), tripod and wire gauze, test tubes and beakers, wood clamp; ice, naphthalene (N), or stearic acid.

Procedure: (a) The temperature of an ice–water mixture is measured. The mixture is heated one minute in the beaker, after a good stirring, the temperature is read off again.

(b) A test tube is filled to a quarter with naphthalene and is heated with the Bunsen burner until all substance is melted; the melt is stirred with the thermometer and monitored. While stirring all the time the temperature is recorded every 30 s until the substance is condensing to a solid and completely solid again.

Observation: As long as the ice melts, the temperature remains constant at 0°C. The naphthalene smells strongly of moth balls (since moth balls contain this substance). As long as there is a mixture of liquid and solid naphthalene, the temperature remains constant at 80°C, when the substance is completely solid, the temperature sinks down to room temperature.

Disposal: The test tubes with the solid naphthalene are provided with a stopper and kept in the laboratory until the experiment is performed again.

E2.2 Boiling Temperature of Water [16]

Problem: Students often know only the scientifically reduced statement: "water boils at 100°C." To show the relation of boiling temperatures and air pressure, the flask for boiling water can be connected to a pump to produce low air pressures and to measure lower boiling temperatures.

It is also possible to replace the air in the boiling flask by water vapor and to condense it by cooling down with a wet cloth (see picture). By condensing of water vapor the pressure in the flask is decreasing, and the measured boiling temperature sinks to nearly 70°C. This results in the motivational incongruence for the students that the boiling temperature of water is not, as usual, reached by heating, but by "cooling."

Material: Round flask (500 mL) with side tube and valve, plug with thermometer (thermal sensor), water jet pump, tripod; water, boiling chips.

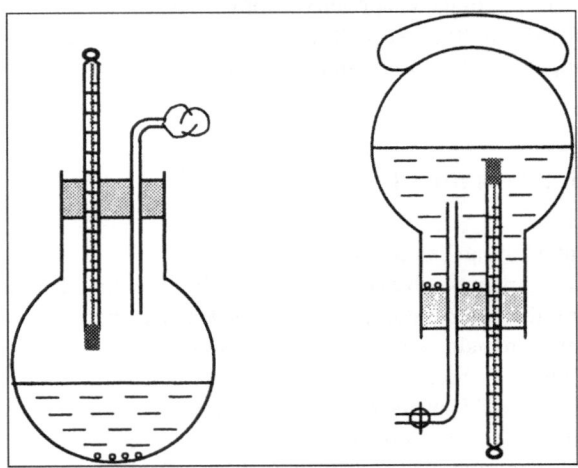

Procedure: (a) The flask is filled with water to a quarter, it is brought to boiling (use boiling chips!) and the boiling temperature is measured at atmospheric pressure. The water pump is connected and the boiling temperature is measured again during the air extraction (caution vacuum, wear goggles!).

(b) The water in the flask is heated to boiling for 1 min until the air has been completely replaced by water vapor (use boiling chips). The burner is taken away, the valve closed, the flask rotated 180° (see picture). A wet cloth is placed on the flask and the temperature recorded. The process is repeated several times, finally, the flask is rotated upright again, and the valve is opened carefully to fill the flask with air again (caution, wear goggles!).

Observation: (a) Under reduced pressure thermometers show boiling temperatures below 100°C. (b) The water shows first the boiling temperature of 100°C. By cooling the flask the water starts to boil again, temperatures decrease down to 70°C. Finally, whistling air penetrates into the flask, and the flask is filled with air again.

E2.3. Same Mineral Tablets, Different Gas Volumes

Problem: Students know that carbon dioxide is dissolved in mineral water and that the same gas is released in small gas bubbles, when mineral tablets are dissolved in water. By collecting the gas of one dissolving tablet (see picture) students are not realizing that only a certain portion of the released gas occurs in the cylinder, the other part dissolves in water to saturation. If in addition a second tablet is dissolved in the presence of this saturated solution, the volume of carbon dioxide is nearly two times bigger than by the first tablet. This incongruity motivates students to reflect and to interpret these phenomena with the solubility of gases in pure water.

Material: Graduated cylinder (250 mL) and matching stopper, glass bowl; mineral tablets (type "carbonate/citric acid").

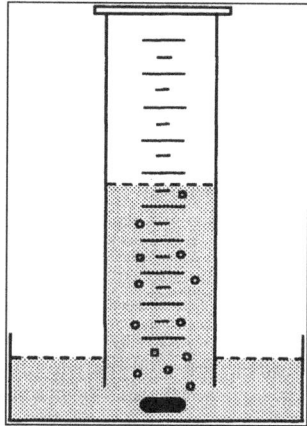

Procedure: The cylinder is completely filled with water and with the help of the stopper placed pneumatically in the glass bowl half filled with water. Under the opened cylinder one mineral tablet is placed, and the produced volume of gas is marked or recorded. A second tablet is placed in the same way and the resulting gas volume is also recorded.

Observation: The gas volume of the first tablet may result in about 70 mL, of the second tablet 70 + 130 mL: a total of 200 mL is observed (the actual volumes differ according to the brand of the tablets, they should be tested prior to use in the demonstration).

E2.4. Extinguishing Fires of Fat and Metals

Problem: Most fires in everyday life are extinguished with water. Therefore, students are usually unaware that it is not possible to extinguish burning metals, e.g., magnesium or burning fat with water – on the contrary, treating these fires with water causes terrible accidents. The experiments can be demonstrated either for a discussion of these safety aspects as well as a way for students to create a cognitive conflict and to solve it. In the experiments you can see that the burning fat at high temperatures of 300°C or more evaporates the added water instantly and an explosive mixture of vaporized fat and air burns with a big yellow flame. In the case of burning magnesium, the vigorous reaction of the metal with water forms metal hydroxide and hydrogen: this gas burns with a bright white flash. Instead of extinguishing both flames they are increasing to big fires – you have to eliminate the air from both fires, with sand in case of burning metals, and by closing the container in case of the burning fat.

Material: Tripod and wire gauze; tea light with aluminum container, magnesium turnings (F), deionized water bottle with water.

Procedure: (a) The wick of a tea light is cut off. The paraffin is heated strongly in the aluminum container until smoke is produced by decomposition of paraffin and the smoke is ignited. A water jet is aimed directly onto the burning paraffin. (b) A spoon of magnesium turnings are put together on the wire gauze and ignited by the strong burner flame. A water jet is aimed onto the burning metal (caution, wear goggles! For protecting the table against burning spots, cover the table with aluminum foil).

Observation: (a) The burning fat creates an up to one meter high yellow flash; (b) the burning magnesium forms a high, bright white flash.

E2.5. Blue Lightning Through Explosions of Gas Mixtures

Problem: A historic show experiment shall demonstrate that striking experiments are usually associated with big emotions: even the Bavarian Queen wished to see the "Blue Lightning" for a second time and Liebig tried to perform the reaction of nitrous oxide and carbon disulfide again (see Sect. 2.2). Instead of nitric oxide he incorrectly used oxygen in the mixture with carbon disulfide and the round flask exploded injuring the king, the queen, and the experimenter himself. Today we use no round flasks for this experiment, but cylinders with parallel glass walls: those explosions cannot destroy the cylinder, the reactions are carried out safely.

Material: Glass cylinder with cover glass, glass bowl, large test tube with a discharge pipe, plastic pipette (5 mL), burner, ammonium nitrate (O), carbon disulfide (F/T), oxygen (O).

Procedure: In the preparation, ammonium nitrate is decomposed in a test tube by gentle heating; the resulting colorless gas of nitrous oxide is introduced pneumatically into the cylinder (under the exhaust hood!). The pipette is filled with 2 mL of carbon disulfide and emptied into the cylinder; the cylinder is covered with the cover glass. After removal of the cover the mixture is ignited by the Bunsen flame. The experiment is repeated with oxygen and carbon disulfide.

Observation: A light blue flash appears and a specific sound is heard reminiscent of a barking dog (the experiment is therefore also called "barking dog"). In the second experiment a white flash is seen, which is accompanied by a loud bang.

E2.6. Diet Coke Is Lighter Than Coca Cola

Problem: The concept of density can be introduced by an effect, which surprises most students and therefore challenges them to explain the effect. Both cans are placed into ice-cold water: the can of Coca Cola drops down to the bottom of the container, the can of Diet Coke swims.

You need to weigh both cans, to discuss the higher mass of 330 mL Coca Cola and to interpret this according to the high content of dissolved sugar in Coca Cola. Moreover, students are motivated to develop the idea of density: Coca Cola has the higher density compared to Diet Coke or Cola Light.

Material: Large glass cylinder, balance, areometers (densities around 1.0 g/mL); 330 mL-can "Coca-Cola," 330 mL-can Cola Light or Diet Coke, ice water.

Procedure: The cylinder is filled to three quarters with ice water; both cans are placed into the water. Then both cans are weighed. Both samples of coke are heated to release the carbon dioxide; the densities of both solutions are measured by an areometer.

Observation: The can of Coca Cola sinks to the bottom of the cylinder, the can of Cola Light swims on the surface. The can of Coca Cola weighs about 20 g more than the can of Cola Light. The density of the Coca Cola is a little bigger than 1.0 g/mL.

Note: The production technology of canned Cola causes an air bubble in the can. These bubbles may differ from can to can and therefore the can of Cola Light may also sink in water. Try out some cans and choose the right ones before you show the experiment.

E2.7. Ice Lets a Bottle Burst

Problem: Another density phenomenon is provided by the anomaly of water: ice occupies a bigger volume than a portion of water with the same mass. We are very familiar with the fact that ice floats on water and we do not think about the fact that normally the solids sink in their melts: a candle sinks in its melt, a piece of lead sinks in the molten metal.

To show the density anomaly of water, you can fill a container with cold water, close it and cool it to below the freezing point: the container bursts because of the larger ice volume.

Material: Small glass bottle with screw cap, thermometer (-20–$100°C$); ice water, ice-salt freezing mixture.

Procedure: The ice-salt mixture is produced and the temperature of $-15°C$ is shown to the students. The bottle is filled to the brim with ice water and then is closed. The bottle is placed into the ice-salt mixture, after a few minutes the bottle is taken out of the mixture.

Observation: The water freezes to ice and the bottle bursts into pieces of glass.

E2.8. Black Carbon from White Sugar

Problem: A most amazing phenomenon for students of any age is the reaction of white sugar and colorless sulfuric acid to black carbon: it can be shown that sugar is a carbon containing compound. Starting with the exciting sugar–sulfuric acid reaction, the students may be motivated to study the chemistry of carbon compounds in organic chemistry.

Material: Beaker (100 mL), glass bowl, glass rod; sugar, pure sulfuric acid (C), water.

Procedure: Sugar is added to about 3 cm high in the beaker and stirred with a little amount of water. It is covered with sulfuric acid about 3 cm high and the mixture is stirred briefly with the glass rod; the beaker is put into the glass bowl for protecting the table. Wait and observe.

Observation: The white mixture turns black, water steam is formed with a strong hissing noise and swelling reaction, thereby the mixture heats up very strongly. It forms a black, porous substance in the form of a sausage, which can be up to 10 cm long. A sweet smell from the decomposition of sugar is noticeable.

Disposal: The black substance – mixed with concentrated sulfuric acid – is carefully wrapped in aluminum foil and is disposed of in the container for solid waste. The beaker is rinsed with water and cleaned with paper.

E2.9 Electricity from the Lemon

Problem: To motivate for electrochemistry an experiment with a lemon is possible: two different metal strips are immersed into a lemon; using copper and zinc an electrical voltage of about 1 V can be measured with the voltmeter. If the same metals are used no voltage should be observed. The same phenomena can be shown with metal strips in sodium chloride solution; with different combinations of metals you can establish the redox series of metals. It is possible to show with a 2-V electric motor that electric current is produced by those simple electrochemical cells.

Material: Beaker, knife, voltmeter, 2-V electric motor, cables, and alligator clips; sodium chloride, metal strips of copper, zinc, and magnesium (F), lemon.

Procedure: (a) A lemon is prepared with a knife to insert two metal strips. Strips of copper and magnesium are placed into the lemon, they are connected by cables and alligator clips with the voltmeter, and the voltage is measured. (b) The beaker is filled halfway with a concentrated solution of sodium chloride. Two different metal strips are provided with cables and alligator clips and connected to the voltmeter; the metal strips are dipped into the solution and the voltage is measured. The experiment is repeated with different metal combinations and also with the same metals. The electric motor is connected as well.

Observation: Dipping into the lemon, potentials of about 1.5 V are measured between copper and magnesium. In the salt solution the same voltage is observed, between zinc and magnesium smaller voltage values are measured. In case of the same metals, no voltage appears. For the higher voltages, the electric motor does work and shows the running electric current.

Note: The juice of the lemon is a weak electrolyte solution: the electricity is not enough to run the electric motor. It succeeds, if two or three lemons are put in series.

E2.10 Brass Name plate

Problem: Students are motivated more strongly if they produce something that they can take home. They can produce their own brass name plate as an introduction into the topic "redox reactions with metals and acids." This plate can be used not only at home, but students can also report their families and friends, how they have produced it: they may feel as experts in this area of chemistry, and may be proud of their competence.

Material: Glass bowl, beaker (100 mL), pipette, brass plate or copper plate, candle, iron nail, knife; pure nitric acid (C), boiling chips, gasoline (Xn/F/N), filter paper.

Procedure: One side of the metal plate is coated with heated liquid paraffin from the candle. With the iron nail you scratch a name or a desired figure deep into this layer of paraffin, so that the metal is exposed. You put the plate with the top layer of wax on a few boiling chips into the glass bowl and put it under the exhaust hood. In the beaker 5 mL of water and 10 mL of pure nitric acid are mixed and the mixture is dripped with the pipette onto the scratched parts of the paraffin layer. You wait about 10 min.

Observation: The acidic solution reacts with the metal by forming a brown gas (nitric oxide) and a blue solution (copper nitrate solution). After removal of the remaining paraffin with the knife and gasoline-wet filter paper, the name or label is clearly visible in the metal plate.

Disposal: The name plate and the glass bowl is rinsed with much water, the gasoline-wet filter paper is to be burnt under the exhaust hood.

M2.1 Sphere-packing model of a salt crystal

Problem: The manual production of structural models is highly motivating for many students, especially if they are allowed to take these models home. It is easily possible to illustrate the topic "composition of salts" with the example of the sodium chloride crystal structure, for example, the arrangement of sodium ions and chloride ions in the ratio 1:1 (see also detailed information in Chaps. 6 and 10).

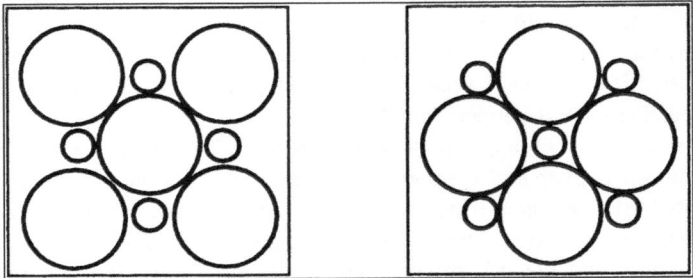

Material: Sodium chloride (rock salt) crystals, cellulose spheres (per student about 18 white spheres (diameter 30 mm) and 18 red spheres (diameter 12 mm)), glue.

Procedure: The balls are glued together to layers as shown (see the picture), two times two layers are put in reverse order together to a square column. It is determined how many small balls touch one large ball and how many big balls touch one little ball. The produced model is compared with the original rock salt crystal. It should be discussed which aspects of the model are suitable, and which limitations of the model exist.

Observation: The layers form a square column and show a closest sphere packing model of the sodium chloride structure. A large sphere is touched in the interior of the packing by six small spheres; a small sphere is touched by six large spheres. The ratio of spheres in the model is 1:1 (see also detailed information in Chaps. 6 and 10).

References

1. Prenzel M, Krapp A (1992) Interessen, Lernen und Leistung. Aschendorff, Münster
2. Schiefele U (1990) Einstellung, Selbstkonsistenz und Verhalten. Hogreve, Göttingen
3. Gräber W, Stork H (1984) Die Entwicklungspsychologie Jean Piagets als Mahnerin und Helferin im naturwissenschaftlichen Unterricht. MNU 37:257
4. Duit R (1996) Lernen als Konzeptwechsel im naturwissenschaftlichen Unterricht. In: Duit R, von Rhoneck C (eds) Lernen in den Naturwissenschaften. IPN, Kiel
5. Heilbronner E, Wyss E (1983) Bild einer Wissenschaft: Chemie. ChiuZ 17:69
6. Barke H-D, Hilbing CH (2000) Image von Chemie und Chemieunterrichts. ChiuZ 34 (2000)
7. Wagenschein M (1971) Die Pädagogische Dimension der Physik. Westermann, Braunschweig
8. Jäckel M, Risch KH (1994) Chemie heute. Schroedel, Hannover
9. Wanjek J, Barke H-D (1998) Einfluss eines alltagsorientierten Chemieunterrichts auf die Entwicklung von Interessen und Einstellungen. In: Behrendt H (ed) Zur Didaktik der Chemie und Physik. Kiel, Leuchtturm
10. Harsch G, Heimann R (1998) Didaktik der Organischen Chemie nach dem PIN-Konzept. Vom Ordnen der Phänomene zum vernetzten Denken. Vieweg, Braunschweig
11. Lind G (1975) Sachbezogene motivation. Weinheim, Beltz

12. Krätz O (1997) Historische chemische Versuche. Aulis, Köln
13. Roesky HW, Möckel K (1994) Chemische Kabinettstücke. VCH, Weinheim
14. Haupt P (1997) Die Chemie im Spiegel einer Tageszeitung. Bände 1–4. Oldenburg 1985–1997
15. Haupt P (2000) Da schmunzelt der Chemiker! NiU-Chemie 11:92
16. Foundation N (1967) Collected experiments. Longmans, London

Further Readings

Ainley MD (1993) Styles of engagement with learning: multidimensional assessment of the relationship with strategy use and school achievement. J Educ Psychol 85:395–405
Ainley M, Hidi S, Berndorff D (2002) Interest, learning and the psychological processes that mediate their relationship. J Educ Psychol 94:545–561
Ames R, Ames C (1991) Motivation and effective teaching. In: Idol L, Jones BF (eds) Educational values and cognitive instruction: implications for reform. Erlbaum, Hillsdale, NJ, pp 247–271
Cohen EG (1994) Restructuring the classroom – conditions for productive small-groups. Rev Educ Res 64:1–35
Deci EL, Koestner R, Ryan RM (2001) Extrinsic rewards and intrinsic motivation in education: reconsidered once again. Rev Educ Res 71:1–27
Deci EL, Ryan RM, Koestner R (2001) The pervasive negative effects of rewards on intrinsic motivation: response Cameron (2001). Rev Educ Res 71:43–51
Gottfried AE, Fleming JS, Gottfried AS (1998) Role of cognitively stimulating home environment in children's academic intrinsic motivation: a longitudinal study. Child Development 69:1448–1460
Hidi S, Harackiewicz JM (2000) Motivating the academically unmotivated: a critical issue for the 21st century. Rev Educ Res 70:151–179
Hidi S, Renninger KA (2006) The four-phase model of interest development. Educ Psychol 41:111–127
Hidi S, Renninger KA, Krapp A (2004) Interest, a motivational variable that combines affective and cognitive functioning. In: Dai D, Sternberg R (eds) Motivation, emotion and cognition: integrative perspectives on intellectual functioning and development. Erlbaum, Hillsdale, NJ, pp 89–115
Iyengar SS, Lepper MR (1999) Rethinking the value of choice: a cultural perspective on intrinsic motivation. J Personal Soc Psychol 76:349–366
Krapp A, Hidi S, Renninger KA (1992) Interest, learning and development. In: Renninger KA, Hidi S, Krapp A (eds) The role of interest in learning and development. Erlbaum, Hillsdale, NJ, pp 3–25
Krapp A, Lewalter D (2001) Development of interests and interest-based motivational orientations: a longitudinal study in vocational school and work settings. In: Volet S, Järvelä S (eds) Motivation in learning contexts: theoretical advances and methodological implications. Pergamon, Amsterdam, pp 209–232
Lepper MR, Henderlong J (2000) Turning "play" into "work" and "work" into "play": 25 years of research on intrinsic versus extrinsic motivation. In: Sansone C, Harackiewicz JM (eds) Intrinsic and extrinsic motivation: the search for optimal motivation and performance. Academic, San Diego, CA, pp 257–307
Meyer DK, Turner JC (2002) Discovering emotion in classroom motivation research. Educ Psychol 37:107–114
Mitchell M (1993) Situational interest: its multifaceted structure in the secondary school mathematics classroom. J Educ Psychol 85:424–436
Nichols JD, Miller RB (1994) Cooperative learning and student motivation. Contemp Educ Psychol 19:167–178

Nolen SB (2003) Learning environment, motivation, and achievement in high school science. J Res Sci Teach 40:347–368

Palmer D (2005) A motivational view of constructivist-informed teaching. Int J Sci Educ 27:1853–1881

Pintrich P (1999) The role of motivation in promoting and sustaining self-regulated learning. Int J Educ Res 31:459–470

Pintrich PR (2000) Multiple goals, multiple pathways: the role of goal orientation in learning and achievement. J Educ Psychol 92:544–555

Pintrich PR (2003) A motivational science perspective on the role of student motivation in learning and teaching contexts. J Educ Psychol 95:667–686

Pintrich PR, Marx RW, Boyle RA (1993) Beyond cold conceptual change: the role of motivational beliefs and classroom contextual factors in the process of conceptual change. Rev Educ Res 63:167–199

Pintrich PR, Schunk DH (2002) Motivation in education: theory, research and applications, 2nd edn. Prentice Hall, Upper Saddle River, NJ

Ryan RM, Deci EL (2000) Intrinsic and extrinsic motivations: classic definitions and new directions. Contemp Educ Psychol 25:54–67

Schraw G, Flowerday T, Lehman S (2001) Increasing situational interest in the classroom. Educ Psychol Rev 13:211–224

Silvia PJ (2005) What is interesting? Exploring the appraisal structure of interest. Emotion 5:89–102

Wentzel KR (1998) Social relationships and motivation in middle school: the role of parents, teachers and peers. J Educ Psychol 90:202–209

Wigfield A, Eccles JS (2002) The development of competence beliefs, expectancies for success and achievement values from childhood through adolescence. In: Wigfield A, Eccles JS (eds) Development of achievement motivation. Academic, San Diego, CA, pp 94–120

Chapter 3
Teaching Aims

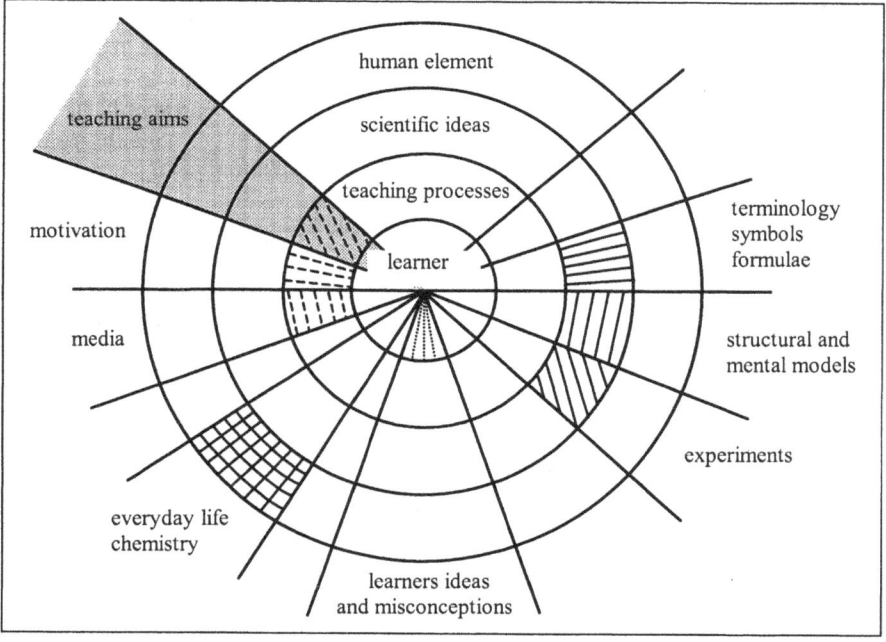

When we ask adults what they still remember from their chemistry lessons at school, we often get the answer: "Uhm, there were these formulae," and they are proud when they still remember the formula H_2SO_4 – not what it means. Most chemistry knowledge did not survive. A senior officer from a school board, a scientist, told me once: Don't talk to me much about the educational value of chemistry... it's only formulae [1]

H.-D. Barke et al., *Essentials of Chemical Education*,
DOI 10.1007/978-3-642-21756-2_3, © Springer-Verlag Berlin Heidelberg 2012

Formulae often dominate our curricula. As long as the learner does not comprehend the information content of these symbols, they also do not understand the
chemistry. They often make fun of their incomprehension of chemistry, and say:
"I never understood chemistry – but look what high level position I could reach
without it" [2].

Apparently, many teachers were not successful in teaching – and in reaching the
goals of science education in the way experts from important professional
associations recommend: "Mathematics and science education are essential parts
of the general knowledge; they support the personality development through
teaching scientific knowledge and methodological competence. They also allow a
basic understanding of technical questions and therefore they are a basis for a
responsible participation in the social discussion of possibilities and limits of
technical development. Mathematics, physics, chemistry and biology contribute
to the scientific knowledge – only a solid basis of science education can support
interdisciplinary learning and the understanding of the complexity and interdependence of natural or technical problems" [3].

To contribute chemistry contents to the general knowledge along those lines, the
goals of chemistry teaching have to be discussed continuously and reformulated
according to current perceptions so that the result is a good understanding of
chemistry for the learner. The following science education and pedagogical fields
have to be considered for a reasonable discussion of these basic questions:

– Goals of teaching and learning, their dimensions and taxonomies
– Goals on the basis of different didactical models and convictions
– Teaching analysis based on those goals

3.1 Scientific Ideas: Teaching Goals and Their Dimensions

The amount of scientific knowledge doubles about every 10 years. Consequently,
it needs to be discussed which attitudes, behaviors, skills, and knowledge have to be
demanded of today's students. German officials therefore ask: "With which objects
and contents should the learner be confronted? In what steps, how and with which
material should they learn? How can the achievements be tested?" [4].

Different levels have to be distinguished for the teaching goals, general headline
goals, or educational goals which cover all fields of education at school or schooling
in general. *Headline goals* are related to specific teaching intentions of the subjects,
and *detailed goals* are related to single learning steps in a certain teaching unit.

Operationalized goals specify the learner's operations, which are to be learned
during a special lesson; they give precise behavioral dispositions, which are being
expected. One example: "The students are to choose three out of five acid–base
indicators and state their colors in the acidic and alkaline range." Since these
concrete operationalizations require constant control by written tests or oral
examinations to evaluate the learner's expected behavior, this kind of goals did
not become accepted.

Teaching goals should connect aspects of content with aspects of behavior. Behavioral changes can happen in three dimensions [5]:

In the area of perception, memory and thinking	*Cognitive* dimension
In the area of interest, attitude, and values	*Affective* dimension
In the area of manual and physical skills	*Psychomotoric* dimension

Teaching goals concerning these three dimensions can be organized hierarchically or transformed into a learning taxonomy. Especially the teaching goals of the cognitive dimension have been organized hierarchically with increasing complexity by Bloom [6]:

1.	Knowledge	Know concrete data, facts, rules, laws, or symbols
2.	Comprehension	Connect facts, interpret and extrapolate data, derive conclusions
3.	Application	Use or transfer knowledge to new situations
4.	Analysis	Divide complex information, analyze data, recognize causal relations or patterns
5.	Synthesis	Put together single bits of information to form a complex, coordinate data, think systematically
6.	Evaluation	Evaluate complex topics, draw conclusions

German experts suggest the following hierarchy [4]:

1.	Reproduction	Reproduce knowledge from memory
2.	Reorganization	Reorganize familiar knowledge for new situations
3.	Transfer	Transfer familiar knowledge to new topics
4.	Problem solving	Solve problems, find new explanations

The different levels of teaching goals can only be evaluated against the background of the student's prior learning and level of knowledge – it is not possible to assign isolated teaching goals to these levels.

Didactical models. Experts have created basically five models of general didactics:

– The educational theoretic didactics according to Klafki [7]
– The learning theoretic didactics according to Schulz [8]
– The curricular didactics according to Möller [9]
– The critical communicative didactics according to Winkel [10]
– The informational theoretic cybernetic didactics according to Cube [11]

Contrary to the normative didactics of past centuries, which focused on giving instructions for reaching educational goals, today's didactical models are oriented towards the actual reality of education. Different didactical models [12] have been developed to illustrate the complex instructional situation at schools – one model alone cannot hold all the essential information. Therefore, the models do not contradict each other, but are complementary: "In didactics we live in a world with a pluralistic perspective. The different didactical models represent single perspectives" [13]. The two first-mentioned models will be explained and compared.

Fig. 3.1 Scheme of steps for planning lectures according to Klafki [14]

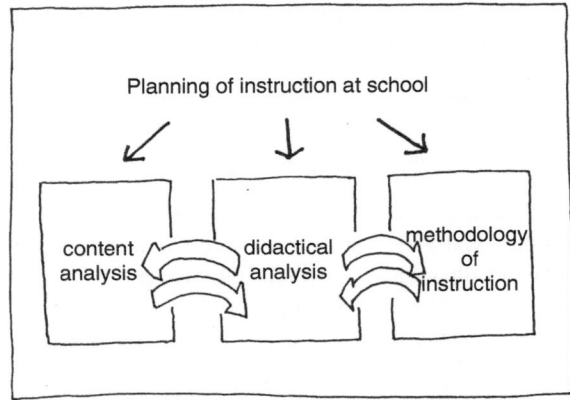

The educational theoretic didactics according to Klafki [7] is based on the "didactical analysis as the core of the teaching preparation and develops a thematic structure over explanations for meanings in the present and the future" [7]. Based on the analysis of the contents in the first step, the didactical analysis follows in the second step, and finally in the third step the methodology of the instruction (see Fig. 3.1).

While the Klafki model [7] uses the primacy of the contents for a topic, the learning-theoretic didactics according to Schulz [8] is based on the interdependency of the decision fields: intentions, contents, methods, and media belong together and are mutually dependent (see Fig. 3.2); the didactical analysis can start from every field [15].

Lesson planning and analysis. Bönsch [16] gives an overview for the lesson planning, which is based on the learning theoretic didactics and the interdependency of requirements, goals, contents, methods, and media (Fig. 3.3). A box with "medium: experiments/models" should be added for the specific issues of chemistry lessons:

– Which experiments/models are planned, which alternatives are possible?
– How can the choice of experiments/models be explained?
– Are the experiments used according to their specific function (introduction, data collection, test of hypothesis, repetition, etc.)?
– Is the experiment planned as a teacher, student, or group experiment?
– Are necessary safety measures provided?

The planning analysis is the basis for the analysis of processes during the lesson (see Table 3.1). A fourth category of questions concerning experimental lessons and a fifth category concerning the use of models are added (see Table 3.1).

In his considerations of the student-oriented lesson planning, Meyer [17] differentiates the teacher's teaching goals and a student's learning goals, especially those for hands-on activities; the factor "student" is taken more into account: "Problems of socialisation due to tracking, the student's everyday consciousness, interest in the class (free choice versus forced assignment), orientation towards performance, work load," are some examples [17].

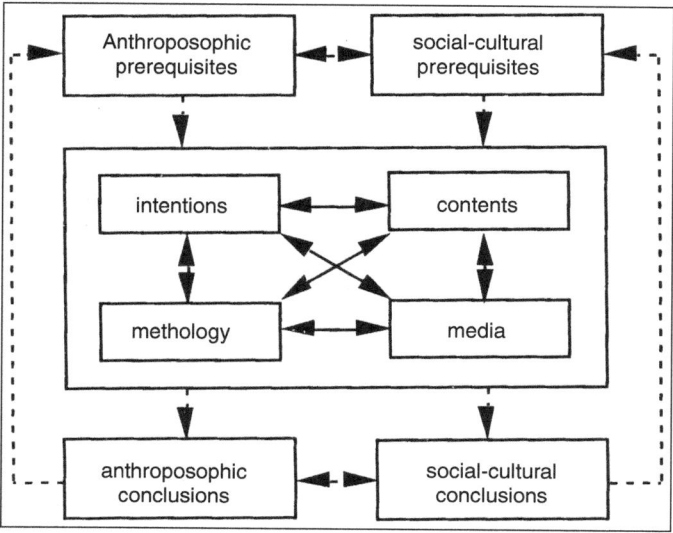

Fig. 3.2   Scheme of steps for planning lectures according to Schulz [8]

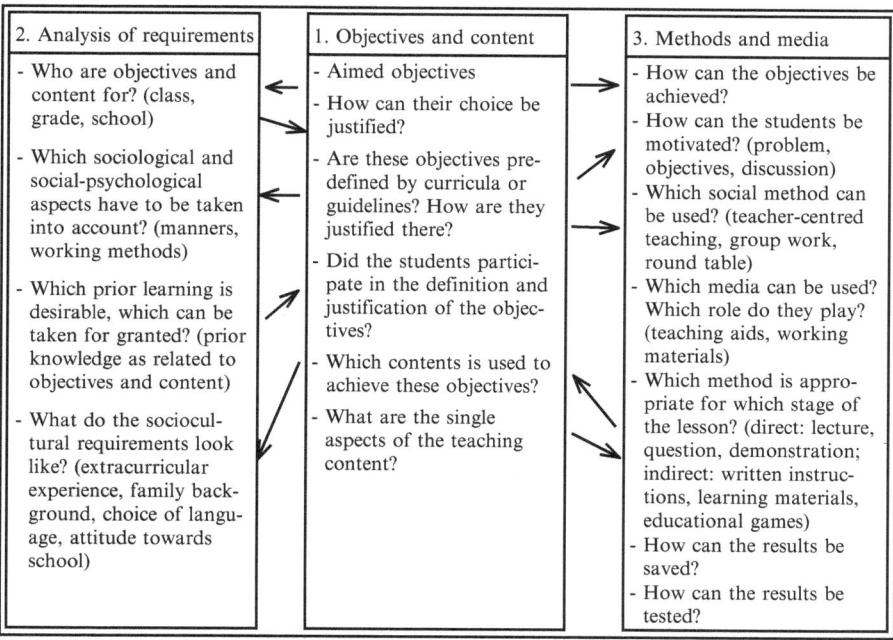

Fig. 3.3   Analysis of lecture planning according to Bönsch [16]

Table 3.1 Analysis of lecture processes according to Bönsch [16]l

1. General: Comparison of the observed lesson with the lesson plan

   Was the lesson carried through according to the lesson plan? If not, why not? Which unexpected events were observed?

   Were the objectives set too high or too low? Was the content already familiar? Were the methods and media appropriate for the objectives/content as well as for the students? What could have been done differently?

   Were the lesson's objectives achieved? Was this tested or can it only be conjectured?

2. Specific: lessons are always complex contacts of people how strong (and annoying) was the teacher's domination? (Proportion of speech, style of leadership)? To what extent were all students involved in the discussion? To what extent were discussions possible? Was the domination of some students annoying or inspiring?

   What did spontaneous aids by the teacher (encouragements, etc.) look like?

   To what extent were students involved in decisions of the definition and achieving of objectives?

   To what extent were independence and participation possible?

3. Specific: Lessons always aim to initiate, control, and create successful learning processes

   How were the students motivated? Which problems were chosen?

   Which individual learning processes could be observed? (general learning processes are not the same as individual learning processes) Was individual learning possible? Were the starting and ending point of the student's learning processes documented and as a result, can individual learning progress be determined?

   Is the learner more an object of organized learning or are they more the subject within the meaning of participation in decisions on learning objectives, contents, methods, media, tests? Which kind of learning processes was intended, which kind was achieved?

   Which kind of media was used, for which purposes and with which effects?

4. Specific: Chemistry lessons are experimental lessons (see Chap. 5 "experiments"). Did the students understand the problem of the experiment? Was the performance of the experiment successful? Did the students draw a picture of the experimental design? Did they understand it? Did the experiment support the problem-solving process? Which questions remained open?

   Were the observations viewable for all students (if applicable using a projector)? Were the instructions for group and student experiments complete and comprehensible? Was the experiment analyzed on the basis of observed phenomena and measured values?

   Did the students have time to write down performance, observation, and analysis of the experiment? Were measurement errors discussed?

5. Specific: Chemistry lessons and the use of models (see Chap. 6 "models")

   Could the students understand philosophy and purpose of the use of models?

   Did they understand all representation ideas of the model? Were all irrelevant additions of the model made clear to the students, for example, through the comparative use of two or three structural models for the same issue?

   Did the students have the chance to comprehend the submicroscopic structure of matter by building structural models, like packing of spheres, space lattices, or molecular models?

   Were structural and bonding models only explained verbally? Or did they become apparent for the student through sketches, space models, or other media? Did the students discuss the media regarding the features of depiction and the features of shortening? Were experiments and models correlated?

3.2 Human Element: Guidelines and Curricula

Different countries developed different basics for teaching and learning goals –
therefore, some examples concerning Germany, the United States of America, and
England are discussed.

3.2.1 Germany

Lesson planning at school and therefore also chemistry lessons are highly dependent on social premises; society exerts influence on the goals of chemistry lessons.
Also environmental problems place a demand on chemistry lessons, political,
ecological, technological, scientific and cultural developments react upon the
lessons.

Teachers should be aware of these relations. They can also be found in
preambles and headline goals of the curricula. The "guidelines and curricula of
chemistry in secondary school" of the state North-Rhine Westphalia say [18]: "The
secondary school teaches a general education with the goal of enabling the student
to a mature and responsible life in a democratic society. It offers them stimulation
and helps to develop their individual abilities and values. Such an education is
being developed dealing with phenomena of nature and society, their structures and
principles, the cultural tradition and the current cultural reality".

The German Association for the Support of Mathematical and Science Education (MNU) has developed educational guidelines in its "Recommendations for the
development of curricula and guidelines for chemistry lessons" [19] and put them
up for discussion:

- Practice complex thinking
- Develop communication skills
- Consider student conceptions

These recommendations [19] also say: "For planning chemistry lessons in
secondary schools these guidelines should be taken into account:

- Methods of working in chemistry
- Substances and properties
- Properties and chemical structure
- Particles between imagination and reality: first atomic and bonding models
- Chemical reaction: changes on the substance, regrouping of atoms, ions and
  molecules, scientific and symbolic language, energy transfer
- Principles of classification for substances and reactions" [19]

Criteria for the conception of curricula for advanced education, in Germany
called *Sekundarstufe II*, other areas, concepts, and contents are discussed (see
Fig. 3.4). Besides criteria of science and chemistry one takes into account special

areas like everyday life, environment, and technologies. Scientific contents are reduced to important basic concepts (see Fig. 3.4): "(a) particle concept: atoms, ions and molecules, (b) structure-property relationships, (c) the donor–acceptor concept, (d) the equilibrium concept, (e) the energy concept. These fundamental concepts should be incorporated into the curriculum whenever possible. In this way the concepts are not only arranged in a regular spiral curriculum, but the rigid structure of chemistry is abandoned in favor of a more open system. This structure allows re-visiting these concepts at different times" [19].

With these basic concepts the idea of competency is involved. With the new shift in Germany towards an outcome orientation, Parchmann and others [20] proposed four important competencies: scientific knowledge, gaining knowledge, communication, and valuation (see Table 3.2). After focusing in the past on the scientific knowledge only, the students should learn more about empirical methods to acquire knowledge, about communication for discussing and exchanging information, and about evaluating chemical facts and reports (see Table 3.2).

Organizing teaching and learning at German schools. In Germany as in many parts of the world, Primary School follows after kindergarten, and it usually lasts

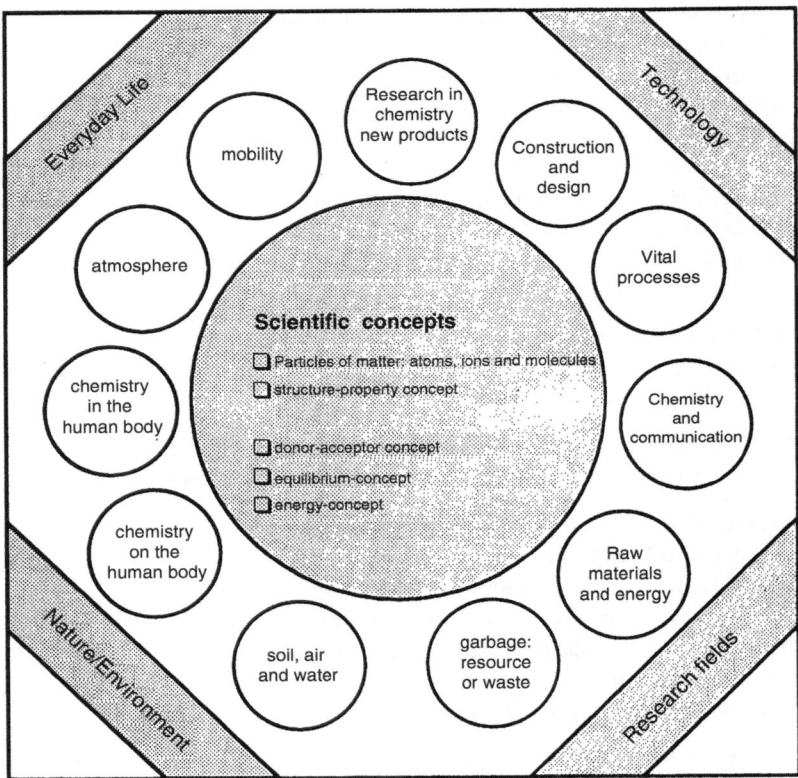

Fig. 3.4 Correlation between basic concepts of chemistry and aspects of the human element [18]

Table 3.2 Competency model for chemistry education [20]

Scientific knowledge	Connecting phenomena, ideas, laws to the basic concepts
Gain of knowledge	Gaining knowledge through experimental methods and models
Communication	Obtaining, reflecting, and exchanging scientific information
Valuation	Evaluating chemical facts in different contexts

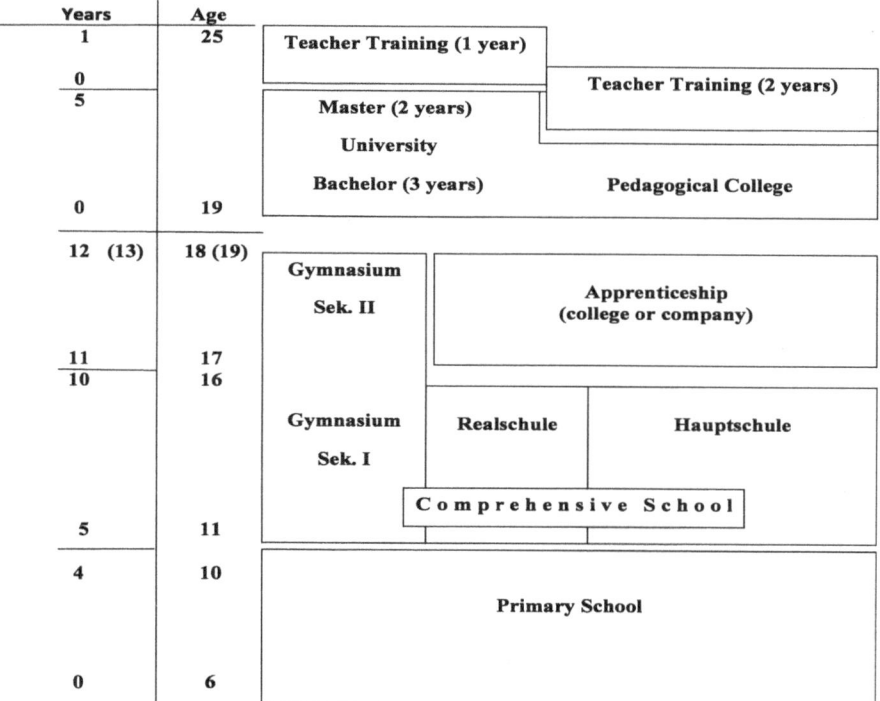

Fig. 3.5 School system and university approach in Germany

4 years (see Fig. 3.5). Some states in Germany are discussing to have 6 years, but most parents want the young students in secondary schools as early as possible. These schools can be gymnasiums from grade 5 to grade 12 and to the graduation (Abitur), they can also be comprehensive schools for 8 or 6 years, parents can also choose types of middle schools (Realschule and Hauptschule). These students can go on to higher education for graduation and for university studies like the students from gymnasiums (see Fig. 3.5), or they can enter an apprenticeship: they are employees in a company or factory with a small salary, but have to visit vocational school 1 day per week to get some theoretical background. After an examination they become full employees with the normal salary – if the company or factory wishes to keep them on.

Primary education. In German elementary schools, social studies and science are taught in an interdisciplinary approach called *"Sachunterricht."* For the sake of completeness, content that is part of primary education will be outlined briefly. Compared with older students 6- to 10-year-olds are highly interested in biological, chemical, physical, or technical phenomena. Their interest should be used to introduce children to many phenomena in the classroom.

The German society of primary education at schools (GDSU) suggests five perspectives [21]:

1. Social and cultural perspective (social science)
2. Spatial perspective (geography)
3. Natural science perspective (biology/chemistry/physics as interdisciplinary teaching)
4. Technological perspective (technics and technology)
5. Historical perspective (history)

The following examples concern chemical issues: "Properties of substances, effects of heat, melting and solidifying, combustion processes, oxygen and respiration, use of electricity, natural powers, wind and water, atmosphere and weather, body and health, hazards to the environment, environment protection, healthy eating" [21]

These topics should be connected to the following inquiry skills: "observe; collect, sort and classify; measure and compare; identify problems and perform experiments to solve them; propose guess and explanation; plan, run, analyze and interpret experiments; create and analyze tables and charts; evaluate explanations; produce skilled drawings" [21].

Hands-on activities play an important role for learning in primary science education. Moeller [22] built special boxes for teachers of primary schools to learn by hands-on experiments more about "swimming and sinking, air and air pressure, transfer of sound, bridges and their stability" [22]. The boxes are made for groups of about six to eight children in class. Teachers can borrow one box for 1 week, carry out the experiments with all the supplied material, give back that box and take another box for the next topic.

The interdisciplinary approach sometimes continues through grades five and six in middle schools and gymnasiums – it differs from state to state in Germany. That subject is often called "world and environment studies" and most topics concern biology. In grade seven or eight, the interdisciplinary approach is abandoned in favor of the sciences: biology, chemistry, and physics. In general, the students have 2 h per week of lectures in every discipline; occasionally teachers offer study groups in the afternoon where students may voluntarily learn more about current questions – especially if more laboratory work is involved!

For advanced studies in grade 11 and 12 (in Germany called "Sekundarstufe II") every student has the choice of one discipline, which they are studying for 3 h per week ("Grundkurs") or 5 h per week ("Leistungskurs").

3.2.2 USA

Curricula and guidelines for chemistry education in the United States are influenced by the National Science Education Standards (NSES) [23]. These standards have been developed by the National Research Council and were published in 1996: government officials, members of business and industry, school administrators, teachers, curriculum developers, publishers, scientists, science educators, and many others were involved in this attempt to improve science education through a set of standards. Until today they are the only nation-wide valid standards for science education from kindergarten to grade 12. The NSES are not a curriculum though. Instead of describing the journey to scientific literacy, i.e., the concrete content and way of teaching, they point out the goals of this journey, i.e., the knowledge and abilities to be achieved by the students. By bringing coordination, consistency, and coherence to the improvement of science education, they form the basis for the development of curricula on the state and local school district level.

A short illustration of the American school system and the role of chemistry for education will help to better understand the concrete description of the NSES. Education is divided into primary (kindergarten (K) until grade 5/6) and secondary education (grade 5/6–12). From K to 5/6 students go to elementary school, from 5/6 to grade 8 students visit middle schools (also called junior high school), and from 9 to 12 students go to high school. Curricula, school laws, and policies vary from state to state due to the federal organization of education in the USA.

Besides mathematics and English, science is one of the key subjects. From K–8, science is taught with an interdisciplinary approach combining biology, physics, chemistry, geology, astronomy, technologies, health science, and environmental science. In middle schools, science is usually subdivided into earth science, life science, and physical science. When entering high school students start the world of individual science disciplines. Besides three mandatory science courses (biology, chemistry, physics) students can choose between a wide range of optional courses like geology, anatomy, forensic science, health science, astronomy, environmental science, and many others depending on the school's options. The long time span of interdisciplinary science teaching offers time for looking at issues from lots of different perspectives. Therefore, it supports a view of the world without discipline borders and deepens the student's understanding of the interconnection and relation of different science disciplines.

Since there neither exist nation-wide curricula nor standards specifically for chemistry education, everything that is said about science education in the NSES applies to chemistry education as well and provides answers to the following three basic questions:

– What should students know, understand, and be able to do?
– What is the way to achieve these goals?
– What kind of assessment enables teachers to test students' attainment of these goals?

Thereby they give guidance to achieve scientific literacy for all students. Scientific literacy is described as the "knowledge and understanding of scientific concepts and processes required for personal decision making, participation in civic and cultural affairs, and economic productivity" [23]. It also includes specific types of abilities and goals for all science teaching: Students should be able to

- "Experience the richness and excitement of knowing and understanding the natural world
- Use appropriate scientific processes and principles in making personal decisions
- Engage intelligently in public discourse and debate about matters of scientific and technological concern
- Increase their economic productivity through the use of the knowledge, understanding, and skills of the scientifically literate person in their careers" [23]

This underlines the importance of scientific literacy as an integral goal of all science teaching. The standards emphasize inquiry as a way of achieving scientific literacy, and they are divided into the following six chapters [23]:

1. Standards for science teaching
2. Standards for professional development of science teachers
3. Standards for assessment in science education
4. Standards for science content
5. Standards for science education programs
6. Standards for science education systems

While Chaps. 2, 3, 5, and 6 are of minor importance here, this section will focus on Chaps. 1 and 4 which concern the standards for science teaching and the standards for science content.

Chapter 1 highlights six standards for teaching science. "Teachers of science

- Plan an inquiry-based science program for their students
- Guide and facilitate learning
- Engage in ongoing assessment of their teaching and of student learning
- Design and manage learning environment that provides students with the time, space, and resources needed for learning science
- Develop communities of science learners that reflect the intellectual rigor of scientific inquiry and the attitudes and social values conducive to science learning
- Actively participate in the ongoing planning and development of the school science program" [23]

According to these standards, inquiry engages students in various hands- and minds-on activities like describing objects and events, asking questions, constructing explanations, testing those explanations, communicating their ideas to others, identifying assumptions, using critical and logical thinking, and considering alternative explanations. Instead of passively learning about science through the teacher, "students actively develop their understanding of science by combining scientific knowledge with reasoning and thinking skills" [23].

Chapter 4 divides science content into eight categories, which represent what students should know, understand, and be able to do:

1. Unifying concepts and processes in science
2. Science as inquiry
3. Physical science
4. Life science
5. Earth and space science
6. Science and technology
7. Science in personal and social perspectives
8. History and nature of science

It becomes apparent that the word "chemistry" does not appear anywhere in these eight categories, but neither do the words "biology," "physics," or "geology." However, this does not mean that chemistry is not part of science education in the United States. It rather means that chemistry ideas and skills are scattered over these eight categories and a thorough study of the content standards shows that all central facts, ideas, and skills of chemistry that students should learn during their chemistry education are included. This also acknowledges chemistry as the "central science" that connects the mostly macro world of biology and the mostly micro world of physics and generally reaches across all natural sciences.

These standards call for a *spiral curriculum* that introduces basic questions and skills of chemistry in K–4, deepens them in 5–8, and reaches sophistication in 9–12. Chemistry content throughout the aforementioned eight categories and from kindergarten through high school will be outlined now.

"Unifying concepts and processes in science" is the only category that is not subdivided for different age groups. Instead it shows the following fundamental concepts and processes for all age groups (in brackets: examples for chemistry education):

– System, order, and organization (system = periodic table; order = classification of matter as solid, liquid, and gas; organization = structure-property connections)
– Evidence, model, and explanation (evidence = experiments leading to new discoveries; model = atomic model; explanation = connecting experiments and models)
– Constancy, change, and measurement (constancy = law of conservation of matter/energy; change = development of atomic theory; measurement = metric system)
– Evolution and equilibrium (evolution = nuclear fusion, creating new elements; equilibrium = acid–base equilibrium, solubility equilibrium)
– Form and function (form = unique properties of water related to its function; function = characteristic reactions of organic molecules related to their functional groups)
– Energy (bond separation and formation, energy change, activation energy)

Since chemistry appears in all age groups and in all categories, the other seven categories can only be outlined briefly.

"Science as inquiry" requires students to understand the principles of scientific inquiry and to develop abilities to conduct scientific inquiry. In K–4, students ask scientific questions, investigate aspects of the world around them, use their observations to construct reasonable explanations, and communicate ideas and explanations. In 5–8, students deepen their knowledge and abilities of scientific inquiry. They systematize their observations, improve the accuracy of their measurements, identify and control variables, develop models, and discover the relationship between evidence and explanation. In 9–12, students sharpen their understanding and abilities of scientific inquiry.

In "physical science," K–4 students learn about substances existing in different states (solid, liquid, gas), they group and sort objects and materials according to their properties. These abilities are precursors for the later introduction of more abstract ideas such as the atomic structure of matter or conservation laws (of matter or energy). In 5–8, students study properties (e.g., density, boiling point, or solubility) and changes of properties through the formation of new substances in simple chemical reactions. The periodic table of the elements is being introduced and students start grouping substances according to their properties (like metals). They make the leap from the description of properties of objects to the properties of substances. In 9–12, students study the structure of matter (atoms, molecules, and ions). They conduct more complex reactions like acid–base reactions or oxidation and reduction reactions. They also investigate interactions of energy and matter (flow of electrons).

"Life science" provides the connection between chemistry and biological processes. In 9–12, students learn that cells are made up of molecules, that cell functions are chemical reactions, that molecules build the DNA and our genetic information. Chemical reactions in humans, animals, and plants are part of the carbon cycle.

Chemistry in "earth and space sciences" lets students in K–4 discover earth materials as sources of fuel and students in 5–8 discover the impact of volcanic eruptions on the water cycle (acid rain) and climate change (carbon dioxide). Beginning in K–4 with balances and thermometers and ending with computer-connected titrations in 9–12, the use of technological devices opens a window for "technological science" in chemistry.

"Science in personal and social perspectives" allows students from all age groups to better understand the chemical background of air pollution and climate change, two issues that have a growing impact on everybody's life.

The "history and nature of science" gives students an understanding of science as a human and interdisciplinary endeavor and fosters an appreciation for science in general. In K–4, students understand that people have learned a lot about the world around us, but that a lot more remains to be understood. In 5–8, students discover that scientists do and have changed their ideas and created better theories about natural phenomena when they found evidence that did not match existing theories. This is a great opportunity for 9–12 students to study biographies of famous scientists like Lavoisier or Dalton who have advanced science by devising new theories about the structure of matter. They also show that science is interdisciplinary, since they contributed to chemistry as well as to other disciplines.

These examples from the NSES show that chemistry is naturally connected to other sciences and that these connections should be highlighted when teaching chemistry at school.

3.2.3   England and the United Kingdom

All teaching and learning in England is based on the National Curriculum (NC), which has been established by the Department for Education and Employment and the Qualifications and Curriculum Development Authority following the Education Reform Act in 1988, and its latest revised edition in 2000 [24].

The NC gives school administrators as well as teachers an idea of the content to teach in science from key stage 1–4 (see explanation below). It sets achievement targets for science learning and determines how student performance can be assessed. It is robust in content yet flexible enough to give teachers the freedom to adapt content to their students' needs. It remains the school's task to include the program of study and the content of the NC into its own curriculum.

A brief description of the English school system supports the understanding of the curriculum. Education is divided into primary (kindergarten – grade 6) and secondary education (grade 7–13). The school system consists of infant school (kindergarten and grades 1–2), junior school (grades 3–6), and secondary school (grades 7–11) with an integrated sixth form (grades 12–13). Compulsory education ends after grade 11. In grade 12 and 13, students prepare for their A-level examinations, which are the entry qualification for higher education. Four key stages are assigned to the different school levels: infant school comprises key stage 1, junior school key stage 2, secondary school key stages 3 and 4 (excluding sixth form). The program of study as well as the achievement targets of the NC is classed according to these four key stages. NC tests assess the student's performance after key stages 1, 2, and 3. Science is taught as an interdisciplinary approach in infant and junior school. In secondary school, science learning takes place in the three disciplines: biology, chemistry, and physics.

While the program of study describes what the students should be taught, the achievement targets represent the expected standards of the students' performance. They include knowledge, skills, and understanding, and they consist of eight levels with increasing difficulty. The NC is laid out as a spiral curriculum, and students are expected to progress from level 1 to higher levels. However, not all students are expected to reach level 8. Descriptions for the content on all levels help teachers to judge students' performance.

The program of study consists of four areas of science:

– Scientific literacy
– Life processes and living things

- Materials and their properties
- Physical processes

Scientific literacy is supposed to be taught through the other three areas. The program of study builds the basis for planning lessons. The NC reduces the amount of facts to be learned by the students and instead emphasizes spiritual, moral, social and cultural development, key skills, and thinking skills. Science, for example, "provides opportunities to promote moral development through helping pupils see the need to draw conclusions using observation and evidence rather than preconception or prejudice, and through the discussion of the implications of the uses of scientific knowledge, including the recognition that such uses can have both beneficial and harmful effects" [24].

According to the NC, students should develop key skills in the following areas through science: communication, working with data, using ICT, team work, autodidactic competence, problem solving.

The way chemistry appears in the science curriculum will be laid out next. During *key stage 1* it is one of the main goals for students to develop skills in scientific inquiry. They learn to observe, explore, and ask questions and they begin to work together to collect evidence. They evaluate evidence and consider whether tests or comparisons are fair, they use reference materials and they share their ideas with their classmates using scientific language, drawings, charts, and tables. In "materials and their properties" students use their inquiry skills to find out about the influence of heat and cold on everyday materials. They group materials according to their properties and learn that properties influence the use of materials.

While improving their inquiry skills in *key stage 2*, students start using models and theories for explanations and discuss the positive and negative effects of scientific and technological development. In "life processes and living things," they discover the role of plants in producing new material for growth. In "materials and their properties," they study the differences between solids, liquids, and gases. Students investigate mixtures of substances, produce solutions, recover solids from solutions by filtering or evaporating, and use their knowledge to decide how mixtures might be separated. At this key stage they also talk about reversible changes (changing the state of matter) and nonreversible changes in substances (chemical reactions like burning).

In addition to a further improvement of their inquiry skills in *key stage 3*, students learn how scientists work together to be more successful in research and how important experimental evidence is to support scientific ideas. In this key stage, chemistry takes over a major part of the curriculum. In "life processes and living things," students learn about the chemical reactions in the photosynthesis process of plants and about elements that are crucial for plant growth like carbon, oxygen, hydrogen, and nitrogen. In "materials and their properties," students deepen their understanding of the states of matter by applying the particle theory. They characterize materials by their melting and boiling point as well as their density. They are being introduced to the periodic table of the elements. They learn that every element consists of identical atoms, that different elements vary in their physical properties, and that elements are combined in compounds. Students learn to represent these compounds and reactions with word equations and formulae. Eventually students take a deeper look into mixtures like air, rocks, and sea water; the methods for separating mixtures are being improved by introducing distillation, chromatography, and others. The topic of chemical reactions is revisited with the law of conservation of mass. The fact that many substances around us are made through chemical reactions establishes a connection to students' everyday life. Possible effects of burning fossil fuels on the environment fit into the current discussion about the pros and cons of fossil fuels and the development of regenerative kinds of energy. Students should also be taught about metals, acids, and bases. The area of "physical processes" lets students learn about particle movement, energy, and the conservation of energy, especially in the context of energy transfer.

Key stage 4 is divided into a single and a double curriculum. The single curriculum is intended for students who focus on other subjects than science (e.g., language or arts). The content in the single curriculum is therefore a reduced version of the double curriculum. The content for chemistry learning in key stage 4 is taken from the double curriculum. The students are able to sophisticate their inquiry skills. They consider the power and limitations of science, especially in social and environmental issues. In "life processes and living things," students put on the "chemistry glasses" to take a look at processes and compositions of substances in the human body like enzymatic reactions, stomach acid, the composition of blood, and the processes in the kidney. The processes of photosynthesis are reinforced as students take a deeper look at reactants and products. In "materials and their properties," students should learn that atoms consist of nuclei and electrons and that the arrangement of electrons has an impact on an atom's properties and chemical bonding.

Models should be introduced to illustrate atoms, ions, and molecules and their arrangement in chemical structures. The students learn that elements are placed in the periodic table according to their masses and properties. They investigate the groups of the noble gases, alkali metals, halogens, and transition metals and their properties, reactions, and uses. They improve their knowledge of chemical reactions by learning about other kinds of reactions like neutralization, oxidation, and reduction as well as thermal decomposition. They are taught that reaction rates vary greatly, but that they can be altered by, for example, varying temperature and concentration. Students learn more about metals, like their extraction from natural resources and electrolysis reactions. Quantitative chemistry (e.g., the chemical equilibrium) and organic chemistry (hydrocarbons, polymers, alkenes, and oil distillation and its products) also play an important role in the key stage 4 curriculum. In "physical processes," students learn about radioactivity, including its beneficial and harmful effects.

Since the curriculum contains a lot of chemical knowledge and skills the abovementioned issues can only be exemplarily. Like in the United States chemistry is taught in a spiral curriculum. Basic ideas of chemistry are taught from key stage 1 on and are taken up and deepened in the higher key stages. The NC tries to put an emphasis on inquiry-based learning and the connection of chemistry to everyday life. Different countries with different education systems have same goals: improving the teaching and learning of chemistry through setting standards and outlining essential contents.

3.3   Learner: Cognitive Development, Preconceptions, Attitudes, Interests

The "student-oriented lesson planning" according to Meyer [17] shows that the learner has to be taken into account for the lesson goals and the lesson planning. Especially the attitudes, interests, and preconceptions of the students have a big impact on the successful planning and conducting of lessons. Developmental–psychological aspects will be outlined first.

Educational goals and developmental psychology. Due to the developmental–psychological conditions, the goals and contents of school chemistry vary on the different levels of school. Piaget distinguished the following four stages of thinking abilities [25]:

Sensorimotor stage	(0–2 years)
Preoperational stage	(2–7 years)
Concrete operational stage	(7–13 years)
Formal operational stage	(>13 years)

The transition from the concrete to the formal stage of thinking is characterized in that the learner begins to think abstractly: they consider variables, derive potential relations, and understand formal mathematical explanations. New studies, however, show that Piaget's age limits are quite arbitrary: they do vary considerably amongst children. Some studies have shown that only 25% of 16-year-old students in grade 10 reach the formal operational stage [26].

Piaget used the terms assimilation, accommodation, and equilibration to explain the development of cognitive structures. The learner takes new information into their existing cognitive structure or compares it to their prior knowledge. As long as new information fits into the existing cognitive structure without changing it, the process is called *assimilation*. If a conflict arises between the new information and prior knowledge, the individual has to change their cognitive structure to work on the new information. This process is called *accommodation*. The learner's tendency is to establish an equilibrium between assimilation and accommodation – therefore the process is called *equilibration*.

Today's learning theories have a constructivist point of view: every learner needs to build their own cognitive structure, knowledge changes are constantly built by the individual. Depending on the kind of knowledge change, constructivists talk about *conceptual growth* and *conceptual change* [27]. Conceptual growth can be compared to Piaget's assimilation, the learning path is continuous. Conceptual change implies the revision of existing cognitive structures and the learning path is discontinuous: often the preconceptions have to be given up and radical restructuring needs to take place. Conceptual growth and conceptual change interact just like assimilation and accommodation.

Preconceptions. At the beginning of chemistry learning the students have cognitive structures, which, according to the constructivist idea, were built on the basis of their lifelong experience. These preconceptions are deep rooted, for example, the assumption of the destruction concept for combustion processes (see Chap. 1). There also exist spontaneous conceptions, which arise during the classroom discussion: they are not deep rooted and easier to correct.

For lesson planning, attention should be paid to those preconceptions that are deeply rooted in the student's cognitive structure: for example, concerning the combustion process, the law of conservation of mass, properties of gases, and the structure of matter (see Chap. 1). The student conceptions may be discussed in class before the scientific idea can be taught with convincing experiments and models. The so-called common-sense explanations or colloquial explanations belong to the deep-rooted preconceptions: often they are "socially accepted misconceptions" [26]. One example is that the sun is "going from the East around the earth to the West during one day" – instead of pointing out that the sun is fixed and the earth is revolving around the sun.

Attitude and interest. They should be taken into account for the lesson goals and lesson planning too. If the learner has a negative attitude towards chemistry or a certain topic in chemistry, it is more difficult to reach set goals in this subject than it is in others, towards which the learner has a positive attitude. If there is a positive attitude towards music or arts, students are highly motivated to follow those lessons.

With pictures drawn by Swiss students in 1983, Heilbronner [28] could show that their attitude towards chemistry and chemistry lessons was absolutely negative: "the vast majority of students felt themselves and the environment to be threatened by chemistry." Around 40% of the pictures showed a destroyed environment, 15% the direct threat to the individual by chemistry and 10% opposed animal testing. Barke and Hilbing [29] picked up on this exercise "draw your picture of chemistry" and repeated the study in 1998: the proportion of negative pictures, which was around 65% in Heilbronner's study, decreased to 40% (see Chap. 8).

Müller-Harbich et al. [30] analyzed secondary school students' attitudes. They found the attitudes to be neutral or negative, without a significant gender-related difference. The place of residence, however, had a significant impact on the attitude: students who live in an area full of industry show a more negative attitude than students who do not live close to industrial locations. When students have a positive attitude towards a subject, they are more likely to develop an interest in contents of this subject. If specific student interests are known, it is helpful to take these interests into account for planning lessons.

Gräber [31] analyzed students' interests and could prove that around half of all participating students are interested in chemistry lessons: "this is remarkable insofar as interest rises with the beginning of chemistry lessons in grade 8 and rises again in grade 10, after a 'hole' in grade 9." He also found differences between boys and girls, when asking for their interest in chemical contexts and different activities (see Fig. 3.6). The item with the highest agreement was "run experiments." Also the items "plan and observe experiments," "watch a film," and "build chemical models" were ranked as interesting.

Barke and Wanjek [32] confirmed this positive interest in student experiments. Instead of using laboratory chemicals for the topic "acids and bases," the students

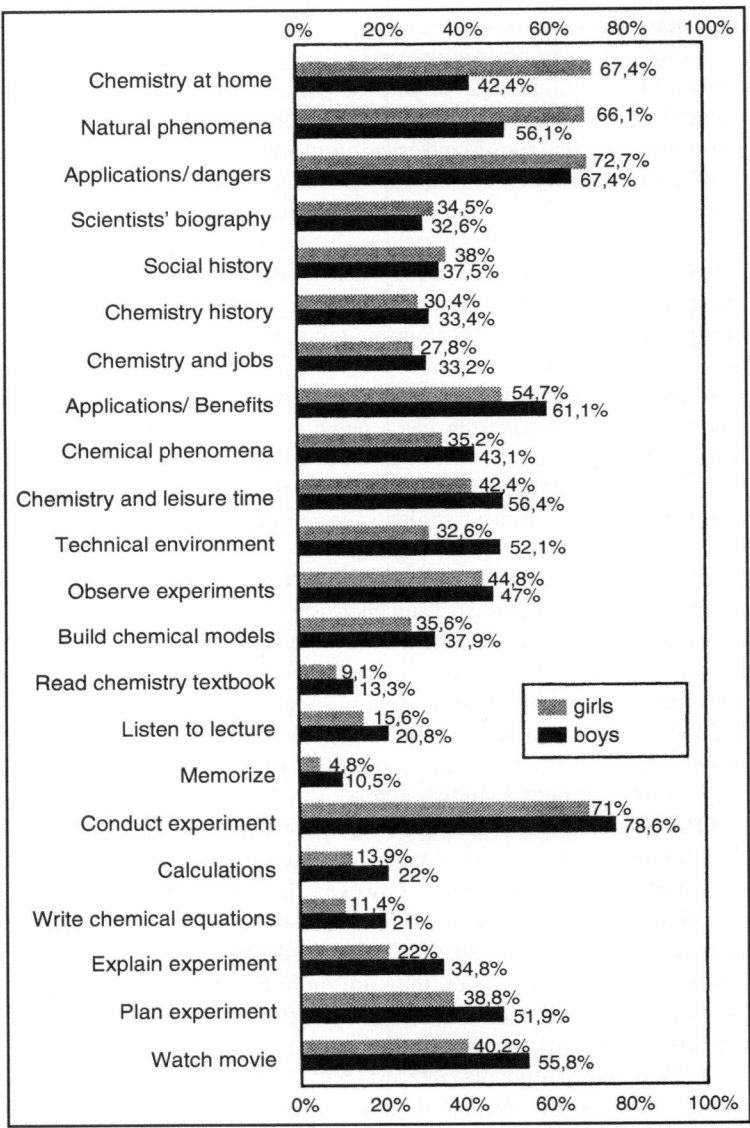

Fig. 3.6 Interest in chemistry topics and related activities [31]

were offered familiar substances from kitchen and bathroom to use in student experiments. This teaching unit was highly accepted by all students. Especially the girls liked this kind of chemistry lessons with everyday chemicals better than the chemistry lessons with regular lab chemicals.

3.4   Teaching Processes: Variety of Methods for Realizing Teaching Goals

There is a full discussion going on about the amount of phenomena and of theoretical consideration students should work on in their chemistry lessons. "Save the phenomena" says one side, "theories and mental models are the true domain of chemistry" says the other side. The truth probably lies in the middle, although phenomena naturally predominate the beginning.

Guidelines and curricula (see Sect. 3.2) therefore put the explanation of phenomena into the focus at the start of chemistry lessons: words are used for description of first substances and chemical reactions, before chemical symbols can be introduced. Harsch and Heimann [33] developed a concept for organic chemistry, which is based on phenomena and on activity orientation. Their curriculum is laid out in a way that it allows the students to classify phenomena and discover connections between phenomena, before molecular models or structural formulae are being introduced. New substances, phenomena, reactions, and formulae are added continuously and integrated into the growing system without problems: "phenomenon-oriented, integrative and interconnected chemistry." This concept will be introduced in Chap. 9.

Every teacher has to decide for his own class to what extent he introduces a simple particle model of matter, to explain changes of state, dissolution processes, or diffusion of gases. This preliminary model is useful and comprehensive as long as mental models of particle arrangements are illustrated by structural models – such as packing of spheres for the structure of metal crystals. If students build these structural models themselves, they are even motivated to think about the arrangement of particles and develop a mental model in their cognitive structure. Studies in grade 3 and 4 of elementary school show that the common pictures of particle arrangements are indeed being understood and can still be reproduced years later.

When the students proceed to the formal operational stage, Dalton's atomic model and terms like element, compound, atom, ion, and molecule can be added to the spiral curriculum of models in chemical education (see Fig. 3.7).

First models of the structure of elements and compounds can be introduced and first chemical reaction of two elements to a certain compound can be interpreted with structural models before and after the reaction. Later atomic, ionic, and molecular symbols; symbols of the structure of molecules or of ionic lattices; and symbols for chemical reactions can be used. Eventually, the topics atomic structure and chemical bonds may follow (see Fig. 3.7).

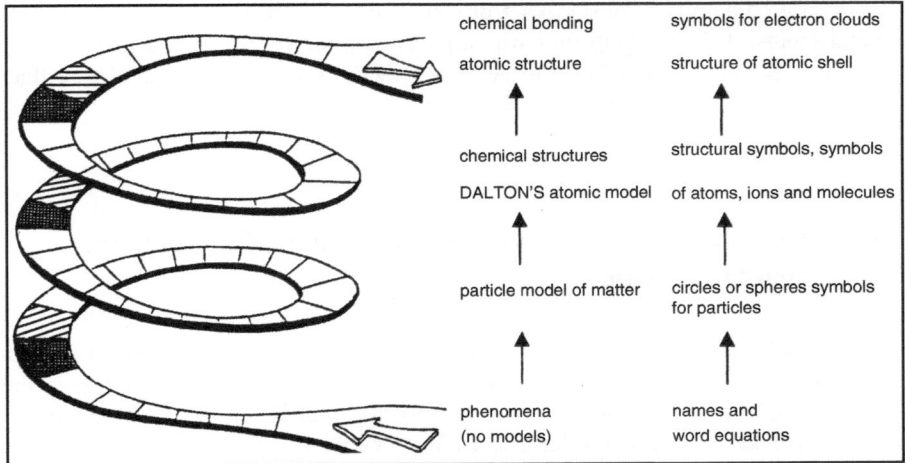

Fig. 3.7 Structural models and chemical symbols in the spiral curriculum

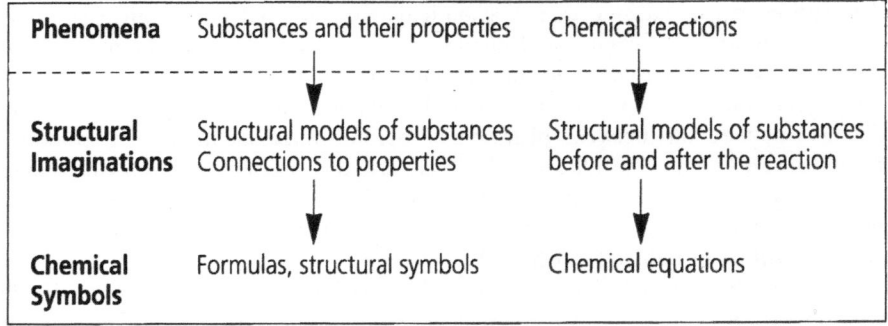

Fig. 3.8 Structural models as a mediator between phenomena and chemical symbols

Models, especially structural models on the basis of Dalton's atomic model are very helpful for the understanding of formulae and reaction symbols – an important goal of chemistry lessons. Structural models – like packing of spheres for the structure of crystals or molecular models for molecular structures – are adequate to teach the structures of involved substances on a medium level of abstraction. Chemical symbols and their different information contents are introduced on the structural level (see Fig. 3.8, also see Chaps. 6 and 7). The structure of matter is continuously taken as a basis in "structure-oriented chemical education" for the interpretation of many phenomena and reactions in chemistry (see Chap. 10).

Planning lessons on the basis of structural models has two big advantages. On one side the models can be arranged in a spiral curriculum and allow for a comprehensive teaching concept. This concept starts with the simple particle model of matter on the concrete operational stage in elementary schools and ends with the nucleus-shell model on an abstract level at the end of secondary school (see Fig. 3.7).

On the other side, the different models and symbols are good for setting approximate teaching goals for certain classes, to decide when these models and symbols are being used and which phenomena and reactions are being conducted and analyzed on the basis of another model. These decisions have to be made depending on the type of school, class level, prior knowledge, and level of the class and have to be reconsidered for every class.

Two other examples for a spiral curricular approach for the terms solubility and acidic solutions can be found (see Fig. 7.4). These approaches begin with the everyday knowledge – the preconceptions – and proceed along different learning paths, until the description reaches the highest abstraction level at school. Schmidkunz and Büttner [34] tried to illustrate the whole school chemistry with the help of a spiral curriculum (see figure 3.9).

Besides all the terms and their spiral curricular arrangement a big goal of chemistry lessons is to teach the scientific cognition process: chemistry knowledge is usually gained with empirical inductive methods. Students will accept this method easier, if they also use it during their chemistry lessons. The empirical inductive method is closely related to the hypothetical deductive method as the following steps show (see also Chap. 5):

– Observe a phenomenon
– Observe other phenomena and empirical results, find a regularity
– Develop a hypothesis to explain the result or regularity
– Test the hypothesis and its conclusions experimentally
– Develop theories and models by combining corroborated hypotheses

Usually not all of these steps can be used on every example in chemistry classes. But it is not reasonable to randomly skip steps and to proceed too fast from a single phenomenon to general coherences. The way from single phenomena proceeding to a hypothesis or empirical regularity is called inductive approach. The way from the theory over the derived hypothesis to the predicted phenomenon is called deductive approach. Both methods are complementary.

Teaching problems. "Lessons have to be considered important, otherwise they are irrelevant. Learning processes have to be built on the student's interest, otherwise they do not affect the student. 'To be boring is the greatest sin of teaching', Herbart wrote in 1806. Therefore teachers have to overcome the rigid teacher-centred teaching in favor of lessons that are based on a variety of learning situations and include an extensive spectrum of teaching methods" [35].

It is necessary to vary the teaching methods, to reach the multiple teaching goals in the most interesting ways. There exist different didactical approaches for teaching chemistry: for example the term-oriented approach, which is based on technical terms and their spiral curricular arrangement. Or the approach, which is oriented towards the scientific way to gain knowledge in chemistry, for example the empirical inductive method, where hypothesizing and testing have priority. Besides the most important teaching and social methods, the above-mentioned teaching methods and chemistry-educational approaches will be presented and discussed to provide an extensive spectrum of teaching methods.

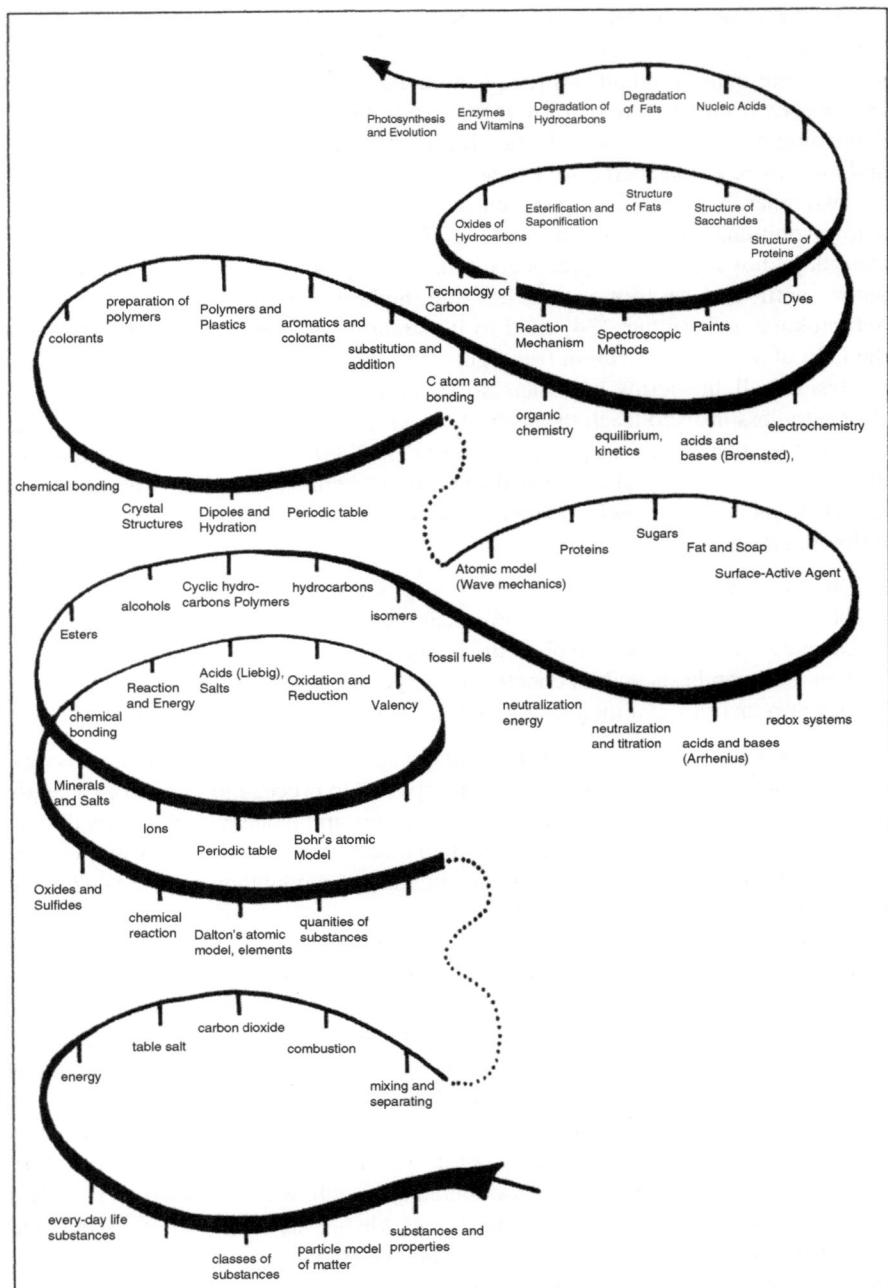

Fig. 3.9 Proposal of a sprial curriculum for chemical education [34]

Didactical approaches. Problems can be presented during the lessons in many ways. Therefore, it is possible to vary the different didactical approaches or to combine them. The following are the main approaches, which are described briefly:

Term oriented	New scientific terms of the structure of chemistry are introduced
Procedure oriented	The way of gaining knowledge by proving hypothesis is shown
Structure oriented	Chemical structure of matter is in the center of teaching
History oriented	History of science is taken as a basis
Application oriented	Applications of student's environment and everyday life
Environment oriented	Environmental pollution of water or air and protection
Activity oriented	Students' activities and hands-on experiments
Project oriented	Interdisciplinarity, activity orientation, cooperation

Problem-oriented lessons are often added to this list. However, problem orientation can only be an umbrella term for all approaches, since every lesson should have a "problem" in the center. It should not always be the teacher, who presents the problem, but also the student, the experiment, or the structural model – true motivation and interest can be awaken and maintained.

The different didactical approaches will be explained with examples from chemistry lessons and connected to teaching and social methods. These will be listed beforehand. Whereas the organizational aspect of teaching is in the spotlight of teaching methods, the communicative aspect is in the spotlight of social methods.

Teaching methods. Teacher-centered teaching is the best known and most common, but also the most controversial teaching method: "international research on school quality agrees that the rigid teacher-centred method is the least favorable for the student" [35]. Therefore, other teaching methods should be favored and are expected in lessons, when teacher students conduct their assessed teaching lessons at the end of their education:

- Problem solving: a single problem, experiment, or structural model starts the lecture
- Inquiry learning: students have to discover a rule or a law after observing experiments
- Differentiated learning: students are learning by themselves in different learning groups
- Programmed learning: learning a special topic by a computer program
- Workshop: learning by hands-on activities supervised through an expert
- Internship: learning from professionals outside of schools in a company or factory
- Excursion: going outside the school and learning in nature, in companies or factories

Social methods. The following forms of communication between teacher and students can be distinguished: teacher lecture, student lecture, teacher demonstration, student demonstration, teacher–class dialogue, discussion, group work, pair work, individual work, silent work, role play, experimental game, simulation, improvisation, free play, etc. Memmert [36] gives an overview of teaching and social methods (see Fig. 3.10).

The possibilities to introduce, for example, organic chemistry or acid–base reactions are shown by combinations of didactical approaches and social methods:

Term oriented	Nomenclature of organic substances (teacher lecture)
Procedure oriented	Composition of methane (group work)
Structure oriented	Structure of butane molecules (pair work)
History oriented	Benzene and Kekulé's idea (student lecture)
Application oriented	Cars do not drive without alkanes (teacher class–dialogue)
Environment oriented	The role of the catalytic converter (student demonstration)
Activity oriented	We produce our own candles (group work)
Project oriented	What are fossil fuels? (pair work)

Possible approaches for the introduction of acid–base reactions:

Term oriented	Brønsted's theory (teacher demonstration)
Procedure oriented	Composition of hydrochloric acid (teacher–class dialogue)
Structure oriented	Structure of sulfuric acid molecules (pair work)
History oriented	Lavoisier and his acid definition (student lecture)

(continued)

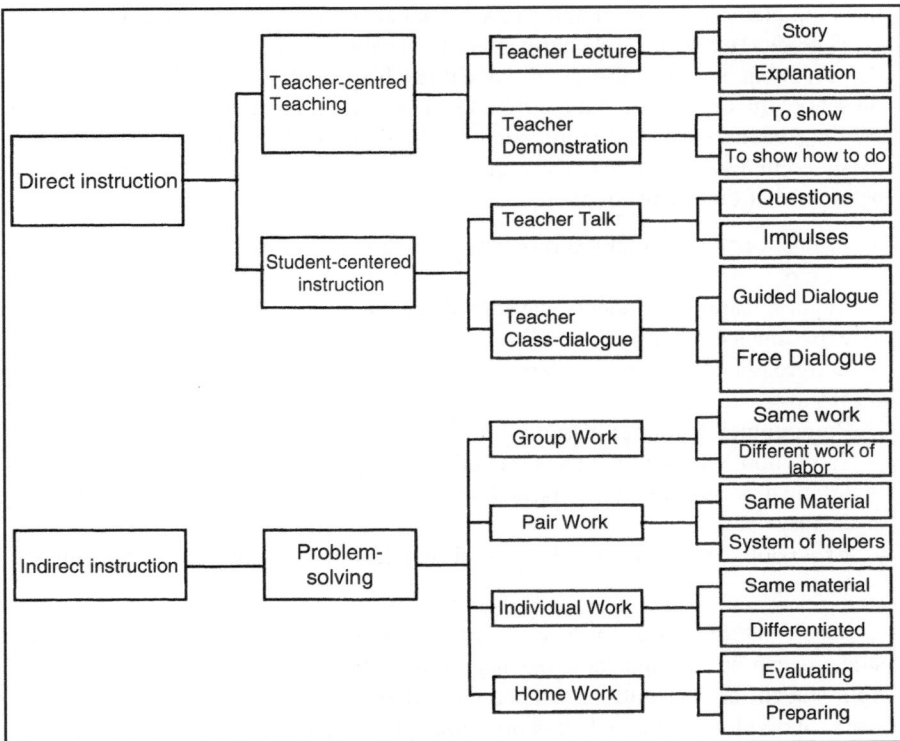

Fig. 3.10 Proposal of teaching and social methods [36]

Application oriented	Limescale remover in the household (group work)
Environment oriented	Sulfuric acid recycling (excursion)
Activity oriented	Producing soft drinks with citric acid (group work)
Project oriented	Acidic rain and environment protection (pair work)

The umbrella term for all didactical approaches should be the problem-oriented idea for lessons. The problem-oriented approach also includes the step sequence according to Schmidkunz-Lindemann [37]. This model for the strategy of problem solving in science lessons distinguishes five steps:

1. Presenting the problem
2. Considerations of the problem solution
3. Attempting a way to solve the problem
4. Abstraction of the gained knowledge
5. Saving the knowledge [37]

Experiments naturally play an important role in this approach: introductory experiments for "presenting the problem," further experiments for "considering and solving the problem" by proving a hypothesis, repeating experiments for "saving the knowledge."

The *history-problem-oriented approach* was developed by Jansen [38] to put a focus on the history of chemistry. This didactical approach assumes that "the history of chemistry should not be regarded as unnecessary ballast that – according to the latest state of science – can be eliminated. The history of chemistry has to be set in relation to the content. The point is that the student has to gain insight into the natural scientific view of the world. The teacher also needs to point out that not all theories are deduced in an empirical inductive way, instead they may have been invented in a speculative way" [38].

At the end we will take a closer look at the *project-oriented approach*. School projects have been discussed since the early twentieth century, culminated in the 1960s, and since then also found entrance into high school education. They can either take place during a whole week (project week) or they are carried out within the regular schedule. With the "basic patterns of the project method," Frey [39] gives planning aid to run a project from the student's initiative to the project close-out in about seven steps (see Fig. 3.11). One example for this kind of project is "water and environment" [40]: this project was realized in a ninth grade class over a period of eight weeks during normal chemistry lessons of 2 h per week.

Münzinger and Frey [41] mention the following characteristics of the project method:

1. Need oriented: students' interests start the project
2. Situation oriented: student's everyday life should be in the center
3. Interdisciplinarity: crossing boarders to other subjects like physics, biology, and others
4. Self-organization: students set objectives, plan and carry out
5. Product oriented: the project is oriented towards a result or a special product
6. Collective realization: division of labor for all participants and social learning

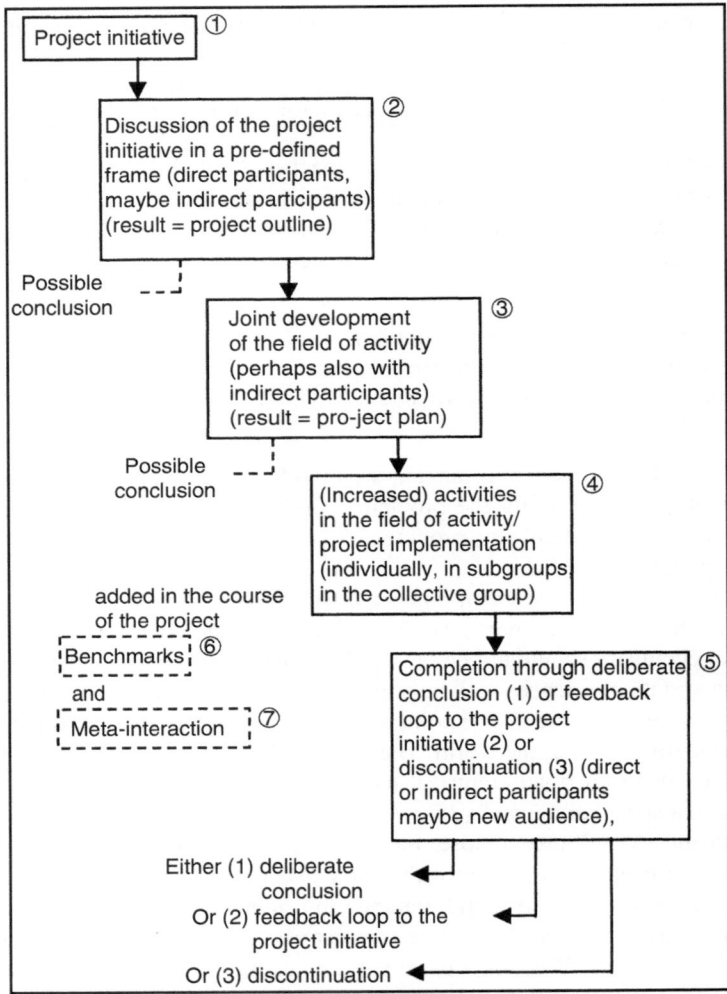

Fig. 3.11 Basic pattern of the project method according to Frey [39]

Bruhn [42] discusses the project-oriented approach in the context of "key problems as fundamental learning content." After lessons in biology, chemistry, and physics, the characteristic "interdisciplinarity" is used as an opportunity to integrate all of these subjects, for example in the last year of high school (see Fig. 3.12): "there exists a lot of positive experience with project-oriented lessons about subproblems of key questions, for example studying the vegetation in urban areas or samples of water, measuring noise exposure at streets or radioactivity inc ventilated or unventilated rooms, energy use, photosmog, questions of armament, waste incineration, nitrogen overfertilization, greenhouse effect, climate change,

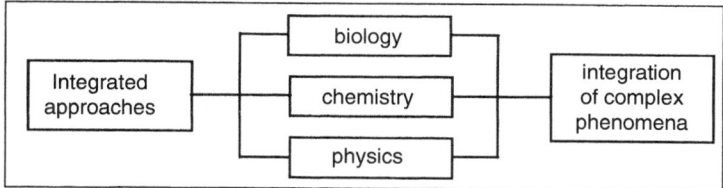

Fig. 3.12 Interdisciplinary approach for the integration of complex phenomena after lessons in the single subjects according to Bruhn [42]

etc. These projects lead to a basic understanding of the key problems of our time and also affect motivation for normal lessons.... Although permanent project-orientation in science lessons is not possible – a project can only complement the lessons in chemistry, physics or biology" [42].

Problems and Exercises

P3.1. Learning objectives can be distinguished according to the cognitive, affective, and psychomotoric dimension. Use examples of your choice to show three different teaching goals for the three dimensions.

P3.2. Operationalized learning objectives give very detailed operations for the student to achieve a very detailed teaching goal. Convert the three written goals of the learning objectives from P3.1 into operationalized learning goals.

P3.3. Learning objectives can be differentiated and hierarchized. Choose an example of learning objectives and hierarchize different teaching goals according to Bloom's taxonomy. Choose another example and differentiate goals according to the taxonomy of German experts.

P3.4. There exist different didactical approaches to not only realize teaching goals in one way, but in many different ways, which make lessons interesting for the student. Name all approaches. Use the example of "redox reactions" to give brief introductions for each approach.

P3.5. There are different schemes to prepare the lessons of 1 or 2 h. Create a lesson plan for a topic and age group of your choice. It should include the didactical discussion of the learning path (see the following "scheme for a lesson plan").

Scheme for a lesson plan (suggestion):

1. Topic, problem, learning objectives
2. Technical bases for the topic
3. Requirements for the students
4. Methodological–didactical considerations of the concept
5. Outline of the lesson (if applicable with the following grid: time, planned teacher behavior, expected student behavior, media/comments).

References

1. Scheible A (1969) Ist unser Chemieunterricht noch zeitgemäß? MNU 22:449
2. Erhart H (1998) Chemie – einer der unbeliebtesten Unterrichtsgegenstände? Chem Sch 4:29
3. MNU, GDCh, GDCP et al. (1998) Mathematische und naturwissenschaftliche Bildung an der Schwelle zu einem neuen Jahrhundert. CHEMKON 5:209
4. Bildungsrat D (1971) Empfehlungen der Bildungskommission. Strukturplan für das Bildungswesen. Beltz, Stuttgart
5. Möller Ch (1973) Technik der Lehrplanung. Weinheim, Beltz
6. Bloom BS (1972) Taxonomie von Lernzielen im kognitiven Bereich. Weinheim, Beltz
7. Klafki W (1980) Die bildungstheoretische Didaktik. WPB 1:32
8. Schulz W (1980) Die lerntheoretische Didaktik. WPB 1:80
9. Möller Ch (1980) Die curriculare Didaktik. WPB 1:164
10. Winkel R (1980) Die kritisch-kommunikative Didaktik. WPB 1:200
11. Cube FV (1980) Die informationstheoretisch-kybernetische Didaktik. WPB 1:120
12. Blankertz H (1973) Theorien und Modelle der Didaktik. Juventa, München
13. Ruprecht H (1976) Modelle grundlegender didaktischer Theorien. Schroedel, Hannover
14. Meyer H-L (2009) Leitfaden Unterrichtsvorbereitung. Cornelsen, Berlin
15. Heimann P, Otto G, Schulz W (1965) Unterricht. Analyse und Planung. Schroedel, Hannover
16. Bönsch M (1976) Unterrichtsanalyse. Erziehung und Unterricht 10:676
17. Meyer H-L (1984) Leitfaden zur Unterrichtsvorbereitung. Scriptor, Frankfurt
18. Kultusminister NRW (1993) Richtlinien und Lehrpläne, Chemie, Gymnasium SI/II. Düsseldorf
19. MNU (2000) Empfehlungen zur Gestaltung von Lehrplänen bzw. Richtlinien für den Chemieunterricht. MNU 53:161
20. Parchmann I, Kaufmann H (2004) Kompetenzen entwickeln. Wie Bildungsstandards zu einer Chance fuer Schulentwicklung werden koennen. Unterricht Chemie 17:4
21. Schmidt M (1972) Didaktik Chemie. Schwann, Düsseldorf
22. Moeller K (2005–2008) Die KiNT-Boxen – Kinder lernen Naturwissenschaft und Technik im Sachunterricht. Spectra, Essen
23. National Committee on Science Education Standards and Assessment, National Research Council (1996) National Science Education Standards. National Academy Press, Washington
24. Department for Education and Employment, Qualifications and Curriculum Authority (2000) The National Curriculum for England – Science. Stationery Office Books, Norwich
25. Piaget J, Inhelder B (1973) Die Psychologie des Kindes. Deutscher Taschenbuch, Freiburg
26. Gräber W, Stork H (1984) Die Entwicklungspsychologie Jean Piagets als Mahnerin und Helferin im naturwissenschaftlichen Unterricht. MNU 37:257
27. Duit R (1996) Lernen als Konzeptwechsel im naturwissenschaftlichen Unterricht. IPN, Kiel
28. Heilbronner E, Wyss E (1983) Bild einer Wissenschaft: Chemie. ChiuZ 17:69
29. Barke H-D, Hilbing CH (2000) Image von Chemie und Chemieunterricht. ChiuZ 34:16
30. Müller-Harbich G et al (1990) Die Einstellung von Realschülern zum Chemieunterricht, zu Umweltproblemen und zur Chemie. Chim did 16:150
31. Gräber W (1992) Untersuchungen zum Schülerinteresse an Chemie und Chemieunterricht. Chem Sch 39:270, 354
32. Wanjek J, Barke H-D (1998) Einfluss eines alltagsorientierten Chemieunterrichts auf die Entwicklung von Interessen und Einstellungen. In: Behrendt H (ed) Zur Didaktik der Physik und Chemie. Leuchtturm, Kiel
33. Harsch G, Heimann R (1998) Didaktik der Organischen Chemie nach dem PIN-Konzept. Vom Ordnen der Phänomene zum vernetzten Denken. Vieweg, Wiesbaden
34. Schmidkunz H, Büttner D (1985) Chemieunterricht im Spiralcurriculum. NiU PC 33:19
35. Winkel R (1993) Langweilig sein, die ärgste Sünde des Unterrichts. DLZ 11, März
36. Memmert W (1977) Didaktik in Graphiken und Tabellen. Klinkhardt, Bad Heilbrunn

37. Schmidkunz H, Lindemann H (1973) Das forschend-entwickelnde Unterrichtsverfahren. München
38. Jansen W et al. (1986) Geschichte der Chemie im Chemieunterricht - das historisch-problemorientierte Unterrichtsverfahren. Teile 1 und 2. MNU 39:321, 391
39. Frey K (1982) Die Projektmethode. Beltz, Weinheim
40. Barke H-D (1999) Wasser und Umwelt. In: Münzinger W, Frey K (eds) Chemie in Projekten. IPN, Kiel
41. Münzinger W, Frey K (1986) Chemie in Projekten. Kiel, IPN
42. Bruhn J (1993) Probleme unserer Zeit als Herausforderung für den naturwissenschaftlichen Unterricht. MNU 46:195

Further readings

Allix NM (2000) The theory of multiple intelligences: a case of missing cognitive matter. Aust J Educ 44:272–288
Ames C (1992) Classrooms: goals, structures and student motivation. J Educ Psychol 84:261–271
Bernal PJ (2006) Addressing the philosophical confusion regarding constructivism in chemical education. J Chem Educ 83:324–326
Beyer S, Riesselmann M, Warren T (2002) Gender differences in the accuracy of self-evaluations on chemistry, English and art questions. In: Paper presented at the Annual Meeting of the American Psychological Society, New Orleans, USA. Available from http://eric.ed.gov
Bodner GM (2003) Problem solving: the difference between what we do and what we tell students to do. Univ Chem Educ 7:37–45, http://www.rsc.org/images/Vol_7_No2_tcm18-7045.pdf
Bodner G, Domin D (2000) Mental models: the role of representations in problem solving in chemistry. Univ Chem Educ 4:24–30, http://www.rsc.org/images/Vol_4_No1_tcm18-7038.pdf
Bucat R (2004) Pedagogical content knowledge as a way forward: applied research in chemistry education. Chem Educ Res Pract 5:215–228
Cardellini L (2004) Philosophical confusion in chemical education research – constructivism and chemical education. J Chem Educ 81:194
Chi MTH, Slotta JD, de Leeuw N (1994) From things to processes: a theory of conceptual change for learning science concepts. Learn Instruct 4:27–43
Diakidoy I-AN, Kendeou P, Ioannides C (2003) Reading about energy: the effects of text structure in science learning and conceptual change. Contemp Educ Psychol 28:335–356
Donnelly JF (2004) Humanizing science education. Sci Educ 88:762–784
Driver R, Asoko H, Leach J, Mortimer E, Scott P (1994) Constructing scientific knowledge in the classroom. Educ Res 23:5–12
Driver R, Easley J (1978) Pupils and paradigms: a review of literature related to concept development in adolescent science students. Stud Sci Educ 5:61–84
Duit R, Treagust DF (2003) Conceptual change: a powerful framework for improving science teaching and learning. Int J Sci Educ 25:671–688
Eybe H, Schmidt H-J (2001) Quality criteria and exemplary papers in chemistry education research. Int J Sci Educ 23:209–225
Eylon BS, Linn MC (1988) Learning and instruction: an examination of four research perspectives in science education. Rev Educ Res 58:251–301
Johnstone AH (1997) Chemistry teaching – science or alchemy? 1996 Brasted lecture. J Chem Educ 74:262–268
Mortimer EF (1995) Conceptual change or conceptual profile change. Sci Educ 4:267–285
Nurrenburn S, Pickering M (1987) Concept learning versus problem solving: is there a difference? J Chem Educ 64:508–510
Shulman LS (1986) Those who understand: knowledge growth in teaching. Educ Res 15:4–14

Strike KA, Posner GJ (1992) A revisionist theory of conceptual change. In: Duschl RA, Hamilton RJ (eds) Philosophy of science, cognitive psychology, and educational theory and practice. SUNY Press, Albany, NY, pp 147–176

Tai RH, Sadler PM (2007) High school chemistry instructional practices and their association with college chemistry grades. J Chem Educ 84:1040–1046

Tyson LM, Venville GJ, Harrison AG, Treagust DF (1997) A multidimensional framework for interpreting conceptual change events in the classroom. Sci Educ 81:387–404

Chapter 4
Media

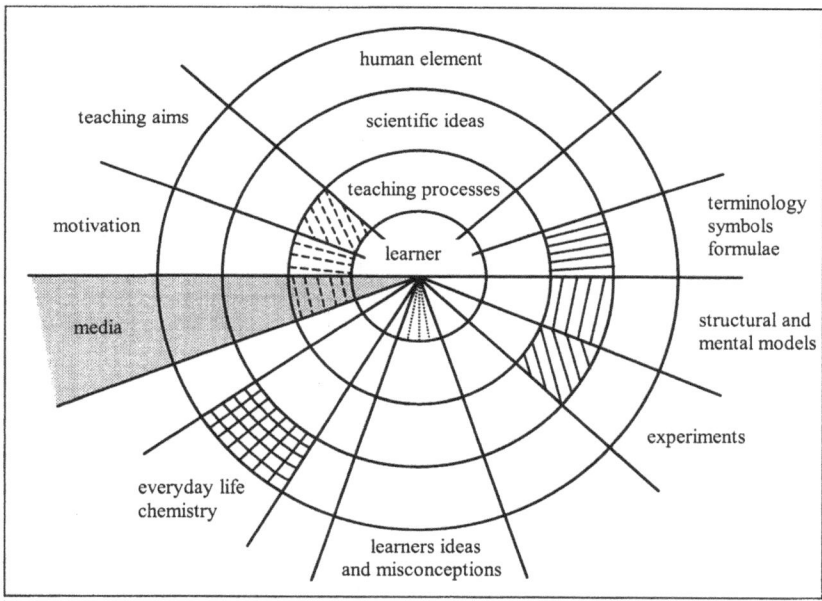

*There is a lot of debate over the use of media in school. On one occasion, a court was
actually appealed to for the question of whether a textbook had to be used in class or not:
"The conflict lasted 8 years already. The teacher addressed a petition to the state parlia-
ment and demanded to not waste money on useless books, but to provide the money for
experiments. His colleagues disagreed with this claim: "The chemistry teacher tried to
impose his ideas without regard for his colleagues and students. Students would not be able
to look up something they learned in class and would have difficulties with a new teacher."
The teacher, who actually went to the Federal Court of Justice, had to commit to using text
books and to record this in class books after his lectures."* [1]

H.-D. Barke et al., *Essentials of Chemical Education*,
DOI 10.1007/978-3-642-21756-2_4, © Springer-Verlag Berlin Heidelberg 2012

Text book and black board are the classical media used in every class and every type of school. The term media can be discussed under the aspects of media education as well as under the aspects of media didactics (see Fig. 4.1). This text will address the basic questions of media didactics: "It deals with the functions and effects of media in teaching and learning processes. Its goal is the advancement of learning processes through a didactically appropriate organization and methodically effective use of media. The choice of media and their use should be appropriate for the teaching goals, contents and methods" [2].

Classification of media. We speak of "personal media as for instance language, gesture, and motion, which are part of the term media. These media, where the teacher is not the carrier of information, are also present in lessons, which are not typically media oriented. They play an important role in every lesson." [3] The text will not take a closer look at the teacher, who represents a medium himself, but instead it will focus on the impersonal media of education at school (see Table 4.1). There is a differentiation between the kind of sensory experiences (visual, audible, audiovisual) and the level of experiences, either primary or secondary experiences.

Becker and Hildebrandt [4] investigated the use of media in the second half of the last century and found with their Fadok data [5] that some media have a long history in education and while others have not been used much (see Table 4.2).

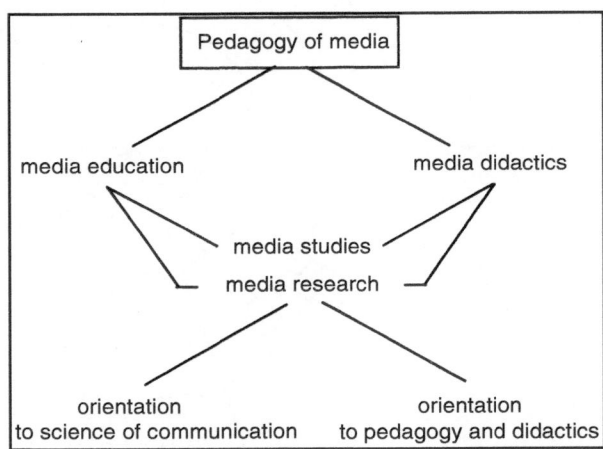

Fig. 4.1 Media didactics as one aspect of media education [2]

Table 4.1 Classification of general media regarding education at school [3]

Media of education				
Sensory experience			Level of experience	
Visual	Audible	Audiovisual	Primary	Secondary
Slide, photo	Radio	Slides with sound	Animals, plants	Preparations
Text book	Audiotape	Movie	Substances, crystals	Models
Transparency	CD	Television	Experiments	Recordings
Black board	Cassette	Video	Starlit sky	Planetarium
Computer	Language lab	Multimedia	Excursion	Movie, video

Table 4.2 Key words in articles about media from 1945 to 1996 [4]

Media/period	1945–1954	1955–1964	1965–1974	1975–1984	1985–1996	Total
Games, playing	0	1	1	39	109	150
Overhead projection	0	0	42	73	74	189
Recording tape	0	4	26	32	3	65
Blackboard	1	6	5	3	13	28
Molecular set (other models)	0 (3)	0 (44)	8 (51)	15 (73)	12 (65)	35 (236)
Text book	19	56	47	64	75	261
Audiovisual	9	65	61	124	94	353
Computer	0	0	7	69	295	371
Work sheets	2	16	13	52	220	303
Applications	0	7	19	17	13	56
Stamps	0	1	2	8	15	26
Media (general)	15	39	107	132	217	386
Total	49	239	389	701	1,215	2,593
Total of investigated articles	581	3,089	3,851	4,706	7,330	19,557

The blackboard with the longest history is not mentioned very often, but is still an important medium. The overhead projector enters into lessons after 1965 like the computer does from 1985. Molecular modeling sets are not mentioned much – it may be that they are present all the time; but the other models of the structure of matter started to make an impact in the 1950s and 1960s and opened the door to a good understanding for 3D structures concerning metals and salts. Work sheets for students in class have been mentioned more and more frequently and have a maximum today.

Considering the number of articles about media in general (which increases dramatically from decade to decade), and comparing it with the total of investigated articles in the past, Becker and Hildebrandt [4] stated that the media for education are claiming more and more attention given the increase from 8% in 1945/54 to 17% in 1985/96. Investigating current articles in journals, the big significance of computers and their software would be confirmed.

Evaluation criteria. Media should be used corresponding to the learning situation:

– Based on the learner's cognitive stage of development
– The learner's prior knowledge and interests
– The intended teaching goals
– The lesson's planned social form
– Aspects of Gestalt and perceptual psychology
– Technical workability and mastery of technical devices

Effects of media. Studies in experimental psychology show that "the human being receives information to

– 78% over the eye, to
– 13% over the ear and to
– 3% over each of the senses of smell, taste, and touch" [3]

Since the capacity for information does not say anything about the retaining of information, relevant studies show the results: "Human beings retain

10% of what they read	50% of what they see and hear
20% of what they hear	80% of what they say themselves
30% of what they see	90% of what they do themselves" [3].

Empirical studies have also been made about "forgetting contents without working follow-up". They show "that the human beings forget what they...

Hear	To 30% after 3 h	To 90% after 3 days
See	To 28% after 3 h	To 80% after 3 days
Hear and see	To 15% after 3 h	To 35% after 3 days" [3]

4.1 Teaching Processes: Variety of Media for Chemistry Lessons

Suitable media for chemistry lessons are those that are scientifically used like experiments and models of the structure of matter, or those that are original in their own way like transparencies, movies, computer programs, etc.

Function of many media. The studies on retaining of information show that the use of media is essential and advantageous. But it has to be considered to what extent a medium can help in chemistry lessons and what difficulties have to be taken into account.

The *functions* can be of the following kind:

– *Motivation*: media can be used to motivate the learner during the chemistry lesson, if there is no primary experience from the learner's environment available
– *Process of perception*: figures or tables of measurements can be used, if it is not possible to run an experiment to test a hypothesis
– *Problem orientation*: pictures, tables, or charts can function as an introduction to a problem; for example, problems of everyday life and environment as well as problems of chemistry history
– *Supply of information*: repetition, learning assessments, etc.

Difficulties of different media may include:

– Transparencies, slides, or movies can cement the teacher-centered instruction and might support inactivity and uncritical acceptance of information on the learner's side
– Media, especially transparencies, that are being used for years and not updated, might include outdated facts

- If the teacher uses ready-made transparencies, tables, or charts and does not develop them during the lesson, students mostly do not have the time to copy or draw the information
- If the teacher shows movies or videos, they often provide information that is far beyond the teaching goal: it has to be decided whether to only show relevant sequences or to show these sequences again after the whole movie

Functions, difficulties, and other aspects will be discussed exemplarily with media on a chemistry didactical basis (see Table 4.3). Important media are textbooks, blackboard, and transparencies.

Text book. A text book has to be approved by the relevant authority to be officially used in class. A commission examines to what extent the book conforms to the guidelines of the curriculum (the book can include contents that exceeds the guidelines). If the result of the commission is positive, the book is put on the list of approved school books.

On the suggestion of the chemistry teachers, the school's committee decides on the introduction of a certain text book. Criteria for the decision can be:

- Structure of the discipline: chemical ideas that represent chemistry in an optimal way, technical correctness, addressee adequacy
- Didactical concept: motivation, problem orientation, use of experiments and models, scientific perception, everyday relatedness of contents, spiral curricular approach, deepening of contents, problems and exercises, etc.
- Methodological concept: two columns for text and pictures, colored photos, experimental instructions, figures of models, tables and charts, phrases and summaries, etc,
- Teacher instruction: a book for teachers accompanying students' text book, an exercise book with (separate) solutions, price, size, weight and others.

Blackboard. A blackboard can be found pretty much in every class room and usually in an exposed position, so that it is visible from every spot of the class room. Therefore, it is being used in nearly every lesson. One should be aware of the importance of its function: often the students copy into their exercise books what is on the blackboard. Therefore, the writing on the blackboard should be carefully planned and well structured (even if the writing is not planned and very spontaneous). If possible, colored chalk may be used for better recognition.

Table 4.3 Classification of media for chemistry lessons

Media for chemistry lessons			
Visual	Audiovisual	For experiments	For models
Text book	Sound movie	Experimental devices	Structural model
Blackboard	Video	Measuring instruments	Model drawing
Transparency	Television	Apparatuses	Model experiment
Slide, photo	Computer	Projection	Functional model
Newspaper	Multimedia	Computer support	Computer model

The blackboard is especially indispensable in lessons with experiments, when the students are supposed to understand all steps from problem raising to problem solving. In these cases, the blackboard can be structured into the following subitems:

− Headline of the lesson
− Problem
− Planning and performance of the experiment
− Observations and measured data
− Analysis and discussion of mistakes

Especially regarding the first subitem: never write on the blackboard without a headline.

Graf [6] gives further arguments for key functions of the blackboard: "The writing on the blackboard is usually being developed with the help of students, which makes sense from an educational psychological perspective. Therefore the content of the board is being sequenced into manageable steps, which makes it easy for the students to understand. If there is a fold-out blackboard available in the chemistry classroom, the outer parts of the blackboard should be used to record the student's questions and ideas or to write down new or familiar technical terms". Graf also lists the pros and cons of the blackboard (see Table 4.4).

Transparencies. Just like a blackboard, you can usually find an overhead projector in every classroom and use transparencies (see Table 4.5). Colorless endless foil or sheets can be used like a blackboard, so what applies to the blackboard also applies to the projected transparency. The following advantages arise:

− The teacher can write with his face toward the students
− The available writing surface is endless
− The transparency can be kept and used again during the next lesson
− Store-bought transparencies can be used as a build-up set of two to five transparencies

The only disadvantages compared to the blackboard are: overhead markers dry out fast so the cap has to be put on every time after writing and the light of the projector can burn out (a back-up light might be available).

Transparencies can either be self-made or store bought (see Table 4.5) and used in addition to the blackboard. If the transparencies are store bought or made before the lesson, the students need enough time to copy all the essential information. Figures that are not appropriate can be covered during the projection or discussed separately. Big companies, especially oil or energy companies, usually offer transparency folders presenting their products and ways of production. They are very helpful for special chemistry lessons.

Newspaper reports. Reading and analyzing current newspaper reports makes lessons on topical questions of everyday life interesting. Often it is also motivating to reveal journalist's factual errors and to resolve these erroneous issues in a class discussion. It can be helpful to start a collection of relevant newspaper articles to be used in class. In particular, with topics concerning the environment and climate

Table 4.4 Advantages and disadvantages of the blackboard [6]

Some advantages of the black board

— Present in every class room
— Quick to use (no dimming of the light or power connection needed)
— Adaptation of the writings to the student's stage of development
— Differentiated adjustment of the writings according to the educational progress
— Involving the students step by step in the development of the writing
— Highlighting important elements or emphases of the lesson without great effort
— Successive additions of the writings can be made
— Multitude of possibilities for creating the writings on the blackboard: outline of the
  lesson, creating mind-maps, experiment planning, observations and interpretations, drawing
  models, etc.
— Student participation in developing the writing on the blackboard
— Corrections can be made easily, also by students
— High in contrast, almost every color can be used (exemption: dark colors)
— Concentration of the student's attention on one point
— Inspires the students to participate actively and to rework

Some disadvantages of the black board

— Limited space without cleaning the blackboard
— Wet black board after cleaning
— Writings at the blackboard are only available for the short time of the lecture
— Writings need to be thoroughly planned (proportions, relationship of picture and text)
— Elaborated experimental setups and technical facilities are hard to draw
— Drawing on the black board has to be learned
— While the teacher draws on the black board, they turn their back on the class and cover
  part of the writings, even when they stand partly sideways towards the blackboard
— Black board needs to be cleaned before the lesson, it also needs to be learned how to
  clean a black board (the cleaning takes time and diligence)
— A big black board tempts to writing too much
— Copying the blackboard takes a lot of time for some students
— When students write on or draw on the blackboard, they do not have the text or the
  drawing in their own notebooks
— Writings which has been developed during the lesson over the entire width of the
  black board, makes it considerably difficult for the student to copy it into their notebooks
— Chalk makes dirty hands and clothes (when writing with chalk and even by cleaning
  the board)

Table 4.5 Different kinds of transparencies for chemistry lectures

Transparencies					
Self-made transparencies			Store-bought transparencies		
With color markers (water-soluble or permanent)	By photocopy (black/white or colored)	By computer printout	Transparency folder	Build-up transp.	Stereo transp.

change like greenhouse effect and ozone layer students need scientific explanations from the teacher, because they mostly do not understand the information and mix up phenomena of greenhouse effect and holes of the ozone layer.

Slides, photos. Today one can either obtain photos on transparencies or print a colored picture on a transparency by oneself. Good slides shown in a darkened room though remain a special experience, especially if aesthetic aspects play a role, for example, showing colored crystals. Slides are a good way to illustrate chemical procedures or technologies during the lesson. Since there is usually no sound available for the slides, teachers can customize their comments to the learners' needs. They also decide the speed in which the slides change.

In the age of digital cameras, it is also possible to show digital pictures via computer and beamer. Students are able to upload pictures to their computers and edit them to their needs. New methodical challenges arise as it is now possible to work on pictures during the lesson.

Sound movie, video. In the near future, old fashioned sound movies will only be available on rare occasions; most of them will be transferred to videos. Companies supplying learning materials offer videos for sale, big chemical companies give them away for free, and educational movie hire services lend them. There are many videos available about chemical engineering, everyday life, and environment, biochemistry, and history of chemistry. These videos do not only offer moving pictures and comments but also models and animations of moving particles or of the regrouping and transformation of atoms, ions, and molecules during reactions. The teacher should watch the video in advance to decide which pictures and animations are appropriate to the students' prior knowledge and which are not appropriate. In any case, models and their animations should be critically evaluated.

Television. The different television channels often show good scientific documentations or critical reports about ecological damage or environmental protection. There also exist movies about certain discoveries or events in the history of science. They can illustrate researchers and important issues in a realistic way for the students. Such movies should be recorded to show them in class at an appropriate point of time.

Computer. The PC has turned into a universal medium, which can do pretty much everything (see Table 4.6): texts can be scanned and processed, tables and charts can be created, pictures or photos can be uploaded and edited. If the right programs are available it is possible to create simulations of real processes, to record, save, and process measured data during experiments, to develop programs in coordination with the information and communication technologies teacher. With a connection to the internet, teachers and students are able to gather information and data about almost every substance, production processes and environmental questions from all over the world, get in touch with other institutions, or present projects on a self-created homepage.

Table 4.6 Different possibilities for computer use in class

Use of the computer		
Common programs on the hard drive	CD-ROM drive	Internet connection, network programs
Text processing	Simulations	Databases
Tables, charts	Measured value acquisition	Email
Drawings, photos	Programming	Homepage

Computer simulations are probably the outstanding ability of the new programs. They allow the students to understand complicated chemical issues that cannot be shown in an experiment. On the other hand, the use of simulation programs should not replace experiments with real substances and apparatuses: the much-lamented loss of reality could occur.

Another way to use the computer is the white board: it combines the ability to use features of the computer and to do corrections or supplements by an interactive pencil.

Multimedia. Educational software has the big advantage that combines the use of texts in picture and sound, pictures with or without comments, film sequences, or model animations. It can switch from one application to another in a second, restart an application, or skip another. On the one hand, it allows every student to interactively work on the computer according to their learning progress. On the other hand, it can be used for the whole class during a lesson via beamer. The teacher should make a selection of the right educational software, since the wide range of additional information can lead to disorientation. It remains to be proven by studies as to what extent this educational software can be used successfully without accompanying lessons.

Experiments. Experimental devices, measuring instruments, and apparatuses (see Table 4.3) are pictured in the experimental instructions of teaching and text books. They can also be found, compared, and ordered in the catalogues of teaching supply companies. Certain laws of Gestalt psychology have to be followed when building apparatuses for experiments (see Sect. 5.2). Apart from that we refer to Chap. 5, which focuses on the function of experiments.

Since the projection of experiments is specifically medial, relevant information will be given at this point. There exist special attachments for using overhead or slide projectors to show certain phenomena.

The overhead projector plays an important role in the projection of experiments, since it is available in almost every class room:

- The overhead projector can be used to illuminate an apparatus, for example, to visualize the development of gas by a gas developer for all students (E4.1). It is also possible to only illuminate part of an apparatus
- The projector can also be used to visualize effects for all students in the room, for example, the model experiment of the equilibrium (E4.2)
- Glass or Petri dishes are also good for the projection on the screen, the electrolysis of zinc bromide solution, for example, can very well be observed with the projector (E4.3)
- Lab supply companies offer tripartite Petri dishes, which allow the comparison of three solutions; for example, an indicator in neutral, acidic, or alkaline solution (E4.4)
- Teaching supply companies sell projection attachments for overhead projectors, which either work with cuvettes for projection experiments (E4.5) or glass cells for electrolysis experiments (E4.6). Full shows a lot of other wonderful applications [7].

There exist attachments for slide projectors, which can project cuvettes or test tubes: all kinds of color reactions in solutions or precipitation reactions are highly enlarged. The colors of the reactions are projected in a realistic way, the ones of the precipitates are not; they always appear black on the screen.

Television cameras and attached monitors offer another kind of magnification for experiments. A gooseneck camera is handy and small enough to get close to every part of the apparatus, and phenomena can be highly magnified. This form of projection especially preserves the real colors (E4.7). Roesky [8] describes this "chemistry in miniature" with many suggestions for experiments. General practice of experiments "in miniature" would cut costs for chemicals and minimize waste – but with the loss of parts of reality.

The use of computers for experiments offers a lot of other advantages. This does not mean to replace experiments by pictures on the computer screen: the real experiment should be run either as a demonstration experiment or as a student experiment, whenever possible. But as soon as data have to be collected and analyzed, tables and charts have to be created or measured curves have to be compared, this can all be done with the help of the computer. Therefore, the use of computers in chemistry lessons should better be called computer-supported measuring. It can be used in multiple ways:

– Measurements can be shown in big numbers on the screen. This way the whole class is able to see them during the measuring process; for example, the mass decrease of a highly volatile solvent during the evaporation on a scale (E4.8)
– Very slow and very fast reactions can be better illustrated on a PC than in the traditional way. The very fast increase of pH-values during the neutralization of acidic solutions, for example, can be shown on the screen (E4.9),
– Series of measured data of one and the same issue can be realized with little effort, the results of the measurements can be compared. The analysis of water and air pollutants at different times of the day is just one example for this use of the computer
– Measured data, which are collected during a real experiment, can be processed in tables and charts, calculation of averages, or variations. Compensating curves or calculations of mistakes can also be illustrated
– It is possible to use the PC for control and automation during experiments. However, good programming skills are needed for the installation and operation of a controlling circuit

Apart from that, it is motivating and fascinating for the students to program an experimental problem in cooperation with information and computer technologies lessons. Hardware and software skills are also needed for computer-supported experimentation. Besides computer and monitor, an analog-to-digital converter (AD-converter), which converts for example the analog data of voltages of a regular measuring device into digital data, is also part of the hardware. Appropriate software is needed for the reception and recording of digital data, for the analysis of data, and for further calculations regarding the average or compensating curve.

Models. Concrete models of the structure of matter, model drawings, mental models, model experiments, or functional models are important media for chemistry teaching. These kinds of models and their discussion form the contents of Chap. 6.

Especially first models of the structure of matter according to the special arrangement of smallest particles (particle model of matter, see Fig. 6.11) have to be introduced. Later, the arrangement of atoms in molecules or ions in ionic structures will be introduced (Dalton's atomic model, see Fig. 6.12). On the base of DALTON's atomic model, it is very important to describe chemical reactions and the rearrangement of atoms in molecules, for example, the reaction of H_2 molecules and O_2 molecules to form H_2O molecules (see Fig. 7.6), or the rearrangement of ions to explain the precipitation from salt solutions (see Fig. 7.8).

Using molecular model sets or closest sphere packing means to take nice colored spheres of wood, cellulose, or plastics, and to take glue or sticks to combine those spheres. Because students like to take these colorful media as originals, they may think that S atoms are yellow and C atoms are black. These misconceptions should be avoided by discussing the "irrelevant items" of the concrete models (see Fig. 6.1): the atoms or ions have no color because no one can see them; there are no sticks or glue between the atoms or ions because they are attracted by electric forces, the material of the models is irrelevant, etc. The models have to only show the spatial arrangement of atoms or ions (the chemical structure), different spheres in sphere packing may show the ratio of diameters of atoms or ions, they may also show the ratio of the numbers of atoms or ions in their arrangement.

Models can also be shown on the computer screen: these representations are even more accepted as models because there are no real spheres, sticks, or glue. We have different programs to create molecular models or crystal structures, and these programs exist in such big numbers that it is nearly impossible to offer detailed information.

4.2 Scientific Ideas: Appropriateness of Media

Educational supply companies as well as educational publishing companies offer a lot of transparencies, slides, videos, drawings, and pictures; often those are different media for one and the same issue. Their scientific appropriateness should be evaluated before use in class based on the following criteria.

Validity. Before the use of any medium in class, it should be checked if it is correct from the view of the scientist or if it can be brought up to date in scientific terms.

There are – for example – many visualizations or transparencies of the atomic structure according to the nucleus-shell model, which uses isolated spheres as models for electrons or circles for their movement around the nucleus. This requires much additional information, which starts from the wave-particle duality and brings terms like energy levels into the discussion. Since protons, neutrons, and electrons can hardly be illustrated by spheres or dots, it should be avoided to use these ambiguous media trying to explain the structure of atoms, ions, or molecules.

Didactical reduction. Many topics have to be didactically reduced to match the class's learning conditions or students' cognitive development. In this context, Reiners [9] talks about special knowledge transformation or "didactics as a transformational science."

In any case it needs to be tested, if such reductions are acceptable in terms of content. Some decades ago, there were pictures of models in school books that were based upon the idea of "elements being made up of atoms and compounds being made up of molecules." (see Fig. 4.2, upper part) The authors of such drawings [10] knew that the element sulfur is made up of S_8 molecules and they also knew about the ionic structure of iron sulfide: nevertheless, they thought that their drawings concerning "FeS molecules" were didactically acceptable reductions. In addition, the statement that one molecule shows the "smallest portion of a compound" is totally incorrect from a scientific point of view and therefore these reductions are not acceptable.

Today didactic reductions are allowed as long as the content stays correct. It is acceptable to describe the reaction of copper oxide and hydrogen with models of lattices and molecules (see Fig. 4.2, lower part). The model shows the rearrangement of atoms [11], in subsequent lessons copper oxide can be explained with ions; and the size of atoms compared to those of the ions may be corrected.

Real objects, real processes. If the opportunity arises to use real objects instead of pictures or transparencies in class, the real object should always be favored. However, they should be accompanied by photos or model drawings, to produce

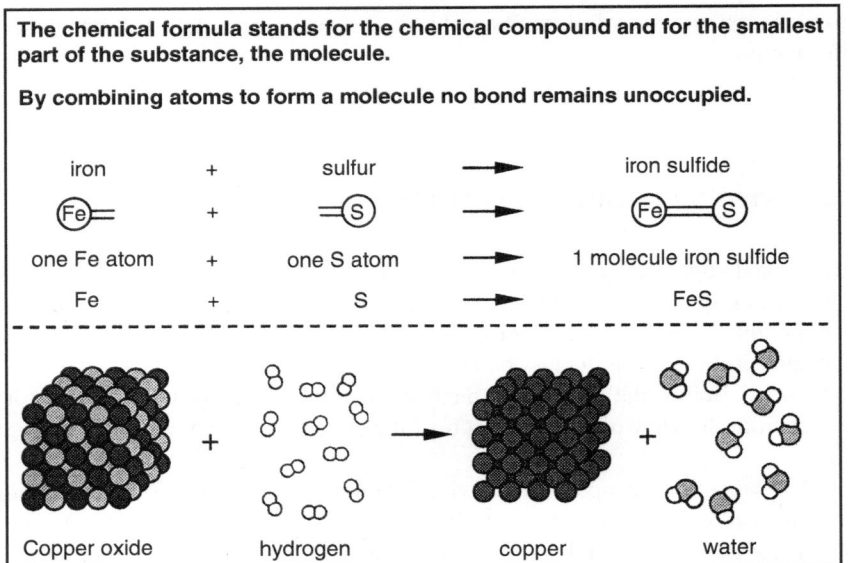

Fig. 4.2 Didactical reductions in the 1950s [10] and today [11]

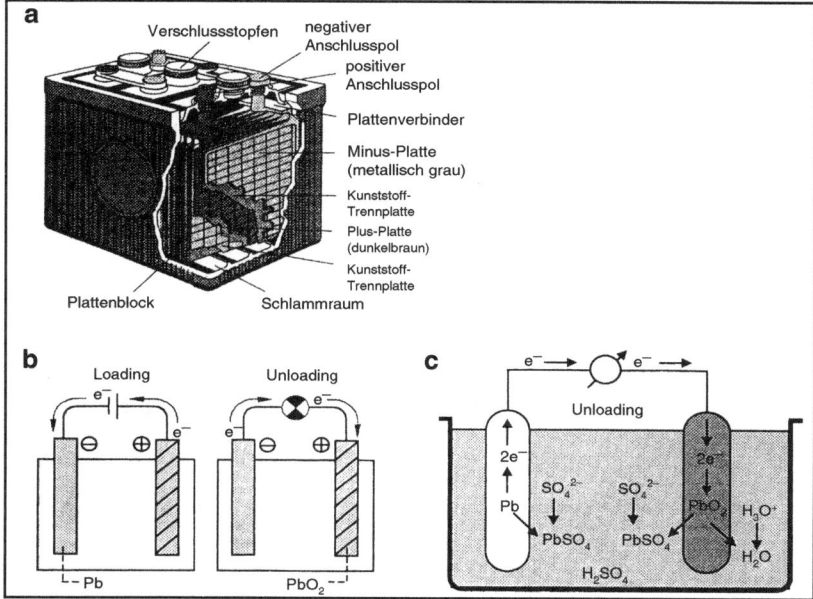

Fig. 4.3 Car battery as a real object, related experiments and transparencies [14]

optimal learning effects. Talking about "car batteries" in class does not only offer the opportunity to show a real car battery but also to run an experiment of the real process (E4.10). After showing the real object and experiment, transparencies and pictures [12] can be used for explanations (see Fig. 4.3).

Daoutsali [13] took a real car catalyst as an object to motivate students to think about its function. From different empirical studies in Greece and Germany [14], she knew that students are thinking more of a filter or storage for toxic gases like carbon monoxide or nitrogen oxides. In order to challenge these misconceptions, she conducted chemistry lectures with experiments using platinum crystals and their surfaces as catalysts, to split molecules like CO and NO and form new molecules like CO_2 and N_2. With the real car catalyst and the possibility to look through the very narrow tubes of the inner material, the students were able to understand it.

4.3   Learner: Media and the Ability for Abstract Thinking

For successful mediation of information, a certain degree of abstraction is naturally assigned to many media. Every medium, which is planned to be used in class, has to match the student's stage of development, especially their ability for abstract thinking. For the use of a specific movie or transparency, it needs to be discussed whether the learner's age group and stage of development is taken into account or whether the media could create or confirm certain misconceptions.

Levels of abstraction. Beginning with real objects and processes (see Chap. 4.2), it should not be a single step to the highest level of abstraction – instead it is better to choose appropriate media on an intermediate level of abstraction first. During the introduction of chemical symbols, for example, which requires a high abstraction on the submicroscopic level, an iconic illustration can help the learner to understand the symbols (see Fig. 4.4).

Starting from the real object, a cubical salt crystal, it is possible to discuss the packing of spheres and the unit cell as media with an intermediate level of abstraction. After that, symbols like $Na_{32}Cl_{32}$, Na_4Cl_4, Na_1Cl_1, or NaCl or the ionic symbols like $(Na^+)_4(Cl^-)_4$, $(Na^+)_1(Cl^-)_1$, or Na^+Cl^- should be derived from these concrete models. Chapters 6 and 7 delve into this problem.

In this way, learners gain an idea of the structure of substances and always have it in mind, when they work with chemical symbols and formulae. When working with iconic illustrations like structural models or model drawings, it is important that the learner understands that these are models, and do not represent an exact copy of reality. This problem can be minimized by working with different concrete models for the same structure, like packing of spheres, crystal lattices, and unit cells (see Fig. 4.4). It is also possible to use red-green drawings of structures that create a three-dimensional impression when seen through 3D glasses. Harsch and Schmidt [15] offer further information about this innovative method.

Spatial ability. Interpreting and understanding those structural models (see Fig. 4.4) is not easy for young students: to imagine the positions of spheres, to count the coordination number 6, or to add the parts of spheres in models of the

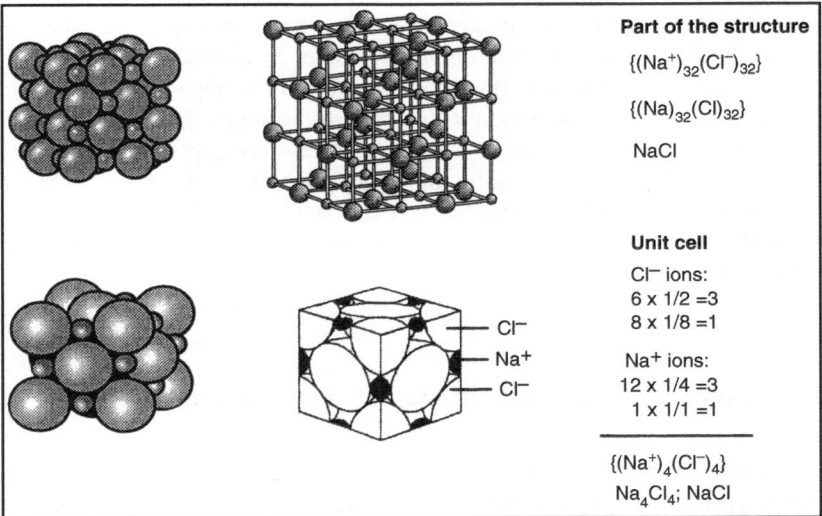

Fig. 4.4 Different models as media for deriving different symbols of sodium chloride

unit cell concerning the sodium chloride structure. Wirbs [16] has introduced those sphere packings and unit cells into chemistry lectures in grade 8 and 9 for deriving formulae from chemical structures – and she could prove that students with little knowledge in fraction calculation are able to add all parts of spheres of the unit cell to find the ratio formula (see Fig. 4.4). On this way, they must apply their spatial ability.

From psychologists, we know that "Space" is a primary factor of intelligence. Bloom stated that "at the age of nine only 50% and at the age of 14 about 80% of the parameter Space have developed" [17]. On this base, a spatial ability test concerning chemical structures was created with the aim to know from which age spatial ability is sufficient to work with two-dimensional pictures of structural models and to interpret them in the three-dimensional way. Five of those questions are represented (see Fig. 4.5).

The results show that in grade 7, the 13–14-year-old students have very different abilities to solve those questions: "Testing the correlation between performance in intelligence and spatial ability the data show that a sufficient correlation does not exist for seventh graders. Conversely, all correlation coefficients in grades 8 and 9 appear to be significantly different from zero. It is striking that on the average the girls' correlation coefficients show a weaker correlation than the boys. In the case of a great number of seventh-grade students, spatial ability has not developed suffi-ciently compared with general intelligence, so that teachers should avoid those problems that require the application of spatial ability." [18]

Like other empirical research of psychologists shows, the boys have performed better than the girls: boys may have a better training of spatial ability by their technical toys than the girls. Temechegn [19] compared results of space tests in Germany and his native land Ethiopia. He found out that for private schools, the same result "boys are better than girls" could be proved, but for government

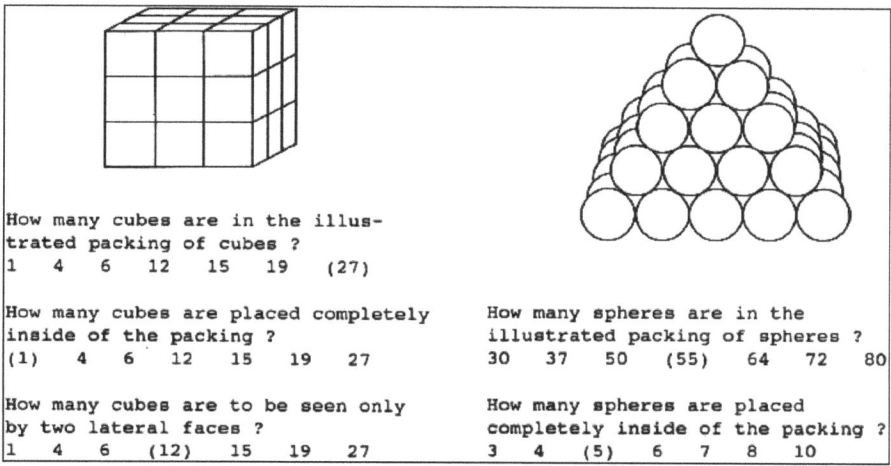

How many cubes are in the illustrated packing of cubes ?
1   4   6   12   15   19   (27)

How many cubes are placed completely
inside of the packing ?
(1)   4   6   12   15   19   27

How many cubes are to be seen only
by two lateral faces ?
1   4   6   (12)   15   19   27

How many spheres are in the
illustrated packing of spheres ?
30   37   50   (55)   64   72   80

How many spheres are placed
completely inside of the packing ?
3   4   (5)   6   7   8   10

Fig. 4.5 Five examples of the space test, correct answers in brackets [18]

schools, this was not the case: those kids have nearly no technical toys, boys and girls have only few chances to develop spatial ability compared to kids of rich parents of private schools.

Sopandi [20] developed for grade 9–12 students a test of chemistry understanding with tasks concerning the structure of matter. The participants should draw their mental models regarding the arrangements of particles before and after well-known chemical reactions, particles in mineral water, or in acidic and basic solutions. Besides this questionnaire, also the evaluated space test [19] was conducted in the same classes and all data have been correlated: the correlation coefficients show high correlation, good chemistry understanding correlates with good spatial ability, spatial ability may be a prerequisite of chemistry understanding [20].

One last aspect of spatial ability will follow. Most curricula in science of grade 5 and 6 want the introduction of the particle model of matter: water particles in ice crystals, in water drops, or in vapor clouds, ethanol particles in ethanol, water, and ethanol particles in an aqueous solution. Schwoeppe [21] developed a space test concerning the understanding of those model drawings and stated big differences in spatial ability. Because the factor Space is just developing during the ages from 10 to 12, the training of this ability seems very important. One way for the training may be building real models concerning the 3D-arrangement of particles on the base of the simple particle model of matter, and by deriving those well-known 2D-model drawings from real 3D-models: they are like many other models very important media in chemistry education.

Prior knowledge. The learners not only bring skills and ideas from their environment into class (see Chap. 1) but they also experience many scientific topics in nonfiction books, children's magazines, or television shows. Teachers should be familiar with these media, to be able to respond adequately, or even to incorporate these media with chemistry lessons

Generally, students start science education with preconcepts or alternative ideas about substances and their changes (see Chap. 1). Asking for example, "what is the mass of the solution when 1 kg of salt is dissolved in 20 kg of water" [22], students like to answer: "the salt is gone and the water is now salty, but has the mass of 20 kg as before" (see Fig. 4.6).

They do not speak about 21 kg of the *solution*, about reclaiming the salt if the water were to be separated by evaporation: students do not have the idea of conservation of mass. So the teacher has to take preconceptions from students and to invoke a conceptual change to the law of conservation of mass.

Concept cartoons. Those cartoons (see Fig. 4.6) are good media to open the discussion about students' different explanations. Temechegn and Sileshi [22] designed concept cartoons for many chemistry topics and lectures. They stated: "Concept cartoons are cartoon-style drawings that put forward alternative conceptions about the science involved. By offering new ways of looking at the situation, they make it problematic and provide a stimulus for developing ideas further. Concept cartoons provoke discussion and stimulate scientific thinking. A typical concept cartoon has the following features: visual representation of scientific ideas, minimal text-in dialogue form, and alternative conceptions in equal status.

Fig. 4.6 Concept cartoon concerning conservation of mass [22]

The organization of a typical lesson using a concept cartoon might be [23]:

- Introduce the topic
- Provide a concept cartoon to focus on a particular situation
- Request a brief period of individual reflection
- Encourage small group discussion and invite groups to see if they can reach consensus
- Give brief feedback to see what range of views is present
- Discuss how to find out which alternative is most acceptable
- Offer small group enquiry
- Share outcomes of enquiry
- Start class discussion, including which alternative seems most acceptable now, why other Alternatives seem less acceptable and what further information we might need to be sure
- Consider how relevant theory applies to the situation
- Draw ideas together and provide an explicit summary of the initial problem, the enquiry, the outcome and what has been learnt from the enquiry

The purpose of relying on this volume on research of the alternative conceptions (misconceptions) is to ensure that the alternatives presented are plausible, so that learners will usually identify directly with some of the alternatives in the concept cartoons. Although many students and teachers find such kind of research on alternative conceptions fascinating, it is *not* always easy to see how it can be used in the classroom. Now we strongly believe that as a strategy of diffusion and use of the research finding the concept cartoons are very valuable and have a profound effect on the quality of teaching, learning and assessment in chemistry education" [22].

Another cartoon (see Fig. 4.7) will show that even after teaching the composition of water, students are not sure about water vapor; it may "contain hydrogen and

Fig. 4.7 Concept cartoon concerning the composition of water vapor [22]

oxygen," or even may be changed into "air" [22]. These kinds of misconceptions acquired in chemistry lectures, are also called "school-made misconceptions" [24].

Experimental kits. There might be some young students that have played with experimental kits before their first chemistry lessons. Parents, who are really interested in science or who comply with their child's wish, buy them an experimental kit. These children do have a considerable prior knowledge and good first experimental skills. Therefore, they can usually contribute a lot to discussions and experiments.

A certain kit of the Kosmos company in Germany [25] contains the material for 250 experiments concerning colors of different indicators, much about separation of mixtures, about water and air, acids and bases, carbon dioxide, lime and gypsum, different metals, food and soap, etc. There are a lot of other kits related to energy, fuel cells, etc. Since most of the experimental kits are very pricy, the number of children, who have one at home, is small.

Media sets. They have been developed to suit the teachers' wishes for giving students the opportunity to work by themselves. Besides substances and reagents these sets contain brochures, slides, transparencies, or videos. One example is the media set "iron and steel" [26] in Germany: the set contains samples of iron ore, videos about the iron furnace and steel production, transparencies, and slides concerning many applications of iron and steel in the society. Further media sets exist for the following topics: soaps and detergents, fats, adhesives, paper production, plastics, colorants, biochemistry, food chemistry.

There is also an environment set [27] to test water or soil samples: it contains battery-powered measuring instruments like pH meter, oxygen measuring electrode, temperature meter, conductometer, luxmeter, photometer; also detection reagents and chemicals for using the photometer, cables for the digital measuring on the computer and an instruction manual are included.

Water analysis kits are an activity-oriented way to introduce the learner to water analysis. The Aquamerck-test set [28] contains tests for total hardness as well as carbonate hardness, pH-value, and for ammonium, nitrate, nitrite, and oxygen contents. Additional individual tests for the analysis of certain metal ions or pollutant particles are available in special stores.

4.4   Human Element: Mass Media

Due to their technical perfection, the influence of media – in the sense of mass media – on the learner is usually stronger than the influence of education media. Mass media like newspapers, magazines, radio, or television shows function as information carriers outside of schools in both the positive and the negative sense.

In the case of chemistry-related content, media can work *positively* by

- Creating interest and curiosity: if the students, for example, find newspaper articles about the relevance of shape memory alloys related to the motor production and for many other technical applications, it might stimulate their interest in the phenomenon of shape memory. The articles can be used as a starting point to discuss this phenomenon in class; further articles from scientific journals can be used as well.
- Motivating a critical reflection: newspapers and television reports about smog alert or about nuclear waste transport can be the motivation for lessons about a critical analysis of these reports and perhaps a correction of journalist's mistakes.
- Presenting content with the help of movies and television: if many students have seen a television report about the discovery and history of dynamite or a movie about Alfred Nobel, it is possible to use student's interest for a digression about nitroglycerin. An experiment concerning the esterification of glycerine with nitrating acid may be planned and conducted; the production of dynamite can be explained.

Mass media can also have a *negative* impact by

- Transferring negative attitudes: one-sided reports about poisonous substances in food or about undesirable side effects of medicines let people assign blame to chemistry. This may affect students to adopt a negative attitude toward chemistry from those journalists. A negative attitude would reduce their learning receptivity for chemistry lessons.
- Delivering wrong conceptions: newspaper or television reports about energy consumption like fuel consumption or power consumption are deepening the belief that energy can "disappear"; the concept of "destruction" – taught by the media – opposes the scientific concept. So the teacher will have a bigger problem to teach the conversion of one form of energy into another form of energy (see Chap. 1).
- Misleading through advertisements and commercials: Many commercials – like the ones for detergents – argue in a magical-animistic way; for example, substances are said "to work miracles." The laundry becomes "whiter than white" or "deep down clean," detergents have the "power of the white giant." If a newspaper says: "organic detergents do not contain chemicals," the reader is totally lead astray. On the one side, this phrase emphasizes the contrast "biology is good vs. chemistry is bad," on the other side, it deliberately conceals that every detergent must contain wash-active substances, therefore chemicals – and the nonactive substances are chemicals, too!

Because of these irritations by the mass media, chemistry education has the big task of educating young boys and girls at school scientifically and to enable them to call newspaper and television reports into question, to interpret them independently, and to draw their own conclusions on the basis of their knowledge and skills. Later in their jobs, they may be able to make the right decisions concerning scientific questions. This task of teachers and schools is one of the very important tasks.

Problems and Exercises

P4.1. Media can have different functions in chemistry lessons. Choose three functions and, by using issues of your choice, explain the use of certain kinds of media according to the chosen function.

P4.2. One kind of media that is equally important for teachers as well as students is the text book. Choose three current text books and decide which book you would use for your lessons. Justify your choice and indicate your criteria.

P4.3. Blackboard and overhead projector are the most important media in the class room. Write down advantages and disadvantages for their use. Explain two different teaching situations in which you use the one and the other medium. Give your reasons.

P4.4. Due to its multiple possible uses, the computer is getting more and more important for school lessons. Explain three different teaching situations, in which you would use the computer in your chemistry lesson.

P4.5. Mass media like television and newspapers have a big impact on the thinking and acting of your students. In which way can this impact be positive for your lessons, in which way can it be negative? Give examples.

Experiments

E4.1. Illumination of apparatuses with the overhead projector
Problem: It is sometimes hard for all of the students to observe a demonstrated experiment clearly. With the help of the mirror system of the overhead projector the bright light can be cast on the apparatus to improve the students' observation.
Material: Syringe, test tube with side tube and stopper, rubber hose; diluted hydrochloric acid (C), magnesium ribbon (F).
Procedure: Syringe and test tube are to be attached to stands and connected, the test tube is to be half-filled with hydrochloric acid. A 5-cm piece of the magnesium ribbon is to be put into the test tube, which is to be closed with the stopper just afterwards. The light of the projector is to be cast on the apparatus.
Observation: The gas generation, the disappearing metal and the increasing volume in the syringe can be easily observed due to the bright light on the apparatus.

E4.2. The overhead projector surface as an experimental table
Problem: The bright projection surface of the overhead projector itself can function as a good experimental table and can be used to run small experiments. The model experiment for the demonstration of the dynamic chemical equilibrium, for example, can be run on the projector surface. This way the experiment can be very well observed from all students.

Material: Two 50 mL-measuring cylinders, two glass tubes of the same lengths (Ø: 8 mm and 6 mm); diluted copper sulfate solution (Xn)

Procedure: One of the cylinders is to be filled with 50 mL of the blue solution, the other cylinder remains empty. Two glass tubes of the same diameter are used to lift and carry the solution from one cylinder to the other, until the volumes are equal in both cylinders. This experiment is to be repeated with glass tubes of different diameters. The volumes and the corresponding number of lifting processes can be illustrated in a chart.

Observation: Equal volumes only result in the first case. In the second case the volume ratio is for example 35:15. The volumes remain on this level even with further lift processes.

E4.3. Projection of electrolysis experiments

Problem: When the illumination of an apparatus does not improve the observation for the student, it is possible to project phenomena with the overhead projector, as long as they take place in transparent solutions. Especially when new substances only originate in small amounts, like during electrolyses, the projection is a good way to magnify the reaction dish and to make the electrolysis observable for all students.

Material: Petri dish, two pieces of platinum wire, transformer and cable; zinc bromide solution.

Procedure: The two pieces of platinum wire (the electrodes) are to be attached to the Petri dish and with the cables connected to the transformer's direct current poles. The Petri dish is to be put on the projector and the bottom is to be filled with zinc bromide solution. A direct current of about 5–10 V is to be set.

Observation: One can directly observe the precipitation of yellow bromine on the positive pole and the creation of zinc crystals on the negative pole. With the right voltage it is possible to see the growth of a "zinc tree". This is better visible with the help of the overhead projection.

E4.4. Experiments in the projected Petri dish

Problem: Many reactions take place in colored solutions; these colors can be made visible if light shines through the solution. Since there is an overhead projector in every chemistry class room, Petri dishes, especially the ones with three compartments, are interesting for showing colored solutions and their reactions. Different reactions can be observed at the same time.

Material: Tripartite Petri dish, pipet; acid-base indicator (F), hydrochloric acid (C), sodium hydroxide solution (C)

Procedure: The three compartments of the Petri dish are to be filled with an indicator solution, for example universal indicator dissolved in tap water: this solution is green. One drop of hydrochloric acid is to be added to the first compartment, one drop of sodium hydroxide solution to the second compartment. One drop of the acidic or basic solution can be used to neutralize the solutions in both compartments of the Petri dish.

Observation: The green indicator solution shows a color change to red and blue. The projector makes it possible to see the streaking and the magnificent play of colors before the colors of the solutions finally change. The colors of the solutions in the first and second compartment can be compared to the color of the green indicator solution in the third compartment of the Petri dish. After neutralization just the green color is to be observed.

E4.5. Special cuvette for the projection on the overhead projector

Problem: Besides the possibility of projecting Petri dish experiments, there exists special equipment for the projection of reactions in test tubes or cuvettes. Due to a mirror system, new equipment projects the content of a test tube upright, older equipment cannot do that: solids precipitate in the projection from bottom to top.

Material: Special equipment for the projection, test tubes; phenolphthalein solution (F), petrol (F/Xn/N), sodium (C/F).

Procedure: The projector is to be switched on. The test tube is to be filled with diluted indicator solution. The indicator solution is to be covered by 2 cm of petrol. A small piece of sodium is to be pitched into the test tube.

Observation: The piece of sodium immerses into the colorless indicator solution, the color of the indicator changes to red. A gas generation can be observed for a short time. The metal piece ascends in the phase of petrol, but drops back into the solution and reacts under the generation of gas. These processes recur until the sodium has dissipated.

E4.6. Measuring cell for the projection on the overhead projector

Problem: Small amounts of gas, especially ones that are to be measured quantitatively, can be better observed, if the apparatus is being projected in a large format. The quantitative water electrolysis is a good example for this kind of experiments.

Material: Measuring cell, transformer, cable, sulfuric acid solution (C).

Procedure: The measuring cell is to be filled with sulfuric acid solution on the projector, the transformer is to be switched on at 5 V, so that a constant gas formation sets in. Hydrogen and oxygen can be confirmed with common tests when the gas volumes are big enough.

Observation: Gas bubbles form on the electrodes. The gas volume on the negative pole is twice as big as the gas volume on the positive pole. Hydrogen occurs at the negative pole, oxygen at the positive pole.

E4.7. Experiments with the help of a gooseneck camera

Problem: Not every reaction in a solution can be projected and observed with the overhead projector. Precipitates, for example, always appear to be black. One possibility to show these and other phenomena true to the original is to use a gooseneck camera with a connected monitor.

Material: Gooseneck camera, monitor, test tubes; iron(II) chloride (Xn), iron(III) chloride (Xn), diluted sodium hydroxide solution (C).

Procedure: One test tube is to be filled with diluted iron(II) chloride solution up to two thirds, the other test tube with the iron(III) chloride solution. These test tubes are to be attached to stands and then enlarged with the help of the camera and the monitor. A few drops of sodium hydroxide solution are to be added to both test tubes.

Observation: The precipitation of green respectively brown iron chloride can be observed. The real colors of both precipitates can be compared on the monitor.

E4.8. Large display of measured values on the computer screen

Problem: Digital measuring instruments with big numbers or with large display are very expensive. The cheap solution for a large display of measured values is the computer screen. An AD-converter and the appropriate software make it possible to connect regular measuring instruments to the computer to display digital measured values.

Material: Computer and AD-converter, software, balance, Petri dish; acetone (F).

Procedure: The balance is to be connected to the computer with the help of the AD-converter. The large display of the measured values is to be prepared with the appropriate software. A few drops of acetone are to be put into the Petri dish, which is placed on the balance.

Observation: The measured values are displayed in a large format on the screen. They show a decreasing mass of liquid acetone.

E4.9. Recording and processing of measured values with the computer

Problem: Some reactions are so fast that they cannot be recorded with traditional methods. A neutralization reaction of an acidic with an alkaline solution and recorded pH values can be realized with a connected computer: one can easily capture the pH jump around the equivalence point. It might be a good suggestion to let the students run some titrations and titration curves in the traditional way, to compare them to the computer's way of recording.

Material: Computer and software, connected AD-converter, calibrated pH-meter with glass electrode, magnetic stirrer, Erlenmeyer flask, burette; 0.1 molar solutions of hydrochloric acid and sodium hydroxide (C).

Procedure: The pH-meter is to be connected to the computer via the AD-converter; the titration is to be prepared with the appropriate software. 50 mL of hydrochloric acid are to be poured into the Erlenmeyer flask, diluted with water, the magnetic stirrer is to be turned on, the electrode is to be immersed and the titration is to be started with the sodium hydroxide solution.

Observation: Before the titration the pH-value is 1.0. During the titration the computer screen draws the titration curve. Adding 50 mL of sodium hydroxide solution the pH value jumps near a pH of 7; after adding a few more milliliters of sodium hydroxide solution pH 13 is almost reached.

E4.10. A model experiment as a medium for the car battery

Problem: The car battery should first be observed in a car and its function may be discussed. The model experiment reproduces the processes of charging and discharging and the reaction of the lead electrodes. Text books or transparencies can be used to explain these processes.

Material: Beaker, transformer, voltmeter, electromotor (2 V), cable and alligator clips; two newly sanded lead plates, sulfuric acid solution (20%; C). Procedure: The beaker is to be half-filled with sulfuric acid solution. Both lead plates are to be positioned and attached in a way that they do not touch each other, they are to be connected to the transformer by cables (see Fig. 4.3). A direct current voltage has to be set so that a generation of gases can be observed for one minute. The transformer is to be removed. The voltage between the two plates is to be measured; the electromotor is to be attached.

Observation: During the current supply a dark brown substance forms on one of the lead plates (as it can be observed on one plate of a battery cell in the car battery). After taking away the current supply a voltage of 2 V can be measured between the two plates. The attached electromotor first spins fast, but becomes slower and slower and finally stands still.

Please note: The sulfuric acid solution is contaminated with lead sulfate and therefore toxic. If possible it should be poured into a labeled container, stored and reused for the same experiment. Otherwise it needs to be disposed off in the container for heavy metal waste.

References

1. NEUE PRESSE vom 1. 12. 93 (1993) Lehrer wollte schlechte Buecher nicht verwenden – er wurde nach Hannover versetzt. Hannover
2. Issing LJ (1987) Medienpädagogik im Informationszeitalter. Studienverlag, Weinheim
3. Stumpf K (1979) Das Lernen mit Medien. CU 10:1
4. Becker HJ, Hildebrandt H (1997) Medien im Chemieunterricht in Artikeln chemiedidaktischer Zeitschriften. NiU-Chemie 8(38):41
5. Becker HJ, Hildebrandt H, Labahn B (1996) Die Literaturdatenbank "Fadok" als Hilfe zur Vorbereitung von Chemieunterricht. PdN-Ch 45(5):34
6. Graf E (1997) Die Wandtafel im Chemieunterricht. NiU-Chemie 38:74
7. Full R (1996) Lichtblicke - Petrischalenexperimente in der overhead-Projektion. ChiuZ 30:286
8. Roesky HW (1998) Chemie en miniature. Wiley-VCH, Weinheim
9. Reiners C (2000) Chemiedidaktik – Quo vadis? Chemkon 7:91
10. Halberstadt E, Waeltermann T (1967) Chemie für Maedchen. Frankfurt, Diesterweg
11. Eisner W et al (1979) Stoff - Teilchen - Reaktion. Klett, Stuttgart
12. Dehnert K et al (2004) Allgemeine Chemie. Schroedel, Braunschweig
13. Daoutsali E, Barke H-D (2010) Der Autoabgaskatalysator im Chemieunterricht. PdN-Ch 59
14. Daoutsali E, Barke H-D (2010) The automotive catalytic converter in chemical education. Poster at the International Conference on Chemical Education ICCE. ICCE, Taipeh/Taiwan
15. Harsch G, Schmidt R (1981) Kristallgeometrie. Packungen und Symmetrie in Stereodarstellungen. Diesterweg, Frankfurt
16. Wirbs H, Barke H-D (2002) Structural units and chemical formulae. CERAPIE 3:185
17. Bloom BS (1964) Stability and change in human characteristics. McGraw-Hill, New York
18. Barke H-D (1993) Chemical education and spatial ability. J Chem Ed 70:968
19. Temechegn E, Barke H-D (2001) Structural chemistry and spatial ability in different cultures. CERAPIE 2:227

20. Sopandi W (2004) Raumvorstellungsvermögen und Chemieverständnis im Chemieunterricht. Schueling, Muenster
21. Schwoeppe C (2007) Das Raumvorstellungsvermögen von Muensteraner Schuelerinnen und Schuelern der Jahrgangsstufen 5/6 im Hinblick auf chemische Inhalte. Schueling, Muenster
22. Temechegn E, Sileshi Y (2004) Concept cartoons as a strategy in learning, teaching and assessment in chemistry. Addis Ababa, Ethiopia
23. Naylor S, Keogh B (2000) Concept cartoons and science education. Millgate, London
24. Barke H-D, Hazari A, Sileshi Y (2009) Misconceptions in chemistry. Springer, Heidelberg
25. Kosmos, (1985) Chemie-Praktikum All-Chemist. Franckh, Stuttgart
26. Fladt R et al (1975) Unterrichtseinheit Eisen und Stahl. Klett, Stuttgart
27. Engler R (1996) Ökologie. Theoretische Grundlagen und Versuche mit dem Umwelt-Messkoffer. Hürth, Leybold
28. Aquamerck, Kompaktlabor fuer die Aquaristik, Sauerstoff-Test. Darmstadt, Merck

Further Readings

Dale E (1969) Audiovisual methods in teaching. New York, 3rd edn. The Dryden Press, Holt, Rinehart and Winston

Elizabeth H (1990) The students laboratory and science curriculum. Mackays of Chatham, UK

Kozma RB (2000) The use of multiple representations and the social construction of understanding in chemistry. In: Kozma MJR (ed) Innovations in science and mathematics education: advance designs for technologies of learning. Erlbaum, Mahwah, NJ

Mayer RE (1989) Systematic thinking fostered by illustrations in scientific text. J Educ Psychol 81:240–246

Mayer RE, Gallini JK (1990) When is an illustration worth ten thousand words? J Educ Psychol 82:715–726

Mayer RE, Sims VK (1994) For whom is a picture worth a thousand words? Extensions of a dual coding theory of multimedia learning. J Educ Psychol 86:389–401

Peeck J (1993) Increasing picture effects in learning from illustrated text. Learn Instruct 3:227–238

Russell JW, Kozma RB, Jones T, Wykoff J, Marx N, Davis J (1997) Use of simultaneous synchronized macroscopic, microscopic, and symbolic representations to enhance the teaching and learning of chemical concepts. J Chem Educ 74(3):330–334

Sadoski M (2001) Resolving the effects of concreteness on interest, comprehension, and learning important ideas from text. Educ Psychol Rev 13:263–281

Sanger MJ (2000) Using particulate drawings to determine and improve students' conceptions of pure substances and mixtures. J Chem Educ 77(6):762–766

Tasker RF, Bucat R, Sleet R, Chia W (1996) The VisChem Project: visualising chemistry with multimedia. Chem Aust 63:395–397

Tasker R, Dalton R (2006) Research into practice: visualization of the molecular world using animations. Chem Educ Res Pract 7(2):141–159

Treagust DF, Chittlebouropugh G, Mamiala TL (2003) The role of submicroscopic and symbolic representations in chemical explanations. Int J Sci Educ 25(11):1353–1368

Williamson VM, Abraham MR (1995) The effects of computer animation on the particulate mental models of College chemistry students. J Res Sci Teach 32(5):521–534

Wu H-K, Krajcik JS, Soloway E (2001) Promoting understanding of chemical representations: student's use of a visualization tool in the classroom. J Res Sci Teach 38(7):821–842

Chapter 5
Experiments

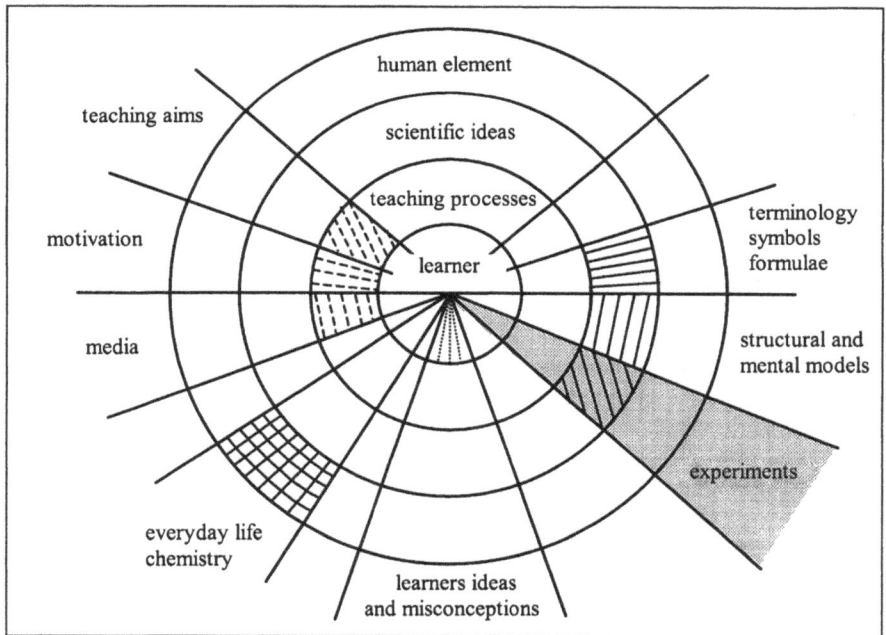

"Many well educated people regard chemistry as a nice experimental way to produce soda and soap, to have better steel, and to deliver good dyes for silk or cotton" [1].

Liebig bemoaned in 1840 already that laymen often believe chemistry to be some kind of trial-and-error approach that, from time to time, delivers a new product accidentally. It is not easily realized that the classical scientific experiment is preceded by a whole philosophy. To define this context, the idea of the "elements" may serve as an example.

H.-D. Barke et al., *Essentials of Chemical Education*,
DOI 10.1007/978-3-642-21756-2_5, © Springer-Verlag Berlin Heidelberg 2012

A preliminary term for the chemical elements already appeared in the philosophic schools of ancient Greece: Empedocles talked about the four roots of all things, which he perceived in a mythical way as deities. These four deities represent earth, fire, air and water, out of which every being is made – the term "element" can be traced back to him. Closer inspection of this idea reveals that the Greek philosophers developed this and other constructs without any experiments: "Indeed their achievements were based on a range of thorough and classified observation in nature, but there can be no talk of experimentation in a scientific way."

The alchemy of medieval times was dominated by the idea of producing gold from base metal: small amounts of gold and silver were set free during the refinement of ore. It was also tried to obtain gold and silver by "metal transmutation" in many ways. During those attempts it was observed that metallic copper deposits, when to vitriol solution – nowadays called a copper sulfate solution – an iron nail is placed. The alchemist van Helmont found that he could also obtain copper from vitriol solution in different other ways and that "copper is somehow contained in the vitriol solution" and just needed to be isolated. The way how "copper is contained" could not be answered.

Boyle took up this thought in his work "The Skeptical Chymist," to arrive at his theory of elements being basic substances, which cannot be decomposed. He pointed out repeatedly that theories about chemical substances cannot be considered valid, unless they are proven by experimental evidence.

However, Lavoisier was the first to successfully topple the phlogiston theory by consequent usage of scales. He showed that pure metals are indecomposable, whereas "metal limes" (today: metal oxides) deliver metal and colorless gases. With his famous mercury oxide decomposition (see Fig. 1.1), he found that red mercury oxide produces silvery shining liquid mercury and colorless gas which ignites a glowing wooden splint. Because of his experiments with dissolved metal and nonmetal oxides in water and proving that nonmetal oxides show acidic reactions, he called that colorless gas "oxygen" and defined the basics of the "oxidation theory."

After many experiences concerning mercury and other pure metals Lavoisier stated that these substances should be *elements*, and the decomposable substances *compounds*. But he also considered that at first indecomposable substances may be later proven to be decomposable with improved technology: "Certain substances are chemically considered to be elements, unless and as long as we do not have an instrument to further decompose them. They are elements for us and our aspects, *pour nous, à notre égard*" [2].

The preliminary ideas of the elements, the old Greek philosophy, the Phlogiston theory, and others were the base for thinking and acting before the time of Lavoisier. With the help of his decomposition hypothesis Lavoisier derived many experiments and proved his hypothesis step by step – through him the experiment got the scientific meaning of an instrument that judges the validity of assumptions, principles, and theories!

According to the chemical elements, a big last step has to appear in science history: the connection of the idea of elements with the idea of atoms as smallest particles. Through his experiments Dalton found this relationship, defined the

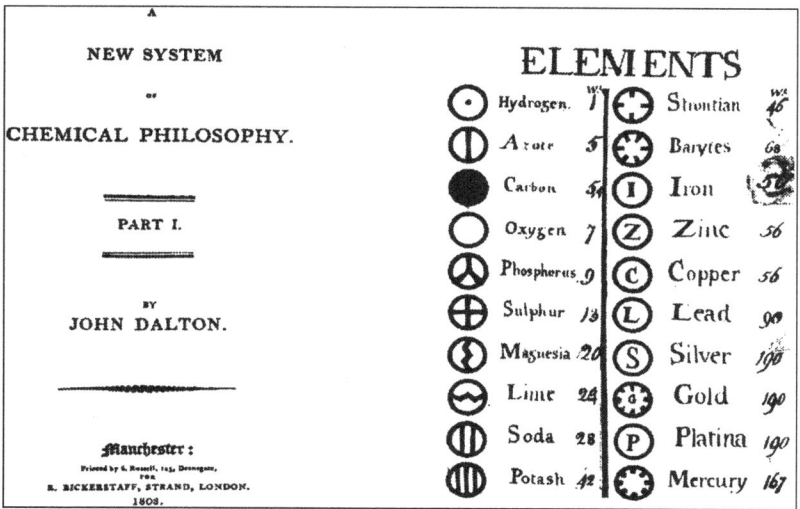

Fig. 5.1 First elements and the atomic mass table of Dalton in 1808 [2]

atomic mass and published in 1808 his famous "New system of chemical philosophy" with the first atomic mass table in science (see Fig. 5.1). With this hypothesis a lot of new research could start, the whole ideas about elements and compounds, about composition of elements and compounds, about formulae, equations and stochiometry were born. But as mentioned Dalton regarded a lot of today's compounds as elements: magnesia, lime, soda, potash, strontian, and baryt – "they were elements for Dalton and his aspects."

5.1 Scientific Ideas: Experiments, Experimental Skills, Safety

As discussed, chemistry developed to an experimental science. Another set of basic questions of chemistry education concerns the role of the experiment in Chemistry, in which way this role also applies to chemistry lessons and which other functions need to be added in this regard. In chemistry education the scientific functions have to apply, but there are also other functions of the experiment like motivation, simulation, or illustration.

Experiments and processes of knowledge construction: One of the most important roles of the experiment in chemistry is to gain scientific and empirical knowledge through experimental proofs of hypotheses under controlled and systematically modified conditions. In the introduction, the evolution of "elements and compounds" is outlined as a fundamental historical example. Three other historical examples will be given. First, the empirical method of perception shall be formalized by outlining the key steps of proving hypothesis (see Table 5.1).

Table 5.1 Steps of the empirical method of proving hypothesis

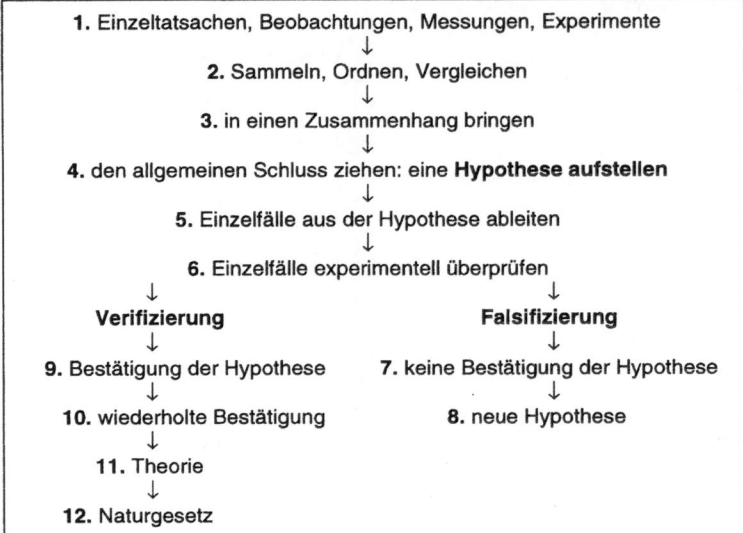

1. Einzeltatsachen, Beobachtungen, Messungen, Experimente
↓
2. Sammeln, Ordnen, Vergleichen
↓
3. in einen Zusammenhang bringen
↓
4. den allgemeinen Schluss ziehen: eine **Hypothese aufstellen**
↓
5. Einzelfälle aus der Hypothese ableiten
↓
6. Einzelfälle experimentell überprüfen
↓   ↓
Verifizierung   **Falsifizierung**
↓   ↓
9. Bestätigung der Hypothese   7. keine Bestätigung der Hypothese
↓   ↓
10. wiederholte Bestätigung   8. neue Hypothese
↓
11. Theorie
↓
12. Naturgesetz

On the basis of considerations about the "horror vacui," Torricelli, together with Galilee, developed the hypothesis of air pressure, and derived the famous mercury experiment in 1648 to prove his hypothesis (see Fig. 1.2). He also verified his hypothesis in additional experiments: air is a substance with a specific density, the surrounding atmosphere shows a certain air pressure, the air pressure decreases with increasing height above earth.

Following the empirical analyses of mass ratios of elements in many organic substances, Kekulé dealt with the expression of these results through formulae, while at the same time he always thought about the structure of corresponding molecules. In the attempt of describing the benzene molecule with his experiences, he failed at first by taking a chain with six C atoms. But after his "famous dream" [3], he established the hypothesis of the cyclic structure of the benzene molecule in 1865 (see Fig. 5.2) and postulated the tetrahedral model of the methane molecule [3].

Kekulé first described the benzene molecule with alternative single and double bonds, but he knew that all six bonds between C atoms should be equivalent. From this hypothesis, he derived many experiments concerning the substitution of H atoms by halogen atoms. He came to the conclusion (see Fig. 5.2) that there is only one mono-substituted molecule (C_6H_5Cl), that there are three doubly substituted molecules ($C_6H_4Cl_2$), and three tri-substituted molecules ($C_6H_3Cl_3$). Many further experiments were necessary to prove the hypothesis of the cyclic structure of the benzene molecule – all students should know this, because most of them are thinking that one single phenomenon may be enough for proving scientific hypotheses.

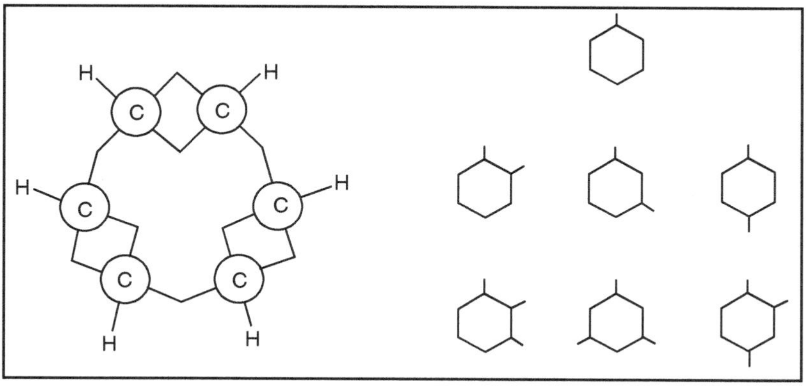

Fig. 5.2 Kekulé's hypothesis for the structure of the benzene molecule and formulae of some substituted derivatives [4]

Since Roentgen's discovery of X-rays in 1895, many scientists worked feverishly on the hypothesis of X-ray diffraction. Laue finally proved it in 1912 and commented on his crucial thoughts: "The discovery of the X-ray interference distinguishes the value of the scientific hypothesis" [5]. Laue arrived at his famous hypothesis because every day he saw in his institute the crystal lattice models of many crystals which were used in Munich by Sohnke and von Groth – and believed in these structures. So he got the idea to take natural crystals like copper sulfate or zinc sulfide as diffraction grids to realize interferences of X-rays by passing them through these crystals (see Fig. 5.3).

Many scientists have sent X-rays through crystals long before, but their observations were confined to the directly passing ray, which did not show anything more than the weakening by the crystal; they did not see the much weaker diffracted rays. The hypothesis of taking crystals as diffraction grids gave the idea to search its surroundings – and they have been found. In 1914 Bragg and his son published the structure of sodium chloride in England, they formulated the Bragg equation as the famous reflection requirement for X-rays and opened the great field of X-ray crystallography.

Data collection: The most important function of the experiment is the validation of hypotheses. Another is to run tests to describe substances and to compare a special substance with other matter. Traditionally density, melting, and boiling temperature are measured to gain a first identification of the substance. Additionally water solubility, solubility in other solvents, refraction index, viscosity, optical activity, thermal conductivity, electrical conductivity, pH-dependency, redox potential, equilibrium and stability constant, spectroscopic behavior, etc. can be tested and the measured values tabulated.

Accordingly chemical processes can be described by determination of mass and volume ratio of reactions, by stoichiometry of the converted amounts of substances, by the number of reacting atoms, ions and molecules, through structural formulae and chemical equations. Furthermore solubility product, equilibrium

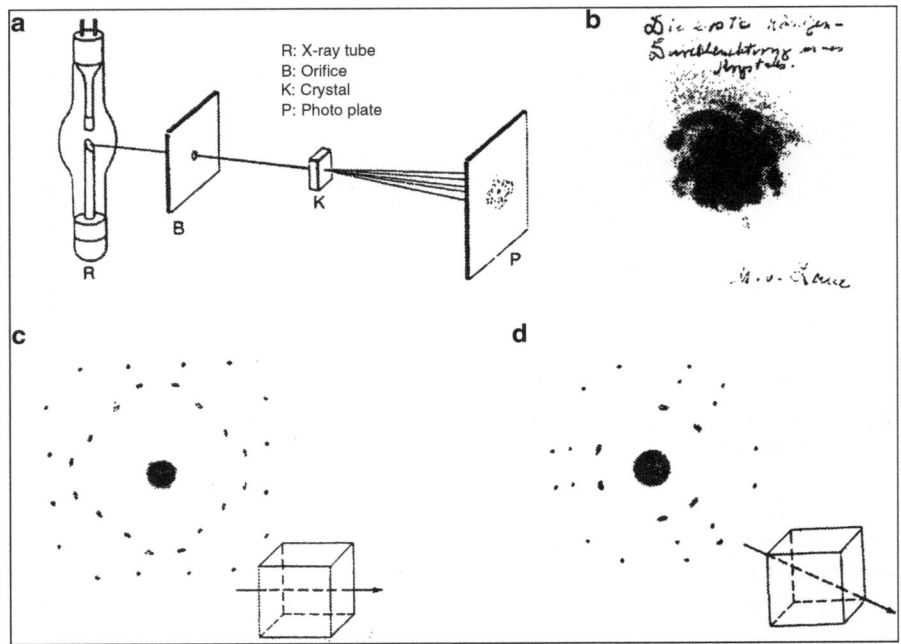

Fig. 5.3 Laue's hypothesis to get diffraction patterns by an X-ray beam through crystals [6]

constant, standard potential, energy turnover, and rate constant identify the chemical reaction.

In present-day laboratories, substances and chemical reactions are described by methods of the instrumental analytics: such as paper-, column- and thin-layer chromatography, gas chromatography (GC), mass spectrometry (MS), combinations of GC and MS, atomic absorption spectroscopy (AAS), infrared spectroscopy (IR), nuclear magnetic resonance spectroscopy (NMR), and X-ray structure analysis.

The required large instruments are not only very expensive, but also need supervision by specially qualified staff. Students become acquainted with the large appliances during their laboratory practicals, while teachers and pupils can inspect them during a visit at a department of chemistry. Smaller instruments that show the method in a clear and didactically reduced way have been developed for school lessons, for example an apparatus for gas chromatography by Wiederholt [7] and one for X-ray structure analysis by Brockmeyer [8]. Corresponding model experiments for gas chromatography (E5.1) and X-ray analysis (E5.2) may demonstrate it.

Synthesis of new substances: Perhaps the oldest function of experimentation is the trial-and-error approach for the production of substances, which cannot be found in pure form in nature. Table 5.2 shows several substances with the approximate date of their first description according to Schwedt [9].

Table 5.2 Extraction or synthesis of some substances in the history of mankind [9]

8000 BC	Ceramics	1648	Hydrochloric acid, nitric acid
3700 BC	Copper, silver and gold	1669	White phosphorus
3000 BC	Lead (Babylonia), bronze (Egypt)	1671	Litmus as an indicator
2900 BC	Glass (Egypt)	1710	Porcelain (Meissen/Germany)
2400 BC	Indigo (Egypt)	1727	Silver nitrate for first photos
2000 BC.	Sulfur from hot springs	1746	Sulfuric acid
1200 BC	Tin and zinc (India)	1747	Sugar (beets)
1000 BC	Iron	1766	Hydrogen
500 BC	Soda, potash, gypsum, mortar (Roma)	1773	Oxygen, nitrogen
400 BC	Mercury (Greece)	1808	Sodium, potassium, magnesium, calcium, strontium, barium
20–80	Soap and mineral colors (Rome)	1810	Chlorine
500	Borax and sodium nitrate (India)	1827	Aluminum
600	Porcelain (China)	1855	Lithium
850	Ammonia, acetic acid, white lead (lead carbonate hydroxide)	1867	Dynamite
900	Paper (Cairo)	1884	Artificial silk
1100	Ink and paint	1894	Argon, noble gases
1227	Spirits of wine as a drug	1898	Polonium, Radium
1230	Gun powder (China)	1901	Indigo
1300	Sulfuric acid	1909	Bakelite
1565	Zinc vitriol (zinc sulfate)	1913	Ammonia
1580	Benzoic acid	1924	Insulin
		1928	Penicillin and others

Experimental skills: Future chemistry teachers not only have to be able to run experiments on their own, but also to do this in school lessons and to put it across to their students. Therefore they possess the skills to plan, run and analyze experiments as well as the manual skills to handle equipment and chemicals in an appropriate way. That is why prospective teachers should not only get the usual experimental training in laboratories during their academic studies, but also a methodological training in experiments on significant topics of chemistry lessons at school.

They especially need experience with the demonstration of experiments that require special safety precaution, such as the handling of:

- Hydrogen, gas mixtures of hydrogen and oxygen and oxyhydrogen reactions
- Alkaline metals, their storage, reactions with water or halogens, as well as safe disposal
- Halogens, the reactions with metals and hydrogen
- The generation of chlorine in gas generators, its storage in a gas jar and reactions
- White and red phosphorus, storage of white phosphorus, and disposal
- Equipment for quantitative experiments such as scale, burette, transfer pipette, aerometer, pH meter, other analytical instruments
- Electric circuits, voltmeter, ammeter and required attachments like battery and transformer
- Electrolysis of salt solutions, Galvanic cells, etc.

The handling of alkaline metals is described in experiments E5.3–E5.10, for reactions with air, water and halogens. Important classroom experiments on essential school topics can be taken from the experimental literature (see [10–14]). In case the students are supposed to run experiments with special effects individually, careful supervision is needed due to the specific safety hazards.

hile working on school experiments the students will get to know a multitude of laboratory equipment. The knowledge of equipment is important for planning experiments for chemistry classes or for running spontaneous experiments, supplementing the classroom discourse. For another thing, it is advantageous to be able to organize the school equipment appropriately. Attaching analytical instruments to a computer makes it possible to collect a series of tests and simultaneously get a graphic representation. With adequate hardware and software data may be collected and analyzed in multiple ways. Since hardware and software develop quite fast, no current literature and equipment list will be given at this stage.

Safety and Disposal: It is self-evident that accidents should not happen during experimentation in laboratory exercises and that carelessness has to be avoided. Laboratory supervisors and teachers have to point out dangers and specific safety measures to all students. A good theoretical and practical-experimental education of chemistry teachers at universities and seminars is the best guarantee for appropriate and accident-free experimentation.

Future teachers need to be familiar with the compilation of chemicals and safety installations, the place and handling of fire extinguishers, fire blanket, sand, safety shower and first aid box. They have to be certain that an adequate disposal of chemical waste is possible or they have to set-up the facilities for it.

The different "Guidelines on Hazardous Substances" define the labeling of all containers for chemicals with standardized danger symbols (see Fig. 5.4) as well as risk and safety statements (R and S phrases). The "Technical Rules for Hazardous Substances" (in Germany: TRGS 450) regulate the handling of hazardous substances in schools. These rules state that the following substances may be used in teacher experiments, but not in student experiments:

– Very toxic substances (T+), e.g. carbon disulfide, nitrobenzene, carbon tetrachloride, white phosphorus, metal cyanides, etc.
– Carcinogenic or teratogenic substances (T), e.g. benzene, chromium (VI) compounds in the form of dusts and aerosols, 1,2-dibromoethane, nickel, cobalt, etc.
– Explosive substances (E), e.g. black powder and mixtures of oxidizable substances with potassium chlorate or sodium chlorate, etc.
– Mixtures of substances that contain pathogens like pathogenic bacteria and fungal cultures, fecal effluent, etc.

Disposal: There are specific instructions for teachers: they have to make themselves familiar with those regulations on hazardous substances and their disposal. They start with considerations of experiments that do not produce hazardous waste at all. These considerations can happen with the students and be part of their environmental education.

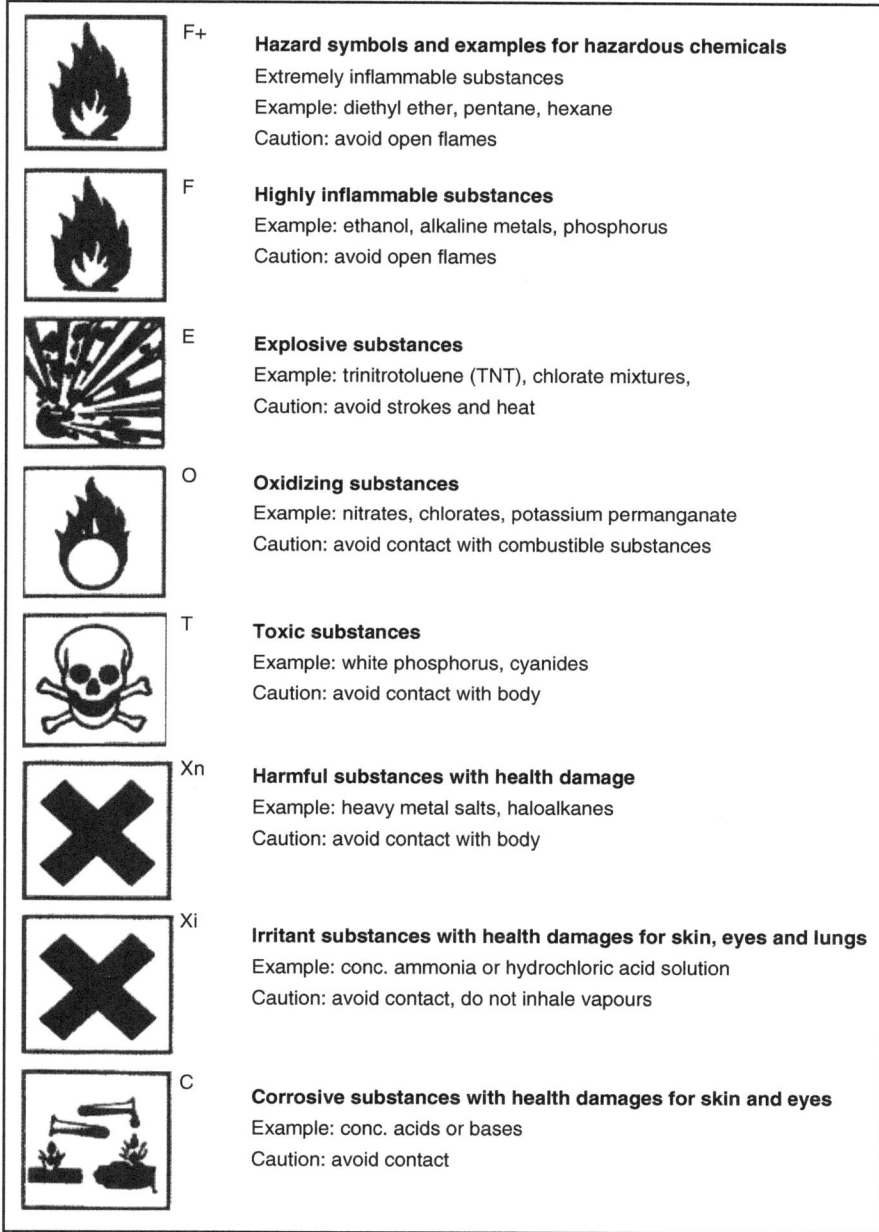

Fig. 5.4 Hazard symbols and examples for hazardous chemicals (experts are developing new symbols with similar meanings, but valid for the whole world)

If an alternative experimentation cannot be realized, an appropriate way of disposal for problematical remainder of chemicals is needed. There are three different ways of disposal:

1. Conversion of hazardous into harmless substances and their disposal in sink: remainder of alkaline metal, for example, can be converted in ethanol, its solution diluted in water and the diluted solution added to waste water into the sink
2. Conversion of hazardous substances and disposal of products in collection vessels: remainder of chromate can be reduced with sodium sulfite solution and produced chromium (III) salt solution can be poured into the heavy-metal salt vessel
3. Immediate disposal in specific collection vessels: hydrocarbons or halogenated hydrocarbons, for example, need to be poured into the appropriate vessels (see Fig. 5.5)

Five collection vessels have proven effective for school laboratories (see Fig. 5.5):

1. For solid waste, wrapped in (filter) paper
2. For concentrated solutions of acids, bases and heavy-metal salts
3. For halogenated hydrocarbons
4. For hydrocarbons and other nonhalogenated organic liquids
5. For polluted mercury waste, intended for purification

From time to time these vessels need to be emptied or deposited at the designated disposal site. The heavy-metal salt solutions of vessel (2) should be made alkaline with bases, prior to the removal: precipitated solid hydroxides can then be filtered and added to collection vessel (1).

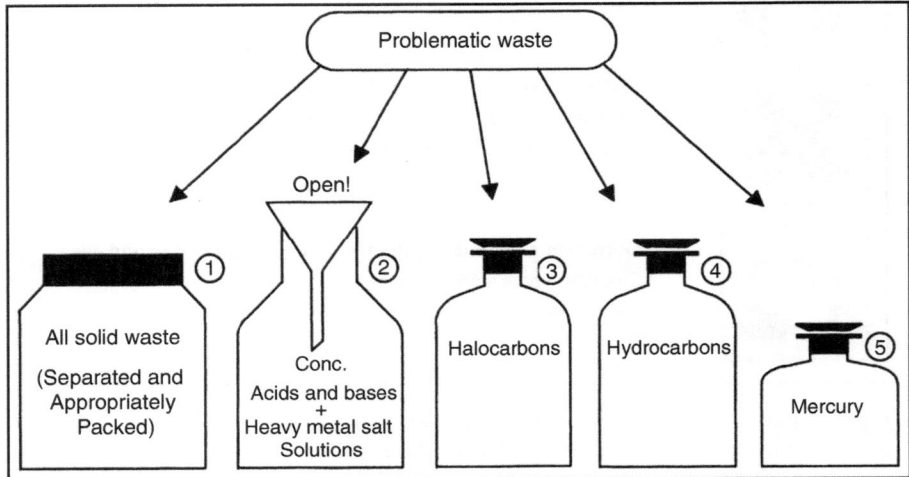

Fig. 5.5 Schematic separation of hazardous substances in schools

5.2   Teaching Processes: Purpose, Selection Criteria, and Forms of Experiments

The scientific purposes of experiments are also educational functions, as for example the process of gaining understanding through setting up and proving hypotheses. If students cannot interpret special unknown facts they can discover the scientific interpretation by the "discovery teaching method": a hypothesis is required, isolated cases are to be deduced. These cases need examination through planning, running, and analyzing experiments (see a detailed description of this inquiry-based teaching by Schmidkunz and Lindemann [15]). In chemical education experiments also have roles, which do not have an equivalent in the scientific field. These purposes (P) will be discussed using examples of experiments (E) with alkaline hydroxides and their solutions.

P1. Motivation.
E5.11: Sodium Hydroxide on the Balance
Students are familiar with weighing from their everyday life: they weigh a chocolate bar to find out that it weighs exactly 100 g and its mass does not change in the earth's gravitation field. In the suggested experiment, students can observe that the mass of a special portion of substance increases while weighing. Regarding this cognitive conflict, students are motivated to explain this phenomenon. Since the second component can only be added from air and since the sodium hydroxide pellets deliquesce aqueous, students might expect aerial water vapor. Sodium hydroxide, indeed, absorbs water vapor from air: this characteristic of hydroxides is called "hygroscopic." But sodium hydroxide also reacts with aerial carbon dioxide to form sodium carbonate. This experiment provides an interesting introduction as well as the motivation to learn more about those reactions.

P2. Stimulate Curiosity.
E5.12: Reactions of Sodium Hydroxide with Air
Surprisingly, the balance shows the increase of the mass of sodium hydroxide (see E5.11). Since the deliquescence of the sample is a hint for water vapor in the air, the question arises, if other components of air are responsible for the mass increase, if nitrogen, oxygen or carbon dioxide might also be responsible. The students are asked to think of an apparatus to verify their suggestion. A simple way is to use test tubes, which are filled with some pellets of sodium hydroxide and a particular gas – connected to a gas syringe, which contains the same gas. The result: only carbon dioxide reacts with solid sodium hydroxide as can be deduced from a volume decrease in the gas syringe.

P3. Testing of Hypotheses.
E5.13: Reactions of Sodium Hydroxide Solution with Carbon Dioxide
Since pure water as well as solid sodium hydroxide react with carbon dioxide gas, the solution in water – sodium hydroxide solution – may also react with carbon dioxide. An hypothesis might be: sodium hydroxide solution dissolves carbon

dioxide. Since sodium hydroxide is easily soluble (a concentrated, approx. 10-M solution exists), the hypothesis might be formulated more precisely: concentrated sodium hydroxide solution dissolves carbon dioxide in a higher amount than the same volume of the diluted solution. Again the students are asked to set-up an apparatus, which tests the hypothesis, again a test tube, filled with few milliliter of solution and connected with a gas syringe, can be used.

The result: 5 mL of concentrated sodium m hydroxide solution react with 100 mL of carbon dioxide completely, while the same amount of highly diluted sodium hydroxide solution only reacts with a few milliliter of carbon dioxide: the hypothesis is verified this way.

P4. Collecting Data.
E5.14: Concentrations and pH-Values of Sodium Hydroxide Solutions
To estimate the concentration of the diluted sodium hydroxide solution used in E5.13, it is possible to measure the pH-value. For a first estimation, universal indicator paper can be dipped into the solution and the pH can be determined by comparison of the obtained color with the colors on a standard scale. For a more precise measurement a pH meter is used: after calibrating the pH-meter for the alkaline range, the glass electrode is to be put into the solution to measure the pH, and the concentration can be calculated from the known pH. The pH-data of other everyday solutions can be measured and collected in a table of pH-values.

P5. Demonstration of a Theoretical Context.
E5.15: Dilution Series for the Illustration of pH-Values
Students might be unable to classify and correlate the pH-values of liquids in E5.14 with the concentration of $OH^-(aq)$ ions in the solution. To explain this connection, a 1-M sodium hydroxide solution is to be presented with pH 14. This solution is to be diluted by the factors 1:10, 1:100 and 1:1,000, and pH-values 13, 12 and 11 will be measured. Now it should be clear that the pH-value changes by one unit, when the concentrations of two solutions vary by the factor 10. Knowing this, the concentrations of $OH^-(aq)$ ions of the solutions in E5.14 can be estimated.

P6. Simulation of Technical Processes.
E5.16: Model Experiment Demonstrating the Mercury Electrode Process
Industry has developed the mercury electrode process to obtain pure, chloride-free sodium hydroxide solution through the electrolysis of cheap rock salt solution. The circulating mercury is charged negatively, sodium ions are reacting and sodium amalgam is formed by the electrolysis. Sodium amalgam decomposes later with water in counterflow, sodium hydroxide solution and hydrogen gas are formed (see Fig. 5.6a). Formation and decomposition of sodium amalgam can be simulated in a model experiment with the electrolysis of sodium chloride solution (see Fig. 5.6b): the U-tube is half filled with mercury, the left tube contains water, the right tube sodium chloride solution. Graphite electrodes are done into the U-tube (see Fig. 5.6b), the electrolysis starts with minus pole on the left side. After 1 min the alkaline reaction on the left side is shown by the dark red color of phenolphthalein indicator, chlorine production can be observed in the right tube.

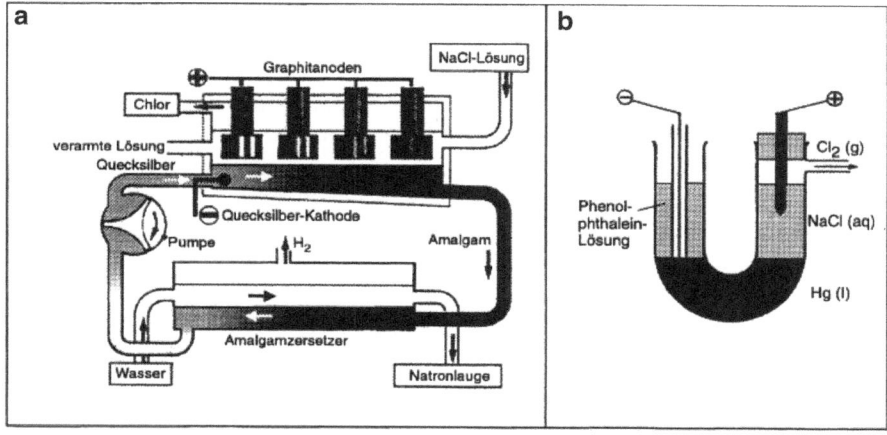

Fig. 5.6 Schematic diagram of the mercury electrode process in industry [16] and school

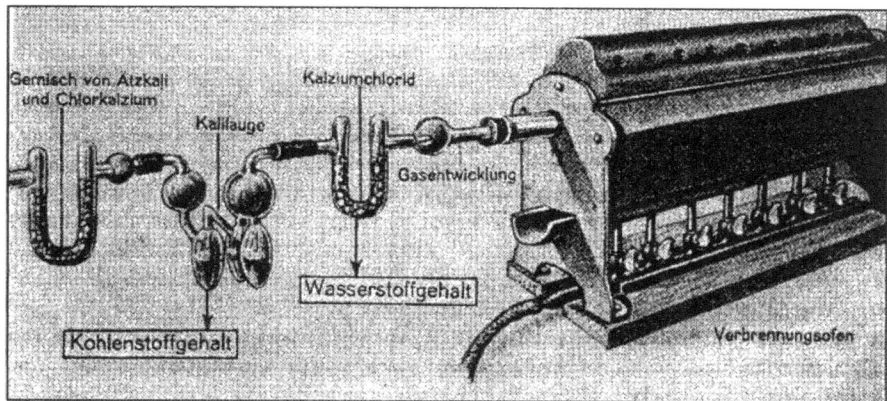

Fig. 5.7 Historical apparatus for the combustion analysis by Liebig [17]

P7. Comprehension of Historical Experiments.

E5.17: Combustion Analysis by Liebig

Sodium as well as potassium hydroxide solution and caustic potash play an important role in the historical combustion analysis by Liebig. He successfully developed the "potash apparatus," which completely absorbs the produced carbon dioxide from the gas stream (see Fig. 5.7). A similar apparatus can be demonstrated schematically for the comprehension of this analysis method. In the version presented in E5.17, however, produced carbon dioxide is being collected quantitatively in a gas syringe, while water is not analyzed.

P8. Repetition and Deepening of Knowledge.

E5.18: Reaction of Tube Cleaner (NaOH–Al-type)

The bathroom chemical "sink cleaner" can be presented and analyzed as repetition and to deepen the previously discussed aspects of substances like alkaline metal hydroxides and their solutions. The label indeed reveals the ingredients sodium hydroxide and aluminum. In the actual substance white sodium hydroxide in grainy form can be observed, and additionally little silver-colored aluminum splinters. When water is added to the mixture, a strong exothermic and alkaline reaction sets in: hot concentrated sodium hydroxide solution is produced. Adding paper to the reaction mixture demonstrates the decomposition reaction of organic substances. With added aluminum and sodium nitrate gaseous ammonia is produced which improves the properties of those cleaners. The formation of gases is being analyzed and their functions are being discussed (see also Chap. 8).

P9. Controlling the Learning Process.
E5.19: Reactions of Alkaline-Earth Metals with Water
The learning process has been successful, if students are able to transfer familiar reactions of alkaline metals with water to similar reactions of magnesium or calcium with water. The formation of alkaline hydroxide solutions and hydrogen gas is familiar, but additionally solid hydroxides form suspensions. Also composition and formulae $M(OH)_2$ (M = Mg or Ca) of these hydroxides are different leading to different chemical equations for the metal–water reactions.

P10. Practicing Experimental Skills.
E5.20: Student Experiments on the Lithium–Water Reaction
Student experiments with hazardous alkaline metals and caustic bases are possible with lithium, as it can be handled much more safely than sodium or potassium. The reaction with water is not only possible in an open beaker, but also quantitatively in a closed apparatus. A piece of lithium can be added to a glass cylinder, which has been filled with water and opened under water in a glass bowl pneumatically (it is not allowed to let sodium and potassium react in this way!).

The student work sheet (see E5.20 at the end of this chapter) exemplifies how work sheets should look in general. It also shows which experiments may be carried out safely and can be interpreted by the students themselves.

Selection criteria for experiments: After the different roles of experiments in chemical education have been experienced, further criteria for the selection of experiments in chemistry classes shall be offered and discussed: The planned experiment should

– Be suited for the age group
– Be based on the student's knowledge
– Be productive scientifically and educationally
– Match the school's chemical collection
– Have a high probability of success
– Not have any safety hazards
– End during an adequate period of time
– End with a clearly recognizable effect
– Be usable as a student experiment

Selection criteria can also be found for specific classroom topics, e.g. for an initial experiment toward the introduction of chemical reactions of different elements (see Table 5.3):

Table 5.3 Fulfillment of criteria 1–12 for choosing initial experiments on chemical reactions

Examples from school books		1	2	3	4	5	6	7	8	9	10	11	12
Formation of	Iron sulfide		x	x	x	x			x		x	x	x	
	Copper sulfide		x	x	x	x	x	x	x	x	x	x	
	Metal oxides	x	x	x	x		x		x	x	x	x	x
	Nonmetal oxides	x	x	x	x		x			x	x	x	x
	Metal chlorides	x		x	x		x			x	x	x	x
	Aluminum bromide		x	x	x	x				x	x	x	
	Silver iodide		x	x	x	x	x	x		x	x	x	x
	Sodium amalgam		x	x	x	x	x	x	x	x	x	x	x

Element A + Element B → Compound AB; energy turnover

1. A, B, AB should be familiar to the students from their everyday life.
2. The reaction should be possible as a student experiment.
3. The reaction should run in a suitable period of time.
4. The experiment should not need too much instrumentation.
5. Substances A, B, AB should be visibly solid or fluid.
6. A, B, AB should be obviously distinguishable by their properties.
7. The mixture of A and B should be easily separable.
8. AB should react back into its elements.
9. The reaction should show a distinct energy turnover.
10. Safety rules must be performed.
11. A and B should be visualized by models (sphere packing or molecular model).
12. AB should be visualized by models (sphere packing or molecular model).

Conduct of experiments: The most frequently used form of experimentation at school is the demonstration experiment. Although the student experiment is considered to be educationally important and valuable, due to lack of time or unsuitable equipment it is not always realized. A compromise between both can be an experiment demonstrated by students: one student or a group of students prepare and demonstrate an experiment to the class (see Table 5.4).

To draw the observer's attention to the essentials and to optimize the perception, a few principles of Gestalt psychology need to be considered for demonstration experiments [18]:

– Parts of an apparatus are to be arranged in such a way that substances flow from left to right for the viewer
– Stands should not disturb the observation of planned effects, they are to be put behind the apparatus from the viewer's perspective
– Apparatuses are to be as easy as possible, needlessly complicated vessels or equipment should be avoided
– Connection tubing should be arranged flat and horizontal, the substance flow should run in a straight line
– Objects or chemicals, which do not immediately belong to the experiment, are to be removed from the table of the experiment
– Planned effects are to be visually, acoustically or olfactorily intensified, if otherwise they cannot be observed sufficiently from the distance

Table 5.4 Some types of experimentation in classes

– Teacher demonstration	– Real experiment
– Student demonstration	– Thought experiment
– Student experiment	– Experiment demonstrated with media
– Division of work	– Qualitatively, quantitatively
– Simultaneous work	– Differentiation by amount of substance:
– Independent work	– Macroscale
	– Halfmicroscale
	– Microscale

Especially the *overhead projector* is a good way to intensify the planned effect by projection and enlargement, for example, of glass dishes with colored indicator solutions and their reactions with acids or bases. Also the bright light of the projector is useful: by adjusting the mirror, the apparatus on the table can be directly illuminated with the light or only the part of the apparatus, which shows the effect (see also Chap. 4). If a *video camera* or a swan's neck camera is available, it is also possible to enlarge effects that are hardly observable from the distance. This method of experimentation gets by with small amounts of chemicals: "chemistry en miniature" [19]. Häusler's semimicro method [20] and Kometz's method of cuvettes [21] also make it possible to run experiments with small amounts of substances and mostly without the use of a hood.

Organization of lab classes: The preparation of lab classes in chemistry is often not limited to a single hour, most of the times a series of experiments of one teaching unit has to be spread over several hours.

Experiments have to be tested in advance, transparencies for the overhead projector or work sheets for student experiments have to be prepared and copied for the lesson. The individual steps can be structured in the following way:

Preparation (well in advance):

– Testing the planned student experiment
– Designing a work sheet for the student experiment
– Designing a power-point presentation
– Preparing chemicals and equipment
– Testing the planned demonstration experiment
– Drawing an overhead transparency of the apparatus
– Preparing the safety features

Procedure (during lecture):

– Presenting the problem of the lecture
– Listening to students questions and discussion
– Explaining the assembly of equipment or apparatus
– Drawing the apparatus on blackboard or transparency
– Giving time to students to copy the drawings
– Explaining the safety features
– Conducting the experiment

- Emphasizing the effects that can be observed
- Making it possible to read the measuring devices
- Finishing the experiment clearly
- Looking for safety

Observation (measured values):

- Gathering and formulating the observations
- Writing down observations (black board, chronological order)
- Repeating parts of the experiment if necessary
- Presenting the measured values tabulary or graphically
- Discussing measurement errors and sources of errors

Explanation:

- Interpreting the particular observations by students
- Repeating experimental parts if necessary
- Offering help for explanation and discussion
- Referring observations to previous knowledge
- Discussing and explaining the overall observations
- Developing models of the structure of involved substances
- Draw beaker models containing symbols of atoms, ions, molecules
- Finding formulae and reaction equations
- Formulating the experimental report on the black board
- Taking the report in the exercise book by every student
- Write-up the experimental report as homework, if required

Follow-up (next lesson):

- Answering questions to the experiment or experimental report
- Completing the experimental report
- Getting back to the presentation of the problem
- Completing the problem and its solution
- Deducing a new topic from the results if applicable

5.3 Learner: Play Instinct, Curiosity, Experimental Skills

A special relevance belongs to the experiment in chemistry classes for young students: due to play instinct and curiosity, high receptiveness for experiments can be expected. Easy phenomena and measurements of density, solubility, melting, and boiling temperatures of substances are most welcome among young students and recommended for student experiments. Often very interested students already have a chemistry set (see Chap. 4) and know a lot of simple experiments: these students especially stand out, because of their knowledge and experimental skills. The following aspects have to be reflected for the learner.

Fundamental manual skills and handling of equipment: In science classes of primary education initial experiments on air and combustion or on water and solutions are to run; mostly without glassware from the chemical lab, but rather with everyday objects from the kitchen. Therefore, the appropriate handling of equipment and chemicals as well as safety precautions have to be practiced. Step-by-step skills for problem-oriented and independent experimentation will be gained. Attention should be paid: boys like to push themselves to the fore, run the experiments and give the girls the part of observation and recording. Teachers should try to compensate this or to work in single-sex groups.

Adaptation of exact observations: Random observation has to be changed to directed observation, so observations have to be planned before starting the experiment. For quantitative measuring, measurement parameters as well as their units have to be made clear, the equipment has to be presented and thoroughly demonstrated, before being deployed for independent experiments by the students.

Recording steps of thinking: Experimental skills will be optimized, if students create experimental protocols during class or as homework. Not only the chronology of individual steps of thinking and experimentation should be discussed, but also measured values displayed in tables or graphs. This last skill is especially difficult for beginners and should be practiced on easily evaluated measurements.

5.4   Human Element: Environmental, Everyday Life, and Historical Aspects

Students are familiar with substances and their reactions from their everyday life and know, for example through questions regarding waste sorting – about environmental hazards of problematic substances. Regarding the human element, the following topics can be discussed in class and reflected upon with the help of experiments.

Experiments on environmental protection: As soon as a topic (e.g. air, water, soil, ecology) touches the problem "environmental protection," additional experiments concerning this matter should be demonstrated and discussed. Regarding a real environmental education, ways for planning and performing experiments without producing harmful substances or with an unproblematic waste disposal can be considered together with the students (see Chap. 8).

Experiments on everyday life and industry: If it is possible to draw a connection to everyday life and to show related experiments, it should be done. This connection can also be used for intrinsic motivation (see Chap. 2). In case of industrial production of certain substances (e.g. sugar, fertilizer, coal, metals) or chemical industry close to the area of the school, experiments should relate to these industrial processes (see Chap. 8). If possible excursions and visits of industrial plants can give insight into problems of mutual influence of laboratory experiments and large-scale technical realization (e.g. wine distillation in class and visit of industrial brandy production, or galvanic cells in class and visit of industrial battery production).

Historical experimental development: Students should learn about the impact of production processes of many substances on social life (e.g. in the Stone, Bronze and Iron Age). They should also learn how certain substances changed social life (e.g. Liebig's research and the relevance of mineral fertilizer for agriculture and food production).

War, in some way, accelerated the development and production of special substances: e.g. sugar made of sugar beets after Napoleon's Continental import embargo of sugarcane; ammonia synthesis for the production of explosives for World War I; synthesis of gasoline and rubber for World War II; development of uranium for nuclear weapons and the first atomic bombing at the end of World War II. These connections should not be embezzled in class – in fact this kind of knowledge can lead to a discussion about threatening developments on account of the war.

Group dynamic processes for solving experimental tasks: Student experiments may help to develop social behavior in a group of boys and girls. Especially, the coordination of boys and girls within a group can affect the cooperation positively and reduce potential prejudices. Confident experimentation brings up group discussions, supports the ecologically responsible use of chemicals and thus develops environmental consciousness and skills for environmentally suitable behavior.

Expositions of experimental results collected by students (e.g. poster displays for results of the examination of air, soil, river or drinking water quality in their home town) can initiate task-oriented discussions of "experts" with schoolmates and contacts with visitors from outside school. This form of education for expertise, critical thinking and democratic open-mindedness of our students is, without a doubt, an important goal of chemistry classes!

Appendix A. Problems and Exercises

P5.1: Experiments in chemistry classes usually fulfill certain roles. Name five of these functions and sketch one experiment for each function and its context for chemistry lessons.

P5.2: Two forms of experimentation are common for student experiments: work-sharing or simultaneous work. Explain with three classroom examples of your choice.

P5.3: Hypothesis testing serves as a way of gaining knowledge in natural sciences as well as in chemistry classes. Describe a historic scientific hypothesis and its testing. Plan a situation in a chemistry lesson, which leads to a hypothesis, and sketch the way of possible experiments for its verification.

P5.4: Experiments for quantitative measurements are of importance. Give five kinds of quantities and their units that are measurable in chemistry classes and write down how to conduct the experiments. Describe corresponding measuring apparatus and draw them schematically.

P5.5: Safety rules and waste disposal play an important role in experimental chemistry lessons. Name five important safety rules and for each rule an example of a dangerous experiment. Describe ways of waste disposal and list important collection vessels for chemical waste.

Appendix B. Experiments

E5.1: Gas Chromatography in a School Experiment

Problem: Usual chromatographs for research laboratories are closed instruments, into which you cannot look ("black boxes"). For understanding the analytical process, it is therefore educationally reasonable to demonstrate an apparatus, which uses a vitreous coiled-tube condenser made of glass which shows all functions openly (see figure). With this apparatus propane and both isomers of butane can be separated and detected by a luminous flames.

Material: Gas chromatograph, gas syringe (10 mL), butane burner (F^+); hydrogen (F^+).

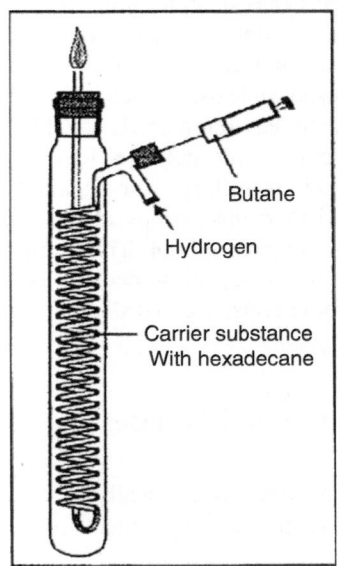

Procedure: The coiled-tube condenser contains a white carrier substance. The surface of the carrier substance is coated with nonevaporating hexadecane ($C_{16}H_{34}$) and therefore it can be used for the separation of hydrocarbons. Hydrogen is being connected and passed through as carrier gas. In case the hydrogen–oxygen reaction is negative, hydrogen can be ignited at the end of the glass tube.

The gas syringe is being filled with gas from the butane burner, and 5 mL has to be injected into the hydrogen stream through the septum. One waits until the colorless hydrogen flame changes for a short time into a yellow flame. The experiment is being finished by turning off the hydrogen stream.

Observation: After about 30 s the colorless hydrogen flame lights up bright yellow and then goes colorless again. That repeats two other times: the first yellow flame shows propane, the second one i-butane, the third one n-butane.

E5.2: X-ray Apparatus for Analyzing Crystals at School

Problem: Most instruments for X-ray analysis are very expensive large-scale devices, which can be seen in research institutes of science departments at universities. The teaching aid industry has developed a school device for the X-ray analysis. Besides the well known shadow pictures of X-rays it can demonstrate the interference patterns of single crystals of salts (Laue patterns), of crystalline powders (Debye–Scherrer pattern) and the glancing angles of crystals (Bragg angles).

Material: X-ray apparatus for schools, equipment such as X-ray film or Polaroid-X-ray-diffraction-cassette; film developer and fixer, lithium fluoride-single crystal (T), sodium chloride powder.

Procedure: The X-ray apparatus can be introduced with X-ray shadows: the bones of an animal, metal pieces in a closed wooden container, etc. Then the apparatus is being converted to take Laue patterns: The single crystal is being adjusted, a photographic plate or a Polaroid-X-ray-diffraction cassette is being placed behind the crystal, a thin X-ray beam is to be separated and directed on the crystal. After an adequate exposure time, the photographic plate is being developed or the Polaroid-picture is being removed from the cassette. The crystal powder is to be handled likewise.

Observation: Besides the primary X-ray, a symmetrical pattern of diffraction points can be seen in the first case. In the second case, there is a pattern of diffraction rings. These phenomena can be explained by the symmetrical order of the ions in the salt crystal; experts are able to calculate the spatial structure of the salt crystal from data of the diffraction pattern.

Precautions for the handling of alkaline metal pieces:

1. Put on safety goggles.
2. Do not touch alkaline metals with bare hands, use tweezers instead.
3. Blot the alkaline metal dry with filter paper.
4. Cut off the crust thoroughly with knife and tweezers.
5. Put the alkaline metal crust and remainder in spirit.
6. Let pea-sized pieces of alkaline metals react.
7. Do not extinguish fires of alkaline metals with water, but with sand.
8. Dilute alkaline solutions with water and wash down the sink.

E5.3: Cut Surfaces of Alkaline Metals

Problem: Alkaline metals are stored in kerosene because oxygen and carbon dioxide of the air or water vapor in the air react with these metals to form oxides, carbonates or hydroxides. To demonstrate it a big piece of alkaline metal is to be cut and the cut surface is to be observed.

Material: 3 watch glasses, tweezers and knife, filter paper; lithium (C/F), sodium (C/F) and potassium (C/F).

Procedure: One piece of metal at a time is to be taken from kerosene, put on filter paper and cut open. The cut surface is to be observed.

Observation: The silvery cut surface starts tarnishing and darkening fast, going from lithium over sodium to potassium.

E5.4: Reaction of Alkaline Metals with Oxygen

Problem: The tarnishing of alkaline metals (see E5.3) is being interpreted by students as a reaction with air or oxygen, respectively. Combustion in air is being assumed and this assumption is to be tested by igniting a piece of the metal.

Material: Tripod with wire gauze, burner, little pieces of three alkaline metals (C/F).

Procedure: Under the hood one piece of metal at a time is to be put on the wire gauze and ignited with the burner's roaring flame.

Observation: Lithium burns with a red, sodium with a yellow and potassium with a violet flame; white smoke ascends, which can be pungent in the nose. White metal oxides remain as combustion products (in case of contamination with kerosene, the product might also appear black).

E5.5: Reaction of Alkaline Metals with Water

Problem: As discussed in E5.4, students might assume that water vapor is the reason for the reaction of the metal cut surface in air. A reaction of alkaline metals with water can be formulated as a hypothesis and tested in experiments.

Material: 3 beakers, 3 watch glasses, universal indicator paper, phenolphthalein indicator solution, big glass dish; little pieces of the three alkaline metals (C/F).

Procedure: Three beakers are half-full with water. One piece of metal is to be put in every beaker, the beaker is to be covered with a watch glass (caution: caustic substance might spurt!). The solutions are to be tested with universal indicator paper after the reactions. A part of the solution is to be concentrated by evaporation, and the resulting white solids are to be tested with indicator paper.

In a second experiment a big glass dish is to be filled with water (about 2 cm high), the glass dish is to be put on the switched on overhead projector. Two or three pieces of sodium are to be put on the water surface; phenolphthalein indicator solution may be added.

Observation: All three metals react with water, the intensity of the reaction increases from lithium over sodium to potassium. During the reaction the piece of sodium melts to a ball, potassium melts and ignites, the burning potassium delivers caustic white smoke.

Three pieces of sodium circulate fizzy on the water surface and leave streaks (well shown in the overhead projection). The streaks are colored dark red in case of the added indicator solution.

All three of the solutions react highly alkaline. White solids remain after the water evaporates. These also react alkaline with wet indicator paper.

Please note: Covering the beaker with a watch-glass serves as safety measure: sodium or potassium balls might stick to the rim of the glass, explode and shoot out of the glass.

Pieces of sodium can also be put on wet filter paper (under the hood!): in this case they ignite with the formation of white smoke and react directly to white solid hydroxide. Caution: the liquid metal may splash around – wear safety goggles!

E5.6: Detection of Hydrogen in the Lithium–Water Reaction

Problem: The sizzling sound and the burning potassium flame show the burning gas that generates on the water surface during the metal–water reaction. The gas can be

collected pneumatically and identified as hydrogen, if the reaction is started under water.
Material: Glass dish, gas jar, cover glass, tweezers, wooden splint; bean-sized pieces of lithium (C/F).

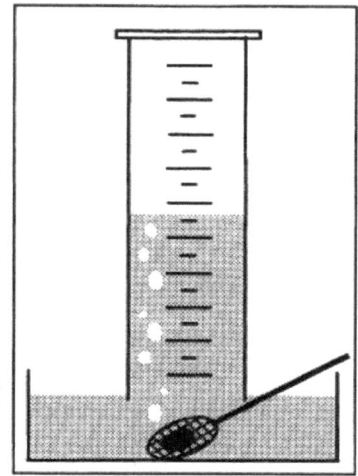

Procedure: The gas jar is to be filled with water and placed pneumatically in the glass dish, which is half-full with water (see figure). A piece of decrusted lithium is to be put with the help of tweezers under the opening of the gas jar. This procedure can be repeated with more pieces of metal until the gas jar is filled with gas. With the help of a cover glass, it is then removed from the glass dish and placed on the working surface vertically. The gas from the gas jar is to be tested with a burning wooden splint. Caution: there might be a little bang.
Notice: this reaction with sodium or potassium is very dangerous – the gas jar may be destroyed during the reaction!
Observation: The vigorous formation of a colorless gas can be observed. The gas burns with a bang at the beginning; the flame is red colored because of the presence of lithium hydroxide solution.

E5.7: Reaction of Sodium with Chlorine
Problem: The reactivity of metals toward oxygen and water can also be assumed regarding halogens. It can be shown exemplarily by the reaction of sodium and chlorine. As an exception the reaction product sodium chloride can be tasted carefully. Interestingly the product of two aggressive and toxic substances is sodium chloride – an essential nutrient!
Material: Gas generator, gas syringe with connected glass tube, test tube; potassium permanganate (O/Xn), conc. hydrochloric acid (C), pea-sized piece of sodium (F/C).
Procedure: Chlorine is to be generated from potassium permanganate and hydrochloric acid in the gas generator under the hood; the gas syringe is to be filled with chlorine and connected with a glass tube. The test tube is to be fixed (stand),

a crust- and kerosene-free piece of sodium is to be heated in the test tube until
sodium melts and lights up (see figure).

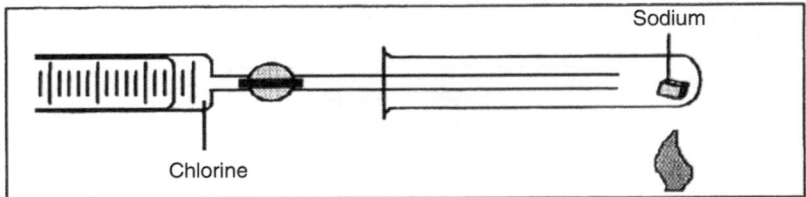

The chlorine is to be passed over sodium slowly with the help of the gas syringe
until a yellow flame appears. Slowly all chlorine of the syringe is to be passed over
the burning sodium. The white substance on the glass tube can be tasted.
Observation: Sodium reacts with chlorine forming a luminous yellow flame; white
powder settles out on the glass tube and inside the test tube: it tastes like table salt.
Please note: The damaged test tube might contain remainder of sodium. Ethanol is
to be added carefully, the test tube is to be rinsed with water and disposed off in the
glass waste.

E5.8: Flame Colors of Alkaline Metal Salts

Problem: When alkaline metals burn in air, the emerging flames show a characteristic
color. These colors can also be observed when the salts of alkaline metals are being evaporated in the hot burner flame. These colorations can be useful for analytical purposes.
Material: 3 watch glasses, cobalt glass; magnesium oxide rods, lithium chloride
(Xn), sodium and potassium chloride.
Procedure: A small amount of lithium chloride is to be put on a watch glass, on
a second glass a small amount of sodium chloride and on a third glass a small amount
of potassium chloride. These samples are to be wet with a little water. The magnesia
rod has to glow in the gas burner flame until the flame does not glow yellow anymore.
Now the magnesia rod is to be dipped into one of the salts and then put into the gas
burner flame. The potassium flame is to be watched through a cobalt glass.
Observation: The burner flame is colored bright red, yellow, and violet by the salts
in the stated order. The last color can be seen easier, if the flame is being watched
through a blue cobalt glass: the yellow flame that is caused by traces of sodium salts
is being absorbed by the blue-colored glass.

E5.9: Electrolysis of Lithium Chloride-Melt

Problem: Solutions of alkaline metal salts can be electrolyzed without difficulty,
but instead of the expected metal, hydrogen forms at the cathode. Therefore this
experiment shows exemplarily that instead of a solution, a melt is to be used to get
the elemental base metal. During this fused-salt electrolysis two technical problems
occur: working at high temperatures and a high current flow.
So a mixture of salts is used to lower the melting temperature; and the direct current
voltage source has to be secured for a high current flow of up to 10 A.

Material: Evaporating dish (Pyrex glass), watch glass, funnel, iron and carbon electrodes, burner, direct current voltage source (10 V, 10 A), tripod with wire gauze; lithium chloride (Xn), potassium chloride.

Procedure: A mixture of 21 g of anhydrous lithium chloride and 7 g of potassium chloride is to be prepared, put in the evaporating dish and – as the figure shows – connected to the electrodes. Set a direct current voltage of 8 V. The mixture is to be melted and electrolyzed on the tripod with wire gauze (under the hood; roaring flame). The electrolysis is to be interrupted after about 10 min to extract the solidified lithium (a layer of chloride protects the metal). The sample can be analyzed after cooling. Especially, the reaction with water and the thereby emerging hydrogen can be demonstrated. Please note: In case of igniting lithium, do not try to extinguish it with water!!! Use sand or sodium chloride!

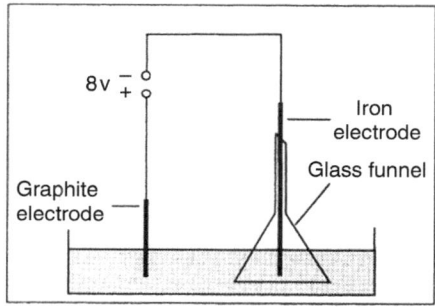

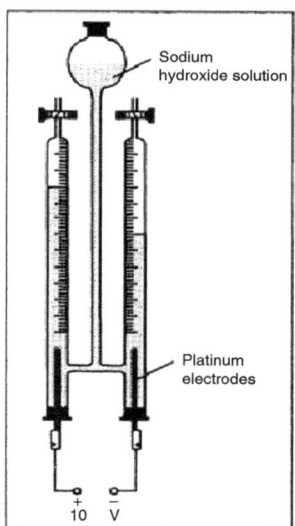

Observation: Lithium deposits at the cathode, chlorine gas forms at the anode. Lithium reacts with water, as known. The emerging hydrogen burns with a red flame.

E5.10: Electrolysis of Sodium Hydroxide Solution

Problem: The electrolysis of melts has been interpreted with the formation of metal in E5.9. In E5.10 it is to be demonstrated that during the electrolysis of sodium hydroxide solution, hydrogen forms at the cathode, instead of sodium metal, and oxygen forms at the anode. In this context you can speak of the decomposition of water and of a special "decomposition apparatus," because hydrogen and oxygen are generated in a volume ratio of 2:1. This apparatus can also be used for all other electrolyses that produce gases at both electrodes.

Material: Decomposition apparatus with platinum electrodes, direct current voltage source (10 V), cable, test tubes, wooden splint; diluted sodium hydroxide solution (C).

Procedure: The apparatus is to be filled with sodium hydroxide solution and connected to the direct current voltage source. The electrolysis is to be run with 5–10 V. The electrolysis is finished, when the cathode cylinder (minus pole) is filled with gas. The volume in the electrode cylinders is to be measured. A test tube is to be put over the cylinder of the minus pole. The tap is to be opened so that gas can escape from the cylinder and be collected in the test tube for a combustibility test. The tap of the plus pole cylinder is to be opened and a glowing wooden splint is to be held above the opening. Afterwards the taps are to be closed again.

Observation: Gases emerge in the volume ratio of 2:1. The first gas burns with a bang: hydrogen. The second gas reignites a glowing wooden splint: oxygen.

E5.11: Sodium Hydroxide on the Balance

Problem: By observing an increasing mass of sodium hydroxide on the balance, students are motivated to explain this unknown phenomenon. Because sodium hydroxide reacts becomes moist on the watch glass, the assumption follows that the reaction with water vapor causes the increasing weight. One has to introduce the term "hygroscopic" for substances that extract water vapor from humid air.

Material: Digital balance with overhead display, watch glass; sodium hydroxide (pellets).

Procedure: The digital balance is to be switched on. The watch glass is to be placed on the balance; the displayed mass does not change. About 20 sodium hydroxide pellets are to be placed on the watch glass, the new mass is to be measured and observed for 5 min.

Observation: The mass of sodium hydroxide increases by a few hundred milligrams. The longer you wait, the higher is the mass; the pellets become moist on the watch glass.

E5.12: Reactions of Sodium Hydroxide with Air

Problem: The reaction of sodium hydroxide with water has been shown in E5.11. But it still needs to be clarified, whether other components of air react as well. Students know the components of air, such as oxygen, nitrogen and carbon dioxide. The question arises, whether these gases as well as water vapor react with sodium hydroxide. The students can create an apparatus independently that makes it possible to answer this question. One way: fill the gases into gas syringes and connect them gas-tight to test tubes containing some sodium hydroxide pellets. A high volume decrease can only be observed in the case of carbon dioxide: sodium hydroxide reacts to form solid sodium hydrogen carbonate in the form of small white crystals.

Material: 3 gas syringes with tap, test tubes, fitting rubber stoppers with fitting hole; sodium hydroxide pearls (C), oxygen, nitrogen, carbon dioxide.

Procedure: The gas syringes are to be filled with one of the stated gases to the 100 mL-mark, closed and connected to the rubber stopper. The three test tubes are to be filled with 10 sodium hydroxide pellets each. One after the other is to be overlayed with one of the mentioned gases: therefore first the gas syringe is to be put loosely on the test tube, the air is to be pushed out of the test tube by the gas until 50 mL remain in the syringe (see figure in E5.13). Then the syringe is to be locked gas-tight to the test tube. The gas volume is to be observed on all three gas syringes for a couple of minutes.

Observation: Neither the oxygen volume nor the nitrogen volume changes, but the carbon dioxide volume decreases: in this case small white crystals form on the sodium hydroxide pellets.

E5.13: Reaction of Sodium Hydroxide Solution with Carbon Dioxide

Problem: From E5.12 the observer concludes that sodium hydroxide reacts with carbon dioxide. They know that water dissolves the gas carbon dioxide – for example in mineral water. The solution of sodium hydroxide should also dissolve the stated gas. The hypothesis may be: sodium hydroxide solution reacts with the gaseous carbon dioxide.

The hypothesis should also include the concentrations of sodium hydroxide solutions: concentrated sodium hydroxide solution dissolves carbon dioxide better than a diluted solution. Again, the students can independently construct an apparatus that verifies the hypothesis. The test series described in E5.12 is one possibility.

Material: 3 gas syringes with tap, test tubes, rubber stoppers with fitting hole; conc. sodium hydroxide solution (approx. 10-M), carbon dioxide.

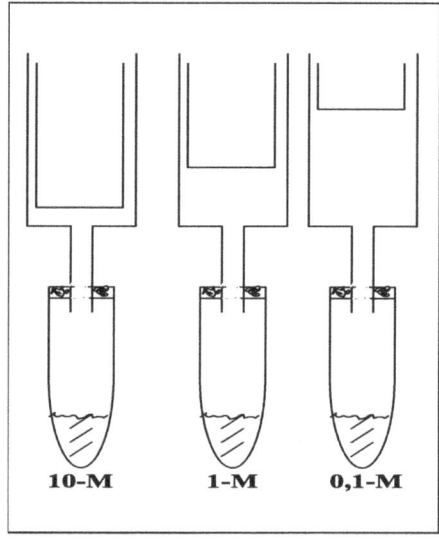

Procedure: A specific volume of the concentrated sodium hydroxide solution is to be diluted at a ratio of 1:10 twice: an approximate 1-M solution and a 0.1-M solution are produced. The gas syringes are to be filled with carbon dioxide to the 100 mL-mark, the tap is to be closed and connected to the rubber stopper. Three test tubes are to be filled with 5 mL of the stated solutions. One after the other is to be overlaid with the gas: the gas syringe has to be put on the test tube first loosely, air has to be pushed out of the tube with 50 mL of gas, 50 mL remain in the syringe. The test tube is then to be closed gas-tight. The gas volume of all three gas syringes is to be observed after a couple of minutes.

Observation: All three carbon dioxide volumes decrease. The concentrated sodium hydroxide solution reacts with the entire amount of gas, a suspension forms which contains a white precipitate. 1-molar solution reacts with half of the gas volume; 0.1-M solution only with a few milliliter of the gas volume.

E5.14: Concentration and pH-Value of Sodium Hydroxide Solution

Problem: Measuring pH-values of acidic and alkaline solutions is the typical form of collecting data. According to the definition of pH, pH-values are a direct measure of the concentration of hydronium or hydroxide ions, respectively. Therefore, these concentrations can be determined easily by measuring the pH. Another way is the titration of a specific volume of alkaline solution with an acid standard solution and recording the changing pH-values to the equivalence point. The result is a table of data, which can be illustrated with a diagram.

Material: pH-meter and glass electrode, beakers, transfer pipette (20 mL), Erlenmeyer flask, magnetic stirrer, burette; buffer solution for alkaline range, diluted sodium hydroxide solutions from E5.13, hydrochloric acid-standard solution 0.1 mol/L.

Procedure: The pH-meter is to be calibrated with the help of two buffer solutions of pH-values in the alkaline range (see instructions for the pH-meter). The pH-values of the two alkaline solutions are to be measured by immersing the glass electrode and then waiting until the pH value stays constant.

With the help of a transfer pipette, 20 mL of the very diluted sodium hydroxide solution are to be poured into the Erlenmeyer flask and stirred on the magnetic stirrer. The glass electrode of the pH-meter is to be dipped into the sodium hydroxide solution to meter the pH-value. After that, it is to be titrated with hydrochloric acid in 1 mL-steps and pH-values are to be measured until they exceed the neutral range. These values are to be recorded in a table and transferred to a diagram; the diagram is to be analyzed in terms of the consumption of hydrochloric acid. The amount of reacting hydrochloric acid to the equivalence point is to be used for a calculation of the concentration of the sodium hydroxide solution.

Observation: At the first step the measured values for both solutions are to be displayed: the pH-values are about 14 and 13. The consumption of hydrochloric acid standard solution is to be observed at the second step and the concentration is to be calculated exactly.

E5.15: Dilution Series for Illustrating pH-Values

Problem: Students might not be able to classify and correlate the pH-value of E5.14 with the concentration of OH^-(aq) ions in the solution. To explain this connection, a 1-M sodium hydroxide-standard solution is to be presented with pH-value 14. This solution is to be diluted with the factors 1:10, 1:100, and 1:1,000. The pH-values 13, 12 and 11 will be measured. Now it is clear that the pH-value changes by one, when the concentration of the solution varies by factor 10. The concentration of OH^-(aq) ions, which matches the metered pH-value of E5.14, can be estimated by this test series.

Material: pH-meter, 3 measuring cylinders (100 mL), transferring pipette (10 mL); special indicator paper (for the estimation of pH-values in the range of 11–14), sodium hydroxide solution ($c = 1$ mol/L).

Procedure: The sodium hydroxide-standard solution is to be tested with the indicator paper. 10 mL of the standard solution are to be poured into the first measuring cylinder, diluted to exactly 100 mL and well mixed by shaking. This solution is to be diluted again by 1:10 and the new solution is to be diluted once more by 1:10. All solutions are to be tested with pH-meter and indicator paper, pH-values and concentrations are to be compared.

Observations: The pH-values are 14, 13, 12, and 11. The pH-values change by one unit, when the solution is diluted by 1:10.

E5.16: Model Experiment Demonstrating the Mercury Electrode Process

Problem: The mercury electrode process is used industrially to obtain pure chloride-free sodium hydroxide solution through the electrolysis of cheap rock salt solution. Sodium amalgam is formed during the electrolysis and decomposed again by adding water. Formation and decomposition of sodium amalgam can be simulated in a model experiment – in this case an experiment functions as a model.

Material: Transformer, iron and graphite electrode, cables and alligator clips, U-tube, glass tube, spill tray; mercury (T), sodium chloride, phenolphthalein solution.

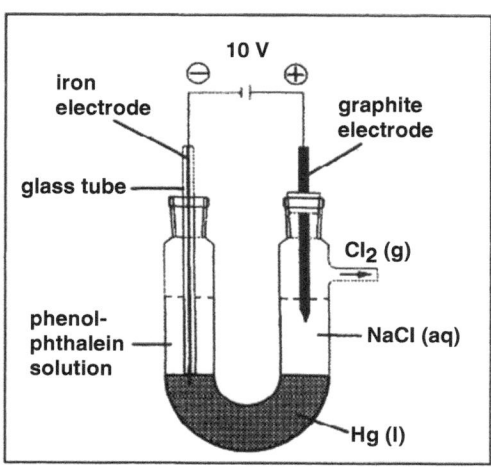

Procedure: Mercury is to be poured into the U-tube (fixed above the spill tray). The iron electrode in the glass tube is to be installed in one leg as a minus pole, water laced with phenolphthalein is to be added. The other leg is to be filled with concentrated sodium chloride solution and the graphite electrode is to be fixed. A direct voltage of about 10 V is to be applied.

Observation: A yellow–green gas with a characteristic smell forms at the plus pole: chlorine. After a couple of minutes the indicator solution turns burgundy at the minus pole, starting from the mercury: sodium hydroxide solution is produced. At the same time gas bubbles rise through the solution on the left side: hydrogen.

Please note: After decanting the aqueous solution, water is to be added to the sodium amalgam and one has to wait until the formation of hydrogen stops. Then the aqueous solution can be decanted and the mercury can be returned to the storage bottle.

E5.17: Combustion Analysis by Liebig

Problem: Sodium hydroxide solution and potassium hydroxide solution play an important role in the historical combustion analysis by Liebig. He successfully used the "potash apparatus," which he developed himself. This apparatus completely absorbs the produced carbon dioxide from the gas stream. A similar apparatus can be demonstrated schematically for the comprehension of this analysis method (see Fig. 5.7): a combustion tube attached to an oxygen bottle on one side and to two U-tubes on the other side, one filled with calcium chloride and weighed for absorbing water vapor, the other filled with potassium hydroxide and weighed for absorbing carbon dioxide. By weighing a specific amount of organic substance and the U-tubes afterwards one is able to estimate the composition of the substance.

In another version of E.17, however, only carbon dioxide is being collected quantitatively in a gas syringe, while the mass of condensed water is not being determined. It can be calculated that 1 C_4H_{10} molecule forms 4 CO_2 molecules: after the reaction of butane with copper oxide the volume of carbon dioxide will be four times larger than the volume of butane before.

Material: 2 gas syringes, combustion tube with copper oxide (wire type) enclosed with quartz wool, connecting rubber tubes; butane gas (F^+), lime water (Xi).

Procedure: One gas syringe is to be filled with 20 mL of butane gas. The apparatus is to be assembled and fixed (see figure). Copper oxide is to be heated strongly with the hot flame. The butane gas is to be pushed or pulled slowly back and forth through the combustion tube – until the volume remains constant. The volume of the produced gas is to be measured. The gas is to be passed from the syringe into a small amount of lime water.

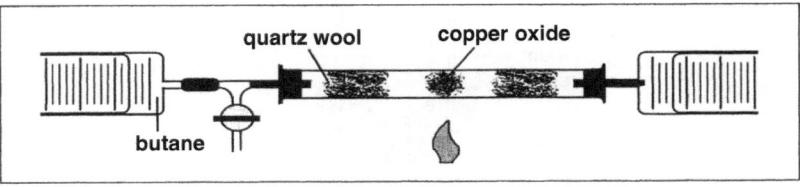

Observation: The copper oxide turns into bright red copper. Eighty milliliter of a colorless gas is formed. As expected, the lime water test shows that the produced gas is carbon dioxide.

E5.18: Reaction of Tube Cleaner (NaOH–Al-type)

Problem: The bathroom chemical "tube cleaner" can be introduced and analyzed for the repetition of discussed facts on substances like alkaline metal hydroxides and bases. The label on a container of tube cleaner reveals the ingredients sodium hydroxide and aluminum. In addition to a white substance in granular form, little silver-colored metal splinters can be found in the tube cleaner. When water is added a strong exothermic and alkaline reaction sets in. Adding paper demonstrates the decomposition by the reaction of hot, concentrated sodium hydroxide solution. Produced ammonia gas may be discussed (see also the explanation of E8.1 in Chap. 8).

Material: Test tubes, tube cleaner (NaOH–Al-type) (C), universal indicator paper, hydrochloric acid (C), wooden splint, pieces of newspaper.

Procedure: The mixture of substances is to be analyzed by visual inspection. The white powder is to be separated from the metal splinters. The powder is to be dissolved in water; the solution is to be tested with indicator paper. The metal is to be put into hydrochloric acid and the emerging gas is to be collected and tested for hydrogen with a burning wooden splint. A small amount of water is to be added to the mixture, little pieces of newspaper are to be added and observed.

Observation: The solution colors the indicator paper deep blue. The metal dissolves with formation of a colorless gas, this gas combusts in air with a whistling noise: hydrogen. The mixture reacts with water to a very hot solution under gas formation; the concentrated solution decomposes pieces of paper.

E5.19: Reaction of Alkaline-Earth Metals with Water

Problem: To estimate the learning success it can be tested whether students can transfer the familiar reactions of alkaline metals with water to similar reactions of magnesium and calcium with water. The formation of hydroxides and hydrogen is well known, but solid hydroxides accumulate as suspensions. The composition $M(OH)_2$, (M = Ca, Mg) has to be given as the formula for calcium hydroxide or magnesium hydroxide, and then reaction equations can be derived.

Material: Glass bowl, glass cylinder with cover glass, outflow pipe, infusible test tube, magnesia groove, test tubes, perforated stopper, glass tube, universal indicator paper; magnesium turnings (F), calcium turnings (F), sand.

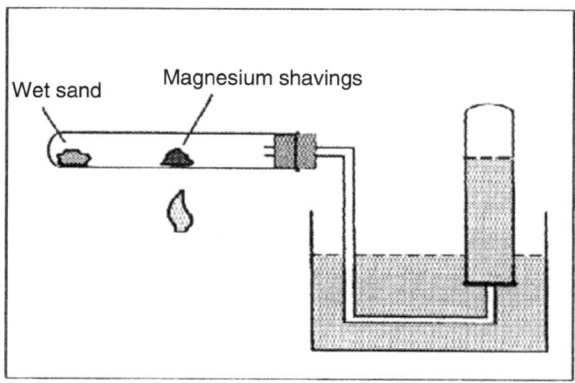

Procedure: The test tube is to be filled with a spoonful of wet sand and fixed horizontally. The magnesia groove filled with magnesium turnings is to be put into the middle of the test tube. Then the test tube is to be closed with stopper and outflow pipe (see figure). The metal is to be heated strongly. By warming the sand, water vapor reacts strongly with magnesium, the produced gas is to be collected in a cylinder, the gas is ignited later. After the reaction the stopper is to be removed, to prevent water from running into the test tube.

A few calcium turnings are to be mixed with water. During the reaction, a second test tube is to be put over the test tube with the opening downwards, this opening is to be put close to a flame. Caution: little bang. The milky suspension in the first test tube is to be tested with indicator paper.

Observation: When heated, magnesium reacts with a very bright light. A rapid gas formation starts, the cylinder fills up with gas. The gas combusts with a whistling noise: hydrogen.

At room temperature calcium reacts to an alkaline solution: calcium hydroxide solution ("lime water"). Solid white calcium hydroxide forms a milky suspension. The observed gas mixture combusts with the familiar whistling noise: hydrogen.

E5.20: Student Experiments on the Lithium–Water Reaction

Student work sheet: "The reaction of lithium with water"

Problem: How can we explain the reaction of alkaline metals with water?

Working material per group: beaker, glass dish, small glass cylinder with cover glass, tweezers; 3 pea-size pieces of lithium (C/F), universal indicator paper, phenolphthalein solution (0.1% in ethanol) (F), matches. Safety goggles must be worn!

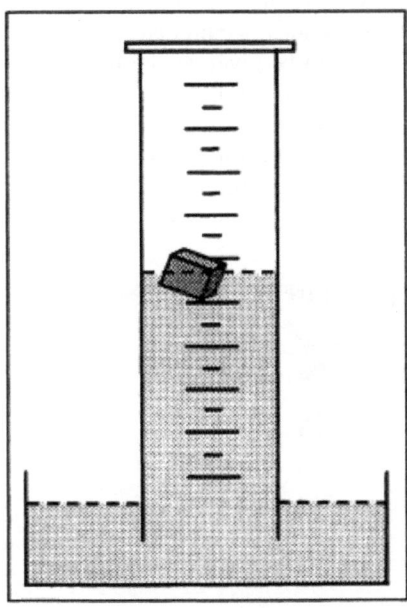

Experiment 1. Fill the glass dish half-full with water and throw in a piece of lithium with the help of tweezers (do not touch it with your fingers!). Record your observations!

Experiment 2. You may notice a gas formation in experiment 1. That gas is presumed to be hydrogen. Try to collect and confirm the hydrogen gas. Therefore, fill the glass dish half-full with water. Then fill up the cylinder with water and close it with the cover glass. Hold on to the cover glass, turn the cylinder around and put it into the glass dish, so that the opening is under the water surface (see figure). Now remove the cover. Take the second piece of lithium with the tweezers, hold it under the opening of the cylinder, and let it rise into the cylinder.

After the reaction cover the cylinder with the cover glass under water (press on tightly!) and take it out of the glass dish. Turn it around again. Ignite the gas! Record your observations!

Experiment 3. The produced solution is called lithium hydroxide solution; this can be shown with indicator paper. Test pure water and the solution with indicator paper by dipping it in briefly. Compare it to the color scale. Note both pH-values.

Experiment 4. Another indicator for lithium hydroxide solution is phenolphthalein solution. Pour a couple of drops into pure water and into the solution. Record your observations.

Experiment 5. Wash out the glass dish and fill it half-full with water and add some drops of phenolphthalein solution. Throw in the third piece of lithium. Record your observations.

Exercise: Explain all observations. Write reaction equations in words and in formulae! Draw your mental models of the reactions.

References

1. Liebig J (1840) Der Zustand der Chemie in Preußen. Ann Chem Pharm 34:97
2. Ströker E (1967) Denkwege der Chemie. Alber, Freiburg
3. Anschuetz R (1929) Leben und Wirken. Baende 1–2. Chemie, Berlin
4. Kekulé A (1866) Lehrbuch der Organischen Chemie. Baende 1–3. Chemie, Erlangen
5. Von Laue M (1961) Mein physikalischer Werdegang. Eine Selbstdarstellung. Vieweg, Braunschweig
6. Von Laue M (1923) Die Interferenz der Roentgenstrahlen. In: Ostwalds Klassiker Nr. 204. Engelmann, Leipzig
7. Wiederholt E (1999) Gaschromatographie – Nachweis von in Wasser gelöstem Sauerstoff und Stickstoff. MNU 52:92
8. Brockmeyer H (1973) Röntgenstrahlen im naturwissenschaftlichen Unterricht. Aulis, Köln
9. Schwedt G (1991) Chemie zwischen Magie und Wissenschaft. VCI, Weinheim
10. Gilbert L et al (1994) Tested demonstrations in chemistry. ACS, Granville
11. Shakhashiri B (1983) Chemical demonstrations. University of Wisconsin, Madison
12. Glöckner W et al (1994) Handbuch der Experimentellen Chemie. Aulis, Köln
13. Arendt R, Dörmer L (1980) Technik der Experimentalchemie. Quelle + Meyer, Heidelberg
14. Meyendorf G et al (1979) Chemische Schulexperimente. 5 Bd. Deutsch, Frankfurt
15. Schmidkunz H, Lindemann H (1992) Das forschend-entwickelnde Unterrichtsverfahren. Westarp, Essen

16. Dehnert K, Jäckel M (1979) Allgemeine Chemie. Schroedel, Hannover
17. Strube W (1981) Der historische Weg der Chemie. Deutscher Verlag, Leipzig
18. Schmidkunz H (1983) Die Gestaltung chemischer Demonstrationsexperimente nach wahrnehmungs-psychologischen Erkenntnissen. NiU-P/C 31:131
19. Roesky H (1997) Chemie en miniature. VCI, Weinheim
20. Häusler KG (1993) Die Halbmikrotechnik. NiU-Chemie 41(2):10
21. Kometz A, Krech K (1998) Küvettentechnik und Mikroglasbaukasten. Chem Sch 45:348

Further Reading

Domin DS (1999) A review of laboratory instructional styles. J Chem Educ 76:543–547
Hart C, Mulhall P, Berry A, Loughran J, Gunstone R (2000) What is the purpose of this experiment? Or can students learn something from doing experiments? J Res Sci Teach 37:655–675
Hawkes SJ (2004) Chemistry is *not* a laboratory science. J Chem Educ 81:1257
Hegarty-Hazel E (ed) (1990) The student laboratory and the science curriculum. Routledge, London
Hodson D (1993) Re-thinking old ways: towards a more critical approach to practical work in school science. Stud Sci Educ 22:85–142
Hofstein A, Lunetta VN (1982) The role of the laboratory in science teaching: neglected aspects of research. Rev Educ Res 52:201–217
Hofstein A, Lunetta VN (2004) The laboratory in science education: foundations for the twenty-first century. Sci Educ 88:28–54
MacNeil J, Volaric L (2003) Incomplete combustion with candle flames: a guided-inquiry experiment in the first-year chemistry lab. J Chem Educ 80:302–304
Milne C, Otieno T (2007) Understanding engagement: science demonstrations and emotional energy. Sci Educ 91:523–553
Morton SD (2005) Response to "Chemistry is *not* a laboratory science". J Chem Educ 82:997
Nakhleh MB, Polles J, Malina E (2002) Learning chemistry in a laboratory environment. In: Gilbert JK, De Jong O, Justi R, Treagust DF, van Driel JH (eds) Chemical education: towards research-based practice. Kluwer Academic, Dordrecht, pp 69–94
Psillos D, Niedderer H (eds) (2002) Teaching and learning in the science laboratory. Kluwer Academic, Dordrecht
Stephens CE (2005) Taking issue with "Chemistry is *not* a laboratory science". J Chem Educ 82:998

Chapter 6
Structural and Mental Models

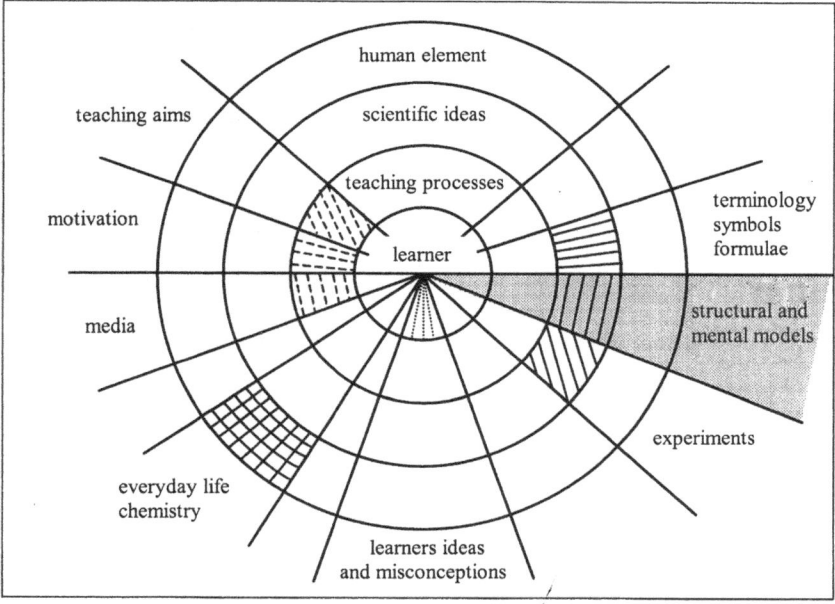

I once was invited to give a slide show about Germany in San Diego/California. One of the photos I found in an illustrated book called "Germany" represented the famous Neuschwanstein Castle near Füssen in Bavaria. I showed this picture first. As soon as it appeared on the screen a woman interrupted: "Beautiful – just like the castle in Disneyland." I tried to correct: "In Disneyland you find a copy of the castle – this photo shows the original in Bavaria" – but the woman did not listen. The difference between original and model was not important to her!

In this example both the original and the model can be observed: the photo of the castle as well as the model in Disneyland can be compared to the Bavarian original in detail. In scientific models for chemical structures the spheres of the model cannot be compared to the submicroscopic original atoms, ions or molecules,

H.-D. Barke et al., *Essentials of Chemical Education*,
DOI 10.1007/978-3-642-21756-2_6, © Springer-Verlag Berlin Heidelberg 2012

because it is impossible to see them – neither with a magnifying glass nor with the best microscope.

Because particles are not visible, one tried all times to develop suitable mental models. Demokritus and other Greek philosophers came up with the idea of the atom more than 2,000 years ago – simply by thinking about their observations concerning matter and trying to find out the composition and structure of matter. Lémery developed a particle model for the effects of acids [1]: "All acids are composed of particles like biting points, all experiences show that acids make pricks, which everyone can feel on the tongue." Lémery had never seen acid particles, but he tried to transfer macroscopic characteristics of acids on the cited mental model.

. Nevertheless structural and mental models are very important for the understanding of science, especially chemistry. One of the key questions of chemistry education deals with the scientific idea of models and how these can be transferred to young learners at school. To get the idea across, general characteristics of models have to be looked at first and later transferred to scientific models.

General characteristics of models. Following an empirical analysis of models in general Stachowiak [2] differentiates three basic characteristics:

– Feature of depiction: models are always a model of something, namely illustrations and therefore representations of certain natural or artificial originals.
– Feature of shortening: models do not cover all characteristics of the original, they represent only the ones that seem to be relevant in a particular context.
– Feature of subjectivity: models fulfil their functions of representation and substitution only for certain subjects, limited to certain mental or real operations, limited to certain time.

Take the picture of the Neuschwanstein Castle another time. It copies the building and the surrounding landscape with fields, trees, roads and mountains in the background downscaled: the feature of depiction is fulfilled. Some of the many shortenings of the original are the missing space dimension, the nonexistent patterns of light and shadow on walls and windows of the castle or the missing motion of trees and branches in the wind: feature of shortening. The specific view of the castle on the picture or the details of the landscape are subjectively chosen by the photographer for a special purpose: feature of subjectivity.

Taking two different maps of the London Underground show that feature 1 is fulfilled: they all show where the underground stations are located. Comparing them one shows the streets close to the stations, the other one not: the shortening is done differently. The differences are due to the users: some need the street maps around the stations, other ones not because they are traveling every day the same stations to work and back at home: feature of subjectivity.

6.1   Scientific Ideas: Thinking in Models

Chemistry found recognition and success, when it departed from trial and error approach of medieval alchemy. Starting with simple laboratory experiments for the description of substances it came to develop first mental models for the structure of matter in the eighteenth and nineteenth century. Some states of perception will be given exemplarily.

Dalton postulated in 1808 that there are as many kinds of atoms as there are known elements. He presented the first atomic mass table, which has been corrected and extended mainly by Berzelius in the next decade. The comparison of experimentally determined mass ratios of elements in compounds with their atomic masses made the empirical analysis possible and led to the knowledge of the composition of many substances as well as the first empirical formulae.

Kekulé deduced the theory of valence from his experience in 1865: He created a first concept of molecular structure with the four-bonding or the tetrahedral model of the carbon atom, the one-bonding model of the hydrogen atom, and the two-bonding model of the oxygen atom. With these mental models it became possible to predict structures of many molecules, verify them in experiments and plan targeted syntheses of new substances.

In 1912 Laue discovered three-dimensional crystal structures through the diffraction of X-ray radiation from salt crystals: he obtained X-ray diffraction patterns, calculated those three-dimensional interference patterns and deduced the structure of investigated salt crystals. Based on these results Bragg solved the crystal structure of sodium chloride in 1914. This destroyed the historical idea of NaCl molecules in table salt: sodium ions and chloride ions with the coordination number 6 form the salt crystals. All following structural analyses were based on Laue's and Bragg's fundamental discoveries, providing the chemical structure of many crystalline substances, in turn making the syntheses of new substances possible.

Models and scientific perception. The idea of models and the process of scientific perception will be explained with a diagram of Steinbuch [3] (see Fig. 6.1): "Any facts of reality, an original, can be transferred to an abstract mental model by taking only the essential and relevant parts of the given context. Certain information, for example generally accepted laws of logic or physics may be added. A mental model for future thinking processes is thereby available. For the purpose of illustration, this abstract mental model can be transferred back into reality by building a concrete display model or even by artistic representation. But these have unavoidable irrelevant components, which the mental model does not have."

This "thinking in models" can be transferred to Laue's way of perception (see Fig. 6.2): The interference pattern (original) that forms from a salt single-crystal in the Laue experiment is let through the filter as "essential." Interferences of light with two-dimensional diffraction gratings and their calculations are familiar (additional information). They are taken as a basis for the calculations of three-dimensional diffraction gratings. The result is a mental model for the three-dimensional symmetrical structure of salt crystals built of ions. The ions in the ionic

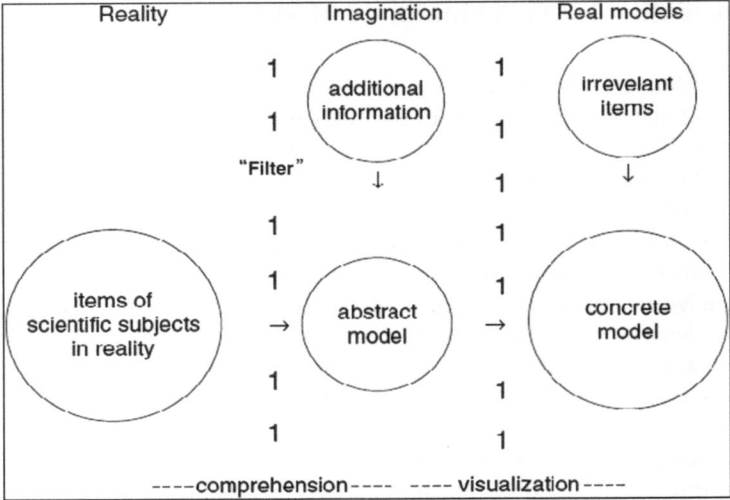

Fig. 6.1 Scheme of "thinking in models" according to Steinbuch [3]

Reality	1	**Imagination**	1	**Real models**
"interference pattern	1		1	
of X-ray reflexes	1	3D-calculation of interferences	1	balls, glue,
of a salt crystal		of waves		sticks
(Laue diagram)"	1		1	
		filter, which only	↓	1	↓
		lets through the		1	
		essentials		1	
	1		1	
	1	3-dimensional	1	
		symmetrically	1	
	1	arranged interference centres	1	
	1	in the salt crystal	1	
	1		1	

Fig. 6.2 Scheme of "thinking in models" applied to Laue's way of X-ray analysis

structure function as diffraction centers for the X-ray radiation beam (abstract mental model).

For the illustration of this mental model, irrelevant components such as balls, sticks and glue can be used for the construction of display models: sphere packing or crystal structure models (see Fig. 6.3). In addition to sphere packing and crystal structure models, the resulting display model for the structure of crystals can also be a unit cell only (see Fig. 6.3).

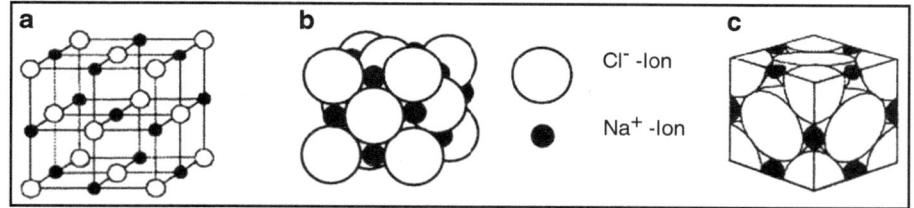

Fig. 6.3 NaCl-crystal structure: *ball-and-stick* model, cubic sphere packing and unit cell

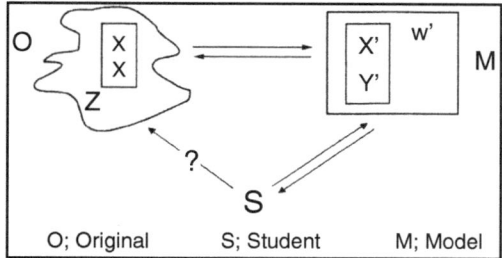

Fig. 6.4 Scheme of the way of perception by models, according to Kircher [4]

The process of perception through models is also described by Kircher [4]. His concept will be explained, using sodium chloride crystals and corresponding models (see Figs. 6.3 and 6.4):

- The original **O** is to be a natural rock salt crystal with a cubic shape, planar surfaces, straight edges and right angles
- The packing of spheres is to be chosen as the model **M** (see Fig. 6.3b). The chloride ions are represented by big balls and the sodium ions are represented by small balls
- The student **S** can now understand the original **O** with the help of the model **M**, the sphere-packing model works as a mediator between the student **S** and the original **O**

1. There are properties x and y of the crystal that have corresponding model characteristics x' and y'. For example, x is chosen for the spatial arrangement of sodium and chloride ions in the crystal, x' is then chosen for the corresponding arrangement of big and small balls with the coordination number 6 in the model. If y depicts the radius ratio of both types of ions in the sodium chloride crystal, then y' depicts the corresponding size ratio of the balls in the sphere-packing model. So x and y represent the parameters in the original, x' and y' those which are displayed in the model. According to Stachowiak x and y are the features of depiction, according to Steinbuch these are "the essentials, which are let through the filter."
2. There exist characteristics z in the original that do not have an analogy in the model. The salty taste or the white color of the crystal, for example, are not

displayed in the model. The model is shortened by this characteristic: Stachowiak calls this "feature of shortening." Similarly the density or melting points of crystals cannot be determined with models. The model builder, however, never had the intention to transfer these original characteristics to the model.

3. There may be characteristics w' in the model that do not have an analogy in the original. The choice of colors – for example white for the big balls and red for the small balls – represents a model characteristic, which is totally irrelevant and arbitrarily chosen by the model builder: Steinbuch calls these irrelevant components. Additional irrelevant components are model materials like wood, cellulose or styrofoam and an adhesive between the balls, such as glue or Velcro tape.

The crystal structure model (see Fig. 6.3a) only shows the cubic arrangement of ions in the crystals, not their size. The option of looking into the model, however, allows a better illustration of the coordination number 6. The connecting sticks between the junctions are only necessary for the stability of the model, they are totally irrelevant components towards the original and do not have a representation function.

The unit cell model (see Fig. 6.3c) shows three features of depiction: the cubic structure of the ionic compound, the size ratio of the ions and the correct 1:1 ratio of the number of ions. If all parts of the big and small balls of the model are summed up, it results in four big and four small balls. Transferred to the original, this shows a unit of four Na^+ ions and four Cl^- ions corresponding to the symbol $\{(Na^+)_4(Cl^-)_4\}$.

The unit cell with this symbol can be regarded as the smallest unit of the NaCl-structure – just like the C_2H_5OH molecule is regarded as the smallest unit of the substance ethanol. Symbols like $\{(Na^+)_1(Cl^-)_1\}$, $(Na^+)(Cl^-)$ or NaCl can be derived from $\{(Na^+)_4(Cl^-)_4\}$ – they are all shortened representations of the sodium chloride structure.

Mental models in chemistry: Due to improving knowledge, scientific models change continuously. Therefore it is not possible to speak of the current atomic model or the current model of chemical bonds.

Quantum mechanical atomic model. The structure of the electron shells of atoms or ions is described by the principal quantum number n ("K shell" or "L shell"), the angular momentum quantum number l (s, p, d and f sub-shell), the magnetic quantum number m and the spin quantum number s. A maximum of two electrons with different spins can form an electron cloud or an orbital (Pauli principle). Starting from the wave–particle duality, wave functions have been developed, which give information on energy levels and electronic orbitals (Schrödinger equation). The calculation of wave functions leads to the description of atoms by atomic orbitals and molecules by molecular orbitals. In this sense, covalent bonds in molecules can be described mathematically.

Historical atomic models. For educational reasons historical models are often used in chemistry education. For example:

– Mass model (Dalton, 1808)
– Mass–charge model (Thomson, 1897)
– Nucleus–shell model (Rutherford, 1911)
– Sub-shell model (Bohr, 1913)
– Electron-pair repulsion model (Gillespie, Kimball, 1966)

Models of chemical bonding. Mental models concerning this matter have to be viewed from two perspectives:

1. The effects of three-dimensional bonding forces are to be displayed with the model material:

   (a) Directed bonding forces that work in designated directions of space are to be displayed by snap fasteners or sticks.
   (b) Undirected bonding forces that work symmetrically around a particle are to be symbolized by spheres, which may be glued together.

2. Bonding forces are difficult to visualize and they are usually described by mathematical models with the distribution of electron densities. The following cases of chemical bonding are to be differentiated:

   – Ionic bonding
   – Covalent bonding
   – Metallic bonding
   – Hydrogen bonding
   – Van der Waals forces (intermolecular forces).

Models of the chemical structure. The mathematical calculation of atomic structure and chemical bonding is mostly a means to get information on the chemical structure. On this basis, it is the aim of many analytical procedures to determine and to describe the atomic or ionic arrangement in given substances, and derive the structural formulae from the determined structures. Different cases of chemical structures can be sketched in the following way:

– Molecular structure (atom types, bond length and bond angle)
– Atomic crystal structure (atom types and lattice constants, unit cell)
– Metal crystal structure (atom types and lattice constants, unit cell)
– Ionic crystal structure (types of ions and lattice constants, unit cell)
– Molecular crystal structure (types of molecules and lattice constants, unit cell)

Models of the chemical reaction. Arrangements of particles in chemical reactions can be described by mental models as well as shortened by reaction equations:

– Rearrangement of atoms in reactions of metals to alloys
– Rearrangement of ions in hydration and precipitation reactions
– Proton transfer in acid–base reactions
– Electron transfer in redox reactions
– Ligand transfer in complex reactions
– Addition, substitution and elimination reactions of organic molecules

Display models in chemistry: Chemists usually work with abstract mental models. For educational reasons appropriate display models are being developed (see Figs. 6.2, 6.3 and 6.5): concrete display models can be build, for example of molecules or crystal structures, in regard to many mental models of different chemical structures.

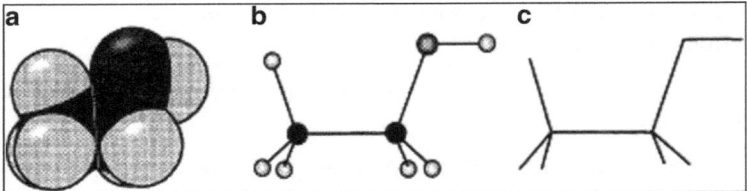

Fig. 6.5 Space-filling, *ball*-and-*stick* and *stick* model of the structure of the C_2H_5OH molecule

Models of molecule structures. The spatial arrangement of atoms in a molecule is described with the help of space coordinates, bond lengths and bond angles, which are to be determined experimentally in the laboratory. It can be illustrated in different ways (see Fig. 6.5):

– Space-filling model (calotte model): the space filling of atoms is being displayed, atom calottes are connected to a molecular model, according to bond lengths and bond angles (a).
– Ball-and-stick model: all balls for the different kinds of atoms have the same size and are specially colored, they are being connected with connection sticks or snap fasteners (b).
– Stick model: instead of balls special sticks are used of appropriate lengths and connected with proper angles (c).

Models of crystal structures. The unit cell is sufficient for the expert to deduce the whole structure. To proceed graphically for the learner, certain parts of the lattice are being chosen to be displayed in models as a crystal structure or a packing of spheres (see Fig. 6.3):

– Crystal structure model: equal spheres, with different colors if applicable, are being connected by connection sticks until the desired part of the structure is being displayed, for example, the part corresponding to the NaCl unit cell (a).
– Packing-of-spheres model: the size ratio of the different atoms or ions is taken into account. The balls are to be glued or piled up according to familiar structure parameters until they display the desired part e.g. the NaCl unit cube (b).
– Unit cell model: it is derived from the corresponding unit cube model and guarantees a proper ratio of atoms or ions (c). Larger structures of any size can be derived by mental translation of unit cell models in all three directions in space.
– 3D-drawing model: red–green drawings of chemical structures are being fixed by both eyes until the three-dimensional picture appears; it can be interpreted spatially [5].

Besides the well-known molecular structure kits there exist a variety of model kits to build packings of spheres:

– "Metal structures" (see Fig. 6.6): this model kit allows to build the three common
  metal structures in the form of a packing of spheres with wooden balls of same
  size (diameter: 3 cm).
– "Model kit for crystal structures" (see Fig. 6.7): a base plate with slots in
  a triangular pattern for stacking colored plastic balls of the same size (diameter:
  1 cm), only the close-packed structures can be built as shown in Fig. 6.6.
– "Solid-state model kit" (see Fig. 6.8): base plate with holes in different patterns
  to hold sticks with pierced glass spheres of different diameters. It is possible
  to build all metal structures with this kit as well as many salt structures with big
  and small glass spheres.

Fig. 6.6   Model kit "Metal Structures" by Geomix [6]

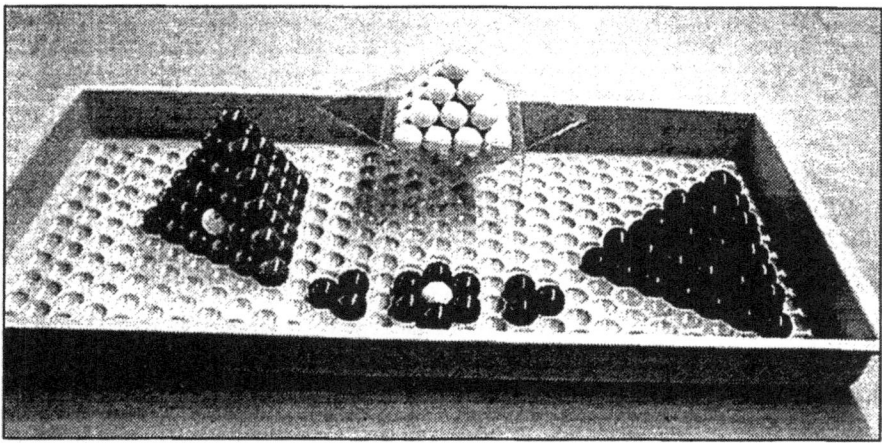

Fig. 6.7   "Crystal-Structure Model Kit" by Leybold [7]

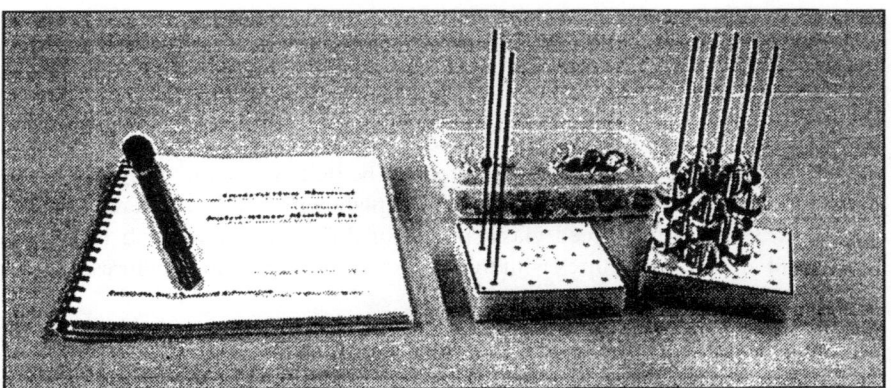

Fig. 6.8 "Solid-State Model Kit" of the Institute for Chemical Education, Univ. of Madison [8]

Dynamic models: Simulation games are dynamic models. They allow the demonstration and interpretation of physicochemical processes on the particle level, unlike the much discussed structural models, which only represent static and time-independent aspects of matter. The processes, which will be simulated on the model level below, can be divided into two categories:

- Processes in which particles (atoms, molecules, ions, photons, etc.) move in space without transforming into other particles. Examples: diffusion, chromatography, distribution equilibria, mixing and separation processes, light absorption.
- Processes in which particles transform into other particles during a chemical reaction. The time behavior of the system, i.e. the kinetics of these reactions are of particular interest. Examples: reactions of elements, consecutive reactions, competing reactions, balanced reactions, oscillating reactions.

The particles are being represented by simplest models in both categories. It is not so much the behavior of the single particle that is of primary interest, but the statistic behavior of a particle arrangement. From this point of view the simulation games can also be called statistic games. The rules for these games are simple, they can be described with the following basic points:

- One can either win tokens from a playboard (board model) or balls from a vessel (urn model).
- A randomly hit particle is considered to be activated – what happens to it next depends on the rules of the game: moving a sphere on the playboard, converting it into another kind of sphere by color change, etc. – the rules should be assigned to the simulated process in a comprehensible way.
- Every event that leads to an activation counts as one time unit, regardless of its specific consequences.
- The kinetics of the statistical process are being recorded by writing down the status of the game depending on the number of time units.

The following examples [9–11] illustrate the mathematical structures of these games as well as their contextualization in complex physicochemical theories (such as Maxwell velocity distribution, Boltzmann distribution, entropy, reaction kinetics). It is not necessary to teach the complex mathematical models as the basis of these games in chemistry lessons. The numerical interpretation of the games is usually enough to illustrate the core of the simulated issue. A mathematical understanding is helpful and even essential for anyone interested in creating computer simulations (see explanations in [9]). The visualization of radioactive decay might be a first qualitative interpretation.

Radioactive decay: Terms like "radioactivity," "the unit Becquerel" or "half-life of isotopes" are part of the public discourse ever since the Chernobyl disaster on 26 April 1986. To enable students to participate in this discourse, it has to be ensured that they understand and use the technical terms in the correct way. This does not replace a further discussion about aims and risks of nuclear power as a basis for consensus building and social interaction.

The following simulation game, which is based on an idea by Eigen and Winkler [12], can be used to understand the process of a decomposition reaction, especially for teaching the term "half-life of isotopes":

- 36 tokens are placed on a board with 6 × 6 squares. These represent the atoms of a radioactive substance, which transforms to inactive atoms with a certain half-life (which is to be determined during the game).
- Coordinates on the board are being determined successively by rolling one dice twice. If there is a white token on the determined field (which will be the case in the beginning), it is to be replaced by a black token (inactive atom). If a field with a black token is being determined during the game, the token remains.
- Every double dice for both coordinates counts as one time unit, regardless of whether a black or a white token is hit.

A typical course of a game is displayed in Fig. 6.9: after about 100 time units almost all the white tokens are replaced by black tokens, i.e. almost all of the radioactive atoms have been transformed to inactive atoms.

The results of at least 10 players are being added up or averaged statistically for the quantitative analysis. An exponential curve progression with a half-life of $t_{1/2} = 25$ time units results. It is important for students to realize that, despite the randomness of the single decay event, a lawful behavior of the overall process results: a curve progression with a constant half-life. If the students know the exponential function from their mathematics classes, the data can also be analyzed analytically (see Fig. 6.10).

Figure 6.10 also shows that the presentation of the data on a logarithmic scale results in a straight line: students can determine the rate constant k from the slope of the line; the value is $k = 0.028$. Since 0.028 is about equal to 1/36, students are able to grasp the relevance of k: 1/36 is the probability with which a single token is taken after rolling the dice. This means that k is the probability with which a single radioactive atom decays per time unit.

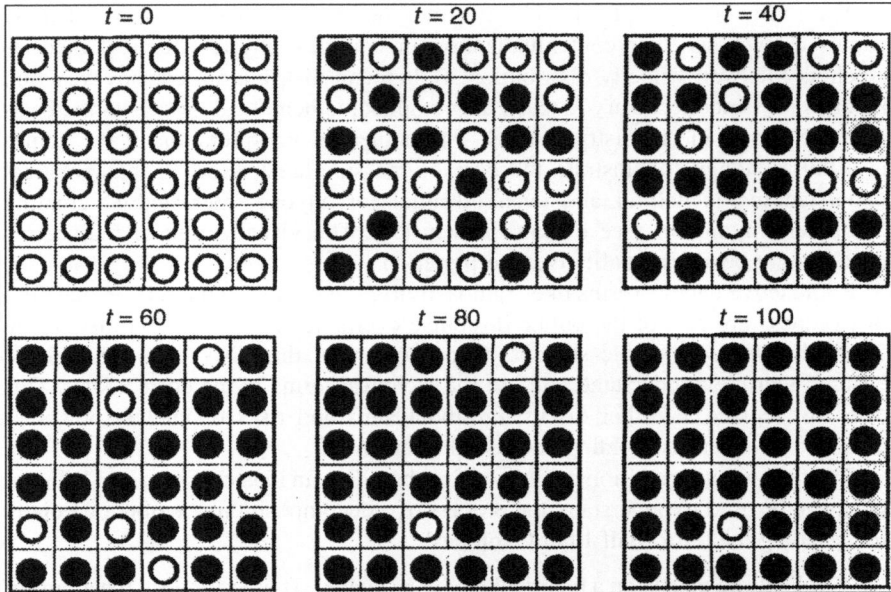

Fig. 6.9 Simulation game for radioactive decay

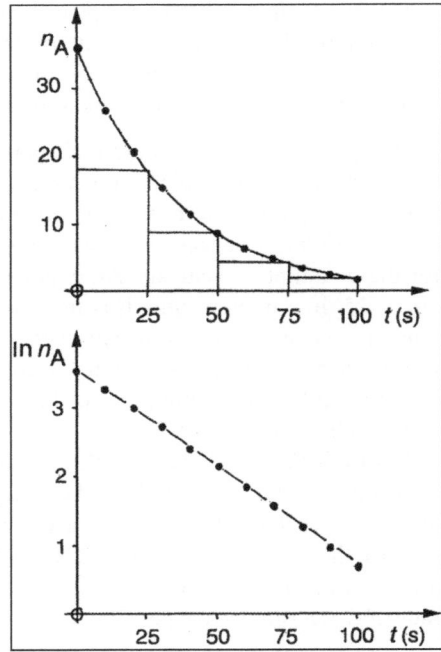

Fig. 6.10 Quantitative analysis of the data in normal and logarithmic presentation (mean values of at least ten games)

In connection with the Chernobyl disaster the caesium isotope Cs-137, which decays with a half-life $t_{1/2} = 30$ years, was discussed in the newspapers. With the help of this game students will be able to grasp the consequences: it takes almost 100 years until the Cs-137 activity decays to 10% of the starting value. Luckily the clouds with radioactive particles spread widely over Western Europe so that experts predicted the dose of Cs-137 in food to be small despite its long half-life.

6.2   Teaching Processes: Models and Their Functions

Students already gained lots of experience in the field of natural phenomena: this field is concrete for them. Therefore they like the elementary lessons in biology, chemistry, geography and physics: they stay in the well-known field of direct illustration and tangible phenomena. As soon as formulae and reaction equations play a role in chemistry lessons, interest in chemistry dwindles: chemistry becomes dry and lifeless, becomes hard to understand. One reason is that formulae and equations belong to abstract mental models (see Chap. 6.1).

Therefore it is important for the education process to find out which display models, for example molecular models, packing of spheres or crystal structures, can be used, before mental models are being introduced. All bonding models or models of the structure of single atoms or ions are abstract mental constructs. They have to stand back until a first understanding of the structure of matter is build up with the help of illustrative structural models. First the chemical structure, then the chemical bond!

Some chemistry and physics educators see the use of particles and ball models for chemical structures with a critical eye. Buck [13], for example, calls the particle model a "nonmodel" and opposes the usual illustration with circles or balls and their arrangement for demonstrating models of the states of matter or aqueous solutions: "Actually we are not allowed to draw spheres, because we are dealing with centers of force. Teachers and authors who should know, accept without problems that atomic and molecular orbitals are asymptotic, can stretch across the whole universe, that limiting lines are drawn arbitrarily, mostly at 85%. To summarize it: the visualisation and illustration by particle pictures is a crucial mistake" [13].

He also recommends initiating "the jump" to the atoms by showing slides of increasing or decreasing complexity and a discussion of system characteristics: "egg $\rightarrow$ hen house $\rightarrow$ farm $\rightarrow$ village $\rightarrow$ country $\rightarrow$ earth $\rightarrow$ universe $\rightarrow$ earth $\rightarrow$ city $\rightarrow$ school $\rightarrow$ students $\rightarrow$ hair $\rightarrow$? What about the next slide? There is no next slide, because such a slide does not exist" [14].

This discussion might be very interesting, for sure. It can take place in class before the introduction of smallest particles. But the consequence of the "nonexistence of the next slide" cannot be to go over to the abstract "centers of force" or "endlessly widespread nucleus–shell systems" right away. For developmental psychological reasons concrete circles, balls, cubes or lego bricks have to be chosen as models of smallest particles. Especially the discussion of form, color or material of the model as "irrelevant components" opens the chance to introduce the idea of the scientific model even on this level. In the beginning the particle model is used as a preliminary mental model. As lessons go along it advances via Dalton's atomic model to the nucleus–shell model. At the end of this topic, "the randomly drawn limiting line" may be discussed with students and the demonstration of balls and circles should be critically evaluated.

Understanding of chemistry through models: Figure 6.1 shows the scheme "thinking in models" and thereby the process of perception in chemistry "from left to right": with the help of additional information a chemist works out a mental model and transfers it to concrete models for visualization. Learners cannot go this way, but they can make their way through the diagram "from right to left," by working first with concrete models: they observe and compare concrete models and develop advanced mental models for relevant issues. It is often the case that mental models interfere with irrelevant issues of display models, but an adequate abstraction happens more and more with time.

After the introduction of first mental models, for example the particle model of matter or the Dalton model of atoms or molecules, chemical facts are to be interpreted with those models, as long as it is possible on this level. Chemistry lessons should now proceed double-tracked, and they should be structure-oriented (see Chap. 10):

Track 1: phenomena and laboratory experience.
Track 2: structural models and mental models (see Table 6.1).

After the decision to introduce the *particle model of matter*, appropriate phenomena and experiments are to be chosen. It is possible to take snow flakes or crystals of sugar and to ask the reason why they have the same shape. The answer should be the symmetrical arrangement of particles: water particles are arranged in snow flakes in a special way, sugar particles in sugar crystals in a different way (see Table 6.1). It makes also sense to choose dissolution processes, for example the dissolution of sugar in water (see Fig. 6.11): the sugar particles are shown by dark spheres, the water particles by light spheres, the dissolving process should be interpreted by moving spheres: in the solution moving sugar and water particles

Table 6.1 The two-tracked approach in chemistry lessons

Track 1:			
Phenomena and laboratory experience	Same shape of crystals, dissolving processes, diffusion, distillation	Chemical reactions of gases, gas laws, Avogadro's law	Redox reactions of metals and salt solutions
Track 2:			
Mental and concrete models	Particle model: arrangement of particles before and after diffusion or dissolution	Dalton's model: models for arrangements of atoms and molecules before and after chemical reactions	Atomic structure: electron transfer from metal atoms to ions in solution, from ions to atoms

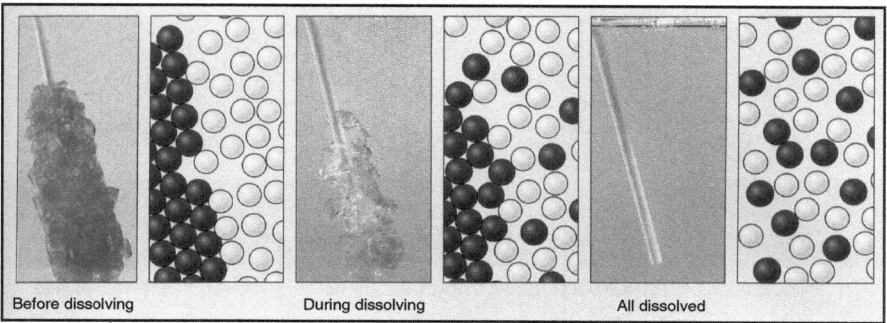

Before dissolving   During dissolving   All dissolved

Fig. 6.11 Dissolving sugar crystals interpreted by the particle model of matter [15]

Fig. 6.12 The
carbon–oxygen reaction
interpreted by Dalton's
atomic model [15]

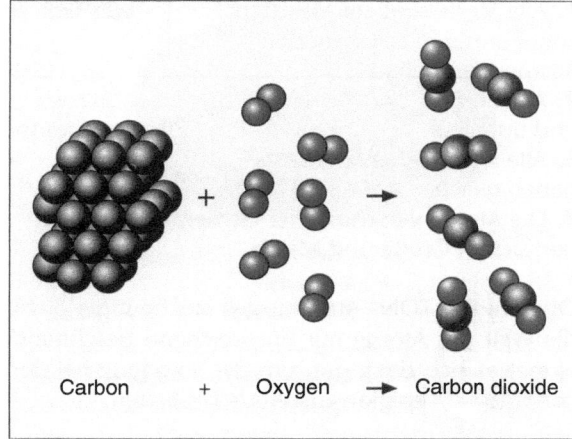

Carbon   +   Oxygen   →   Carbon dioxide

are mixed. The motion can be visualized by shaking mixtures of two kinds of
spheres in a glass bowl on the overhead projector.

If it is decided to introduce atoms and molecules with *Dalton's atomic model*,
reactions of carbon and oxygen may be conducted (see E1.9) and interpreted with
C atoms and O atoms (see Fig. 6.12): C atoms in a carbon crystal are shown by
big spheres in a closest sphere packing; two O atoms in one O_2 molecule are
presented by two linked smaller spheres. Carbon dioxide as the reaction product
can be shown by CO_2 molecules: molecular models consist of two small spheres
with one big sphere in the middle. The reaction is interpreted by the rearrangement
of C and O atoms.

Packing of spheres. Since metals are in the center of beginning chemistry
lessons, it is possible to introduce the particle model (copper particles) or the
Dalton model (Cu atoms) through the structure of metals. The advantage of this
approach is that the three-dimensional arrangement of metal atoms in many metal
crystals can be shown with close-packed spheres (see Fig. 6.13): learners accept

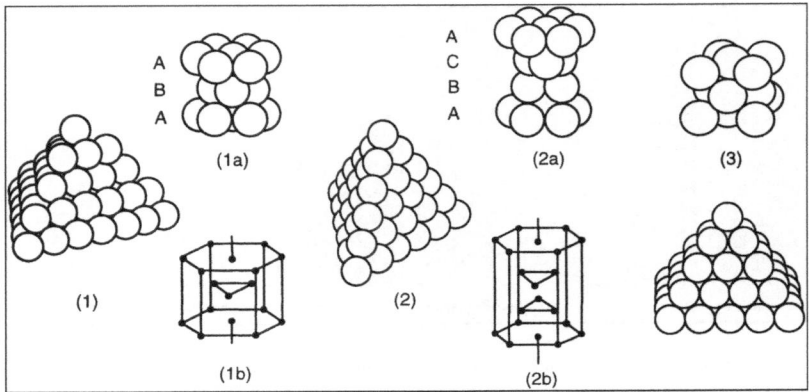

Fig. 6.13 Hexagonal and cubic close-packing as models of different metal structures [17]

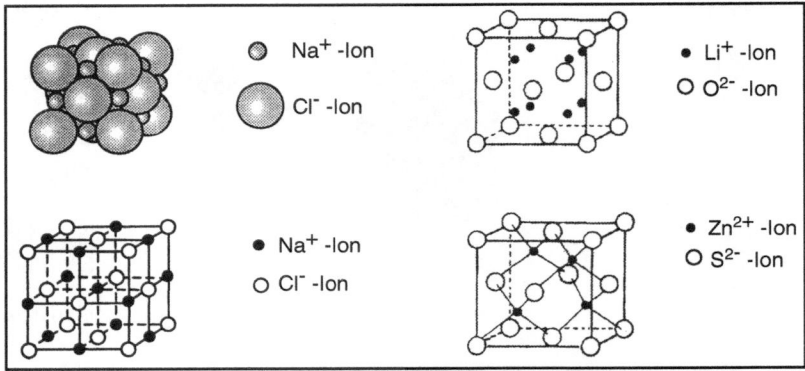

Fig. 6.14 Cubic-close-packing as models for salt crystal structures [18]

sphere packings easily when they have the opportunity to produce these models themselves by stacking spheres. With M6.1–M6.7 (see the "modeling course" at the end of this chapter) the building of sphere packings is suggested for the laboratory. Further information on metal structures can be found elsewhere (see also [5] or [17]).

If the holes in the cubic close-packing of spheres are introduced, one obtains the sodium chloride structure by filling out the octahedral holes with small balls (see Fig. 6.14). If the tetrahedral holes are being filled, the lithium oxide structure results. The zinc sulfide structure results in case that only half of the tetrahedral holes are filled. The construction of these salt structures is suggested for the lab with M6.8–M6.15 (see end of this chapter, also [5] or [18]).

Molecular models. As soon as gases are to be interpreted with the structure-oriented approach, the structures of appropriate molecules are to be demonstrated with molecular models. Since the bonding of nonmetal atoms is usually represented by push buttons on balls in molecular kits, it is fairly easy for the learner to build molecular models independently. Thereby they become acquainted with atomic bonding over time (see Fig. 6.5 and [19]).

If the learner observes the directed sticks from ball to ball in the molecule model, these sticks can be interpreted as *directed bonding* between atoms. Subsequently it is possible to understand the bonding of metal atoms in metals or of ions in salts as *undirected bonding* and to separate this from the different bonding in molecules [16]. It is also possible to differentiate the *finite* arrangement of atoms in molecules from the *infinite* one of atoms or ions in crystals.

If at least two molecule kits are available and two different models are being built for the same molecule, the fixation of one single model with all irrelevant components can be prevented. In a discussion of these models the learner has to find out similarities in both models and recognize them as the feature of depiction. The same applies to sphere packings later on: if two models of different colors and materials are being built for the same metal or salt structure, a one-sided internalization of material or color can be avoided.

In conclusion it should be pointed out that the spatial ability of the learner can be enhanced with these three-dimensional structural models (see Chap. 10). Empirical research has shown [20–22] that spatial ability increases with the construction and discussion of structural models, with building up sphere packings to show metal structures, with the comparison of octahedral or tetrahedral holes and coordination numbers in ionic structures, with the intensive use of three-dimensional models of the structure of matter in chemical education.

Adaptation and extension of models in chemistry lessons: In most cases chemistry lessons start with the simple particle model of matter: one smallest particle is chosen for every pure substance, for the substance copper the copper particle, for water the water particle, for sugar the sugar particle. The assignment of carbon particles causes difficulties since there are two substances: diamond and graphite.

But there are no specific graphite particles and different diamond particles. Carbon particles exist in both substances: they build the diamond structure in a specific way and the graphite structure in another way (see Fig. 6.15). Carbon particles are therefore neither colorless nor black – colors are characteristics of substances and not characteristics of particles. The same applies to densities or melting temperatures: they are not particle characteristics, but substance characteristics!

Applications and limits of models. Table 6.2 shows that a specific model is not fixed for every time: there are spiral curricular changes depending on purpose and level of knowledge. Thereby one can indicate to the learner in chemistry lessons that models and mental models have to be extended according to latest understanding and knowledge – as it was the case historically.

In history the Greek philosophers were already discussing the atoms as "smallest particles of matter." In 1808 Dalton got the big idea that there are as many kinds of atoms as elements, after discovering more and more elements. He brought us the first table of atomic masses and the sphere as the first mental model of one atom. In 1884 Arrhenius postulated ions as smallest particles of salts, acidic and basic

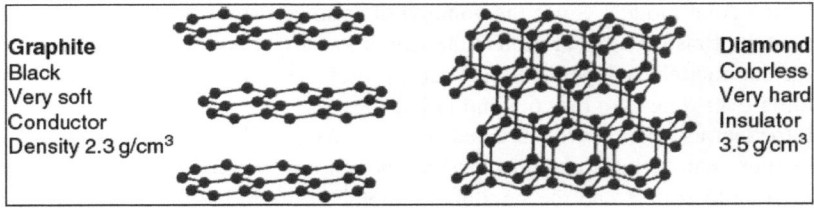

Fig. 6.15 Arrangement of C atoms in the graphite and diamond structure [19]

Table 6.2 Concrete models and mental models, their application and limits

Particle model of matter	Interpretation: state of matter, change of state, kinetic theory of gases, diffusion and solution processes, chemical reaction without change of particles (e.g. formation of alloys), etc. Limits: ↓
Mass-charge model (Dalton)	Atom, atomic mass, element, compound, periodic table, chemical reactions: regrouping of atoms and ions, law of conservation of mass in chemical reactions, ions, ionic symbols, ion lattice, undirected bond, molecule, molecule symbol, molecule structure, directed bond, empirical and structural formula, reaction equation, etc. Limits: ↓
Nucleus and shell model (Rutherford)	Nucleus, protons, neutrons, isotopes, radioactivity, atomic shell, moving electrons, electron density, electron clouds, electrolysis, metal–nonmetal-reaction, redox reaction, electron transfer, etc. Limits: ↓
Nucleus and sub-shell model (Bohr)	Ionic charge number, periodic table and octet rule, chemical bond, ionic bond, covalent bond, hydrogen bond, etc. Limits: ↓
Valence shell electron pair repulsion model (VSEPR, Gillespie)	Electron clouds, repulsion of electron clouds, bond angle, bond length, spatial structure of molecules, etc. Limits: ↓
Atomic orbital model (Schroedinger)	Orbital, hybridization, delocalization, structure of the benzene molecule, calculation and prediction of lattice or molecule structures, molecular design with computers, etc.

solutions – from now atoms and ions should be called the basic particles of all substances, and the ions may be added to Dalton's model. Following Rutherford's discovery of the nucleus of atoms with his famous experiment in 1911, the nucleus–shell model was created. Subsequently Bohr and Schroedinger described and calculated the arrangement of electrons in sub-shells and orbitals.

Further functions of display and mental models: Models have not only the function to understand the structure of matter – there are other functions to understand chemistry better.

Reduction of the learner's anthropomorphic conceptions. Preconceptions like "sunbeams clear the puddle" or "acids eat up metals" have been described in Chap. 1 "students' alternative conceptions." These or similar comments can be discussed and challenged with the help of display models. Despite initial difficulties with scientific models, the learner is open to give up his alternative conception for a new mental model, when concrete models of the structure of matter are consequently used. Advanced mental models can be developed afterwards.

Reduction of complex contexts. The reduction of difficult contexts is a fundamental teacher's task: wherever possible complex contexts have to be reduced without losing the essentials. If the definition of element and compound is being reduced to the mental model of elements being built of atoms and compounds being built of molecules – as it happened in the 1950s – this reduction has to be avoided. If the chemical bonding of atoms in molecules is being reduced to the mental model of atomic valences and one introduces the bonding number 4 for the C atom or the bonding number 1 for the H atom, the reduction is reasonable and easily applicable to the methane molecule CH_4, and other hydrocarbon molecules. This conception is acceptable – it can be extended in lessons on the nucleus–shell model later on to form the mental model of binding electron pairs between the atoms in molecules.

Generalization of facts. Beginning with the use of models, students are often able to generalize. If the CH_4-tetrahedron is being introduced as a model of the methane molecule and the extension to the ethane molecule is demonstrated, then students are able to deduce all homologues of the alkane series by generalization. They develop a spatial ability for the structure of molecules and are more and more able to understand usual structural formulae or even half-structural symbols.

Illustration of reactions. If a chemical reaction is to be interpreted, structural models of the substances before and after the reaction can be demonstrated to the learner (see Chap. 7). Before the reaction to prepare and illustrate ester formation, for example, the structure of the involved acetic acid molecule and alcohol molecule has to be demonstrated with the molecular model kit. The elimination of a water molecule and the reaction equation using half-structural formulae can be written and illustrated with structural models:

$$CH_3COOH + C_2H_5OH \rightarrow CH_3COOC_2H_5 + H_2O$$

No expert would describe the ester formation with empirical formulae only:

$$C_2H_4O_2 + C_2H_6O_1 \rightarrow C_4H_8O_2 + H_2O$$

It is easy for reactions of organic chemistry: molecular structures before and after the reaction have to be taught and half-structural symbols have to be derived – otherwise empirical formulae of the above kind have to be memorized. In inorganic chemistry it is often thought that one can use symbols like Na, Zn or Al for metals and NaCl, ZnS or Al_2O_3 for the corresponding salts. Without illustrating the structure of metals and salts by sphere packings and crystal structure models, however, learners will develop misconceptions of molecules: NaCl molecules or ZnS molecules. The empirical formulae lack any structural information and therefore these formulae are not helpful to understand substances and reactions: mental models of the structure of metals and salts have to be demonstrated (see [16–19, 23] and also Chaps. 7 and 10).

Illustration of mathematical logical facts. Descriptions of many chemical facts are based on abstract mathematical terms. Substantial mathematical reasoning is necessary for the deduction and use of equilibrium constants, for example, to recognize that the equilibrium of a weak acid is almost completely on the side of the molecules. The "glass-tube-cylinder model" (see E4.2) should be introduced, before equilibrium constants come into play. It is to demonstrate that equilibria do not only occur with a ratio of 50:50 (glass tubes of equal size in the model experiment), but also with ratios of 20:80 or 10:90 (glass tubes with different diameters). If the single steps of this experiment are plotted in a graph (x-axis) with the measured volumes in both cylinders (y-axis), two pairs of curves can be observed (see Fig. 6.16): equilibria and corresponding equilibrium constants can be understood with this model experiment. Another statistical game can show the same mathematical relationship.

The following rules regulate the game with A- and B-spheres:

– At the beginning 140 A-spheres are located in a glass bowl, no B-spheres are present
– Per time unit one sphere is taken out of the bowl: in case it is an A-sphere (that is certain at the start), it is changed to a B-sphere
– If it is a B-sphere, it is changed with the probability of 1:2 (throw a coin!) into an A-sphere. In case the coin decision is negative, the B-sphere is put back into the bowl

The result after 400 time units is shown in Fig. 6.16 (top): the model equilibrium is reached after about 300 time units, the number of B-spheres in the bowl is double that of the A-spheres.

If one starts with 140 B-spheres instead of A-spheres and plays with the same rules, the same model equilibrium as before is reached (see bottom of Fig. 6.16): $n(B):n(A) = 2:1$. If equilibrium is reached you can continue and play another 200 time units: the ratio of 1:2 will not change.

The transfer to reality shows the properties of a *dynamic* equilibrium: the forward reaction and the backward reaction are going on and on, but the concentrations of particles stay constant. The mathematical terms for these observations are the equilibrium constants: they are easier to understand with those models than without. Another nice model experiment shows the "Apple fight" of Dickerson and Geis [24] in Fig. 6.17.

Illustration of processes in chemical engineering. If a production process for plastics is to be discussed in class, an experiment takes the part of a model. The experimental production of a nylon thread from the simple starting chemicals adipine acid and hexamethylene diamine, for example, can be demonstrated by pouring some mL acid on top of some mL hexamethylene diamine. A thin film forms at the interface, a thread can be drawn out of the beaker with the help of a glass rod. This model experiment serves as a basis for demonstrating the production of nylon or other polymers in chemical engineering.

Construction of hypotheses. Mental models of molecules allow the prediction of characteristics and specific reactions. If students are already familiar with the

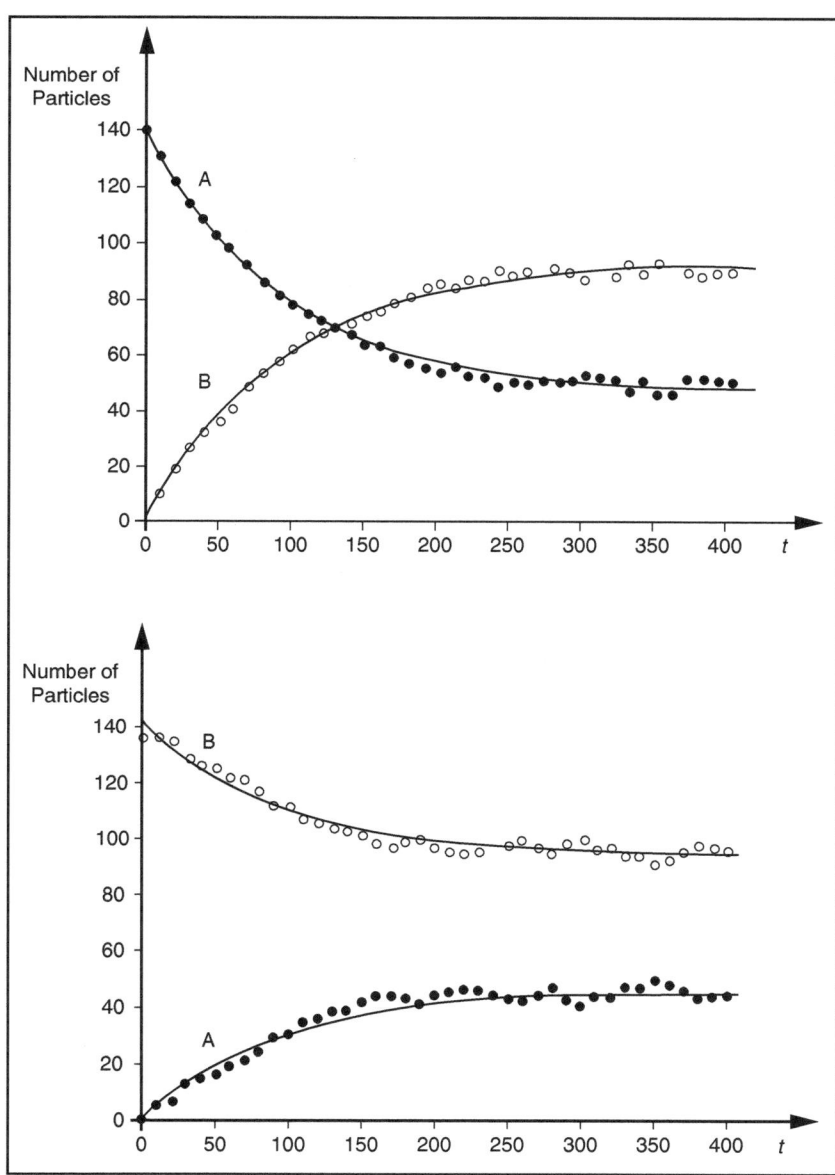

Fig. 6.16 Simulation of an equilibrium [9]

structure of alkane or alkanol homologues, they are able to devise a hypothesis for the solubility of these homologues with the help of appropriate mental models concerning structures of molecules: alcohols with short carbon chains are not soluble in petrol, but in water; alcohols with long carbon chains are not soluble in water, but in petrol. Appropriate experiments can be planned for testing hypotheses.

Fig. 6.17 (**a**) An illustration of the equilibrium with the "Apple fight," Part 1 [24]. Continued.

Science history also offers many examples. Watson [25], to only name one, cut out the shapes of base molecules from cardboard for the prognosis of base-combinations in the assumed DNA double helix: "The purine and pyrimidine models that I needed were not ready in time. So I used the whole afternoon to cut out exact models from thick cardboard. I started to move the base models back and forth and arranged them in pairs in every possible combination. Suddenly I realized that an adenine–thymine pair, which was held together by two hydrogen bonds, had the same shape as a guanine–cytosine pair" [25]. These sentences indicate the importance of very simple

Fig. 6.17 (b) An illustration of the equilibrium with the "Apple fight," Part 2 [24]

cardboard models for the understanding of the structure and function of nucleic acids. Thanks to this model conception, Watson and Crick were able to build a complete structural model of the DNA. They also formulated hypotheses that were in accord with all known facts and looked for new hypotheses, namely for the reduplication of DNA and the processes for the development of life.

6.3   Learner: Experience with Models

Students come to chemistry lessons with their experience of models in three areas: their own toy models, for example, soft toys, car or ship models. They also know concrete models of other school subjects like biology or geography, in mathematics they may even be used to mental models: most graphs can be shown by mathematical terms or formulae.

Toys. Children like to compare their dolls with themselves or others. They find out that many characteristics are the same in model and in the original: for example place and form of mouth, nose, eyes and ears. They also realize that there are functions, which the doll does not have: the doll does not eat or breath.

This experience is contrary to the understanding of scientific models. If young students take a structural model, they cannot compare it to the original: they are not able to look into the salt crystal, to recount the coordination number 6/6 of ions in common salt crystals. They find coordination numbers only in structural models. Due to the learners experience of concrete toy models the scientific model has to be developed and defined step by step.

Fun with models. The fun that children have playing with their dolls or car and ship models can be used for working with models of the structure of matter in chemistry. It has to be connected to an activity-oriented introduction to the scientific world of models, for example by building close-packed sphere models or crystal structure models.

If you show the packing of spheres for the sodium chloride structure to students (see Fig. 6.3b) and ask them to build this model with white and red spheres from cellulose (diameter: 30 mm and 12 mm, respectively), they have a lot of fun and proudly present this model at home. They might even discuss this model with their parents and siblings and might explain to them what it is a model for: now they are the experts, who explain chemical ideas to others! Instructions for model building exercises can be found at the end of this chapter.

If you show a model of the sodium chloride structure (see Fig. 6.3a) and ask them to rebuild this model with tooth picks and soft candy of two colors (for example red and black "strawberries" from sweety producers) they like to do this – and after the hard work and explanations of the coordination number 6 they may eat their model.

Older students, prospective teachers or even teachers at meetings for teacher training also like to work with structural models of all kinds, according to our experience. They convince themselves of the number and kinds of isomers in alkane molecules or special alcohol molecules, by building molecular models with the help of molecular model kits. If you offer even spheres of foam, of styrofoam, or of plexiglass to teachers, they like to build the different sphere packings (see the modeling course at the end of this chapter), and even like to build difficult models of unit cells (see Fig. 6.3c). At the end of the course they can take their self-built models to the school collection and use them in their chemistry lectures.

Models from other subjects. Students have lots of experience with models from other subjects. *Biology*: The school collection normally holds models of eyes, ears or the human skeleton (only rarely an original). Model character, shortenings and irrelevant components of the models are very evident in these cases. These models are very illustrative and motivating due to the relevance for oneself: they can be used for a discussion of the idea of scientific models. But since visible body parts of the original can be compared with the corresponding model, this kind of biological model is not to be transferred to the models of chemical structures: no one is able to directly observe original structures of matter!

Geography: Maps, for example those of their home town or familiar hiking trails, are also models that are easy to understand for the students. Even the globe as a model can be compared to the original: pictures of Earth, seen from space, can be shown to the students today. Before the time of space travel the globe was nearly a mental model that could not be compared to the original directly. Models of Earth's interior cannot be obtained by optical comparison: they are derived from empirical analysis of experiments on Earth's surface.

Mathematics: Geometric drawings can be regarded as models. Certain triangles, for example rectangular triangles or equilateral triangles, are models for different patterns, and students can reproduce them for drawing special parquetry layers on the basis of these triangles. When students build three-dimensional models of cubes, cuboids, octahedra or tetrahedra with the help of given cardboard pattern and try to fill the whole space with them, the transfer of these models to chemical structures is possible and very useful. Structural models can be explained on this mathematical basis in chemistry lessons.

When students start to draw these cubes, tetrahedra and octahedra in three dimensions in their math class, these skills can be transferred to chemistry lectures to develop the ability to also draw three-dimensional models of the structure of matter. Moreover this trains and advances their spatial ability. The advancement of these skills is an important interdisciplinary task for mathematics, but also in chemistry: spatial ability is important for everyone's life!

6.4 Human Element: Interdisciplinary Mental Models

Working with models has an interdisciplinary character and makes a contribution to general education. Besides the understanding of the importance of scientific models, especially for chemistry, one can reflect on the relevance of models in many other fields. Models and mental models play an important role in

Industry: cycles of material
Economy: cycles of money and goods
Sociology: behavior patterns of specific groups of people
Politics: voting behavior of special groups of people
Ecology: cycles of substances in nature and ecological systems

Appendix A. Problems and Exercises

P6.1 You will find models in your chemistry collection at school: NaCl packing of
       spheres, NaCl crystal structure and NaCl unit cell (see Fig. 6.3). Give the
       features of depiction of the three models. Discuss and compare the irrelevant
       components of these models.
P6.2 The NaCl-structure can be described by the cubic face-centered lattice of
       chloride ions where the octahedral holes are filled with sodium ions. Describe
       the Li_2O-structure and the ZnS-structure in a similar way and draw the
       corresponding models (see Fig. 6.13).
P6.3 Planning of chemistry lessons applies two levels: level 1 deals with pheno-
       mena and lab experiences, level 2 with structural and mental models (see
       Fig. 6.11). Describe and draw your mental model of the diffusion of hydrogen
       sulfide (H_2S) in air on the basis of (a) the particle model of matter, (b)
       Dalton's atomic model.
P6.4 Usually mental models are introduced in chemistry lessons from the particle
       model, via Dalton's atomic model to the nucleus–shell model. Chose (a)
       a substance and (b) a chemical reaction and make model drawings on the
       basis of these three models. Discuss the differences in interpretation based on
       those models.
P6.5 The chemical equilibrium can be illustrated with model experiments or
       with every-day experiences. Give one example for each and establish connec-
       tions to an example of a real chemical equilibrium.

Appendix B. Modeling Course: Structures of Metals and Salts

Material: 100 white cellulose balls $d = 30$ mm [26], 50 red cellulose balls $d = 12$ mm
[26], triangular wooden frame ($a = 17.5$ cm), square wooden frame ($a = 15$ cm),
modeling clay, glue, two equilateral ball-triangles (six balls with $d = 30$ mm each).
   Structure the plural: Close-packing of spheres (1 metal atom $\equiv$ 1 sphere) are
structural models to describe the structure of many metal crystals (find drawing
models at the end of this sheet).

M 6.1: Fill up the triangular wooden frame with a closest layer of balls in the
         triangular pattern. Put as many layers of balls on top as possible. Draw the
         layers of balls.
M 6.2: The coordination number stands for the number of balls that touch one ball
         in the middle of the packing. Find out the coordination number of one
         sphere in the close-packing of spheres. Draw three layers of balls to
         visualize this number.
M 6.3: Two different close-packings of spheres with the coordination number 12
         are possible:

(a) With the layer sequence ABCABC..., (b) with the layer sequence ABAB...

Build both packings! Draw the layers of balls with triangular pattern so that (a) and (b) become apparent.

Definition: A layer sequence ABCA... exists, when the fourth layer of balls is congruent with the first layer – seen from above. The layer sequence ABA... exists, when the third layer of balls is already congruent with the first layer (the layers in triangular pattern are meant).

Information: An elementary cube (see picture) can be found in the ABCA-packing of spheres. Therefore this packing is also called *cubic close-packing* of spheres.

M 6.4: Draw the crystal structure model next to the shown packing. Draw a perspective cube: instead of the balls only give the central points of the balls and connect these points.

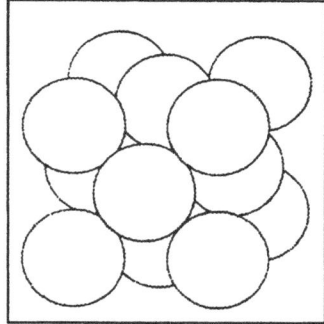

M 6.5: Put together the shown model with the help of the two equilateral 6-spheres triangles and two additional spheres.

Try to build this cube into the closest ABC-packing (M 6.3).

Draw two possible ways for building the elementary cube:

(a) Connect layers in a triangle pattern (1 + 6 + 6 + 1), (b) in a square pattern (5 + 4 + 5).

M 6.6: Take the square wooden frame. Build the cubic close-packing of spheres beginning with the square pattern. Try to get the elementary cube into the packing. Determine the coordination number. Draw the sphere layers so that the coordination number can be determined.

Information: The following models show the structure of metal crystals:

1. The hexagonal close-packing of spheres with the layer sequence ABA: it shows how crystals of magnesium and zinc are built up of atoms. One can say that there are crystals with hexagonal symmetry or crystals of the *Mg-type*.

2. The cubic close-packing of spheres with the layer sequence ABCA in triangular pattern: it shows in which way crystals of copper, silver or gold are built up of atoms. One can say that there are crystals with cubic

symmetry or crystals of the *Cu-type*. The elementary cube has one
sphere in every face center – therefore this structure is also called
face-centered cubic.

3. The name "face-centered cubic" is possible to point out the difference to
the *body-centered* cubic packing of spheres. It is not a close-packing
anymore, the coordination number is 8 (see picture). Metal crystals of
tungsten and alkaline metals have this structure, the *W-type*.

M 6.7: The nine-spheres packing shows the elementary cube of body-centered
cubic metal structures. Draw the crystal structure next to the picture:
draw a perspective cube, instead of the balls only give the central points
of the balls and connect these points. Compare it with M6.4.

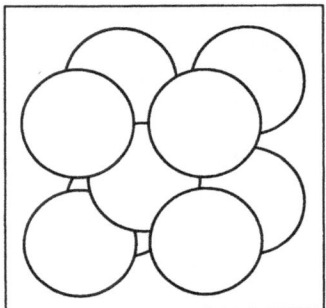

Structures of salt crystals: The structure of many metal crystals can be
illustrated by packings of spheres of *one* kind – the structure of salt crystals
needs *two* kinds of spheres. Models for the sodium chloride structure and
three other salt crystals will be built in the following (drawing models of
the crystal structures are shown at the end of this section).

M 6.8: Na^+ ions in sodium chloride are to be represented with red balls ($d =$
12 mm), Cl^- ions with white balls ($d = 30$ mm). With the help of the
triangle frame build a close-packing of spheres with both kinds of balls.
Draw the layers of balls.

M 6.9: Determine the coordination number for both kinds of balls. Draw the layers
of balls to visualize the coordination number of both kinds of balls.

M 6.10: In the close-packing of spheres there are two different-sized kinds of holes
or gaps. Determine the number of balls that form those two different gaps.
Draw the gap-producing balls for both kinds of holes in the form of the
layers of balls: (a) for big gaps, (b) for small gaps.

Information: Two different types of holes can be found in the close-
packing of spheres. Convince yourself with the help of the model M 6.8:

1. Big holes are formed by six balls with octahedral geometry: octahedral
gaps (OG)
2. Small holes are formed by four balls with tetrahedral geometry: tetrahedral gaps (TG)

3. Spheres, OG and TG exist in a ratio of 1:1:2 in the close-packing of spheres.

> The structure of the sodium chloride crystal can therefore be explained like this: Cl^- ions form a cubic close-packing of spheres, all octahedral gaps are occupied by smaller Na^+ ions. The coordination numbers are 6 and 6, the ion ratio is 1:1, the ratio symbol should be $(Na^+)_1(Cl^-)_1$
>
> Explain the cubic shape of the sodium chloride crystal.

M 6.11: Take the elementary cube from M6.5, look for octahedral gaps and fill them with small red balls.

(a) Complete the model drawing by filling the gaps with small spheres (see picture).

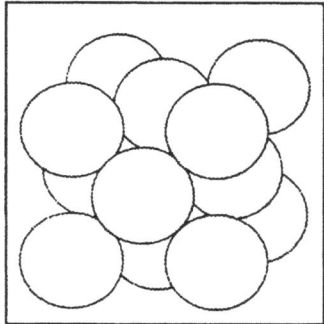

(b) Draw the crystal structure next to the picture. Draw a perspective cube: instead of the balls only give the central points of the balls and connect these points (see M6.4).

M 6.12: Build the cubic close-packing of spheres with the help of the square frame and with both kinds of spheres. Draw the sequence of the 2-spheres layers.

M 6.13: Convince yourself that the elementary cube can be build into the packing of spheres beginning with the triangle pattern (M 6.8) as well as the one beginning with the square pattern (M 6.12). Which position does it take in both models?

Draw the elementary cube with the help of (a) layers of spheres in the triangular pattern, (b) layers of spheres in the square pattern.

M 6.14: A model for the aluminium oxide structure (corundum structure) may be built:

(a) Stick together three layers with 15 white balls each (see picture).
(b) Add ten small balls in the shown pattern (see picture).

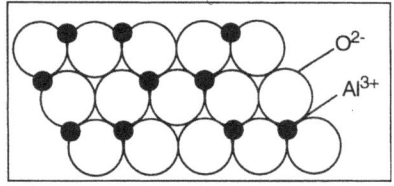

(c) Stack up three layers so that the sequence is ABA. The coordination number of the small balls has to be six, of the big balls has to be four. What is the ratio of the balls?

Information: The O^{2-} ions form a hexagonal close-packing in aluminium oxide crystals, only 2/3 of octahedral gaps are occupied by Al^{3+} ions. The coordination is 6/4, the ratio of ions is 2:3. Therefore the models have to be abbreviated to formulae like $\{(Al^{3+})_2(O^{2-})_3\}$ or Al_2O_3.

M 6.15: Form a few small balls with modeling clay that fit into the tetrahedral gaps of big spheres. With the small and the big spheres build an elementary cube

(a) For the zinc sulfide structure (see picture).
(b) For the lithium oxide structure (see picture).

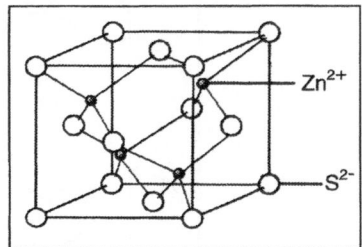

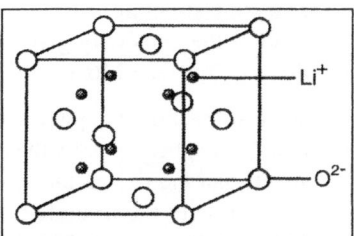

Information: zinc sulfide can be described as a cubic close-packing of S^{2-} ions, half of the tetrahedral gaps are occupied by Zn^{2+} ions. The coordination is 4/4, the formula for the unit cell is $\{(Zn^{2+})_4(S^2)_4\}$, the empirical formula ZnS.

Lithium oxide can be shown as a cubic close-packing of O^{2-} ions, all of the tetrahedron gaps are occupied by Li^+ ions. The coordination is 4/8, the formula for the unit cell is $\{(Li^+)_8(O^{2-})_4\}$, the empirical formula Li_2O.

On the following two pages you will find the expected model drawings for the models M6.13 to M6.15. Compare your own drawings with these.

M 6.1:

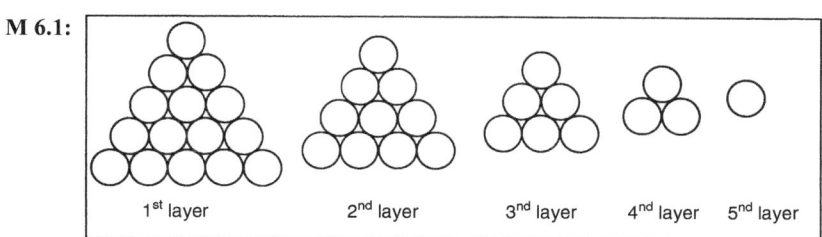

M 6.2:

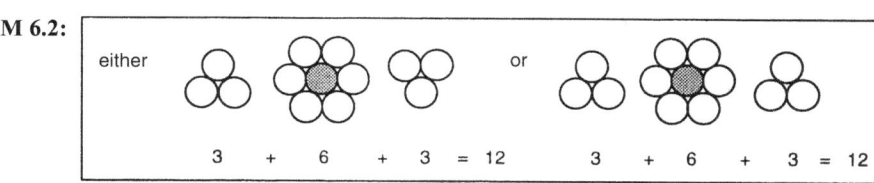

M 6.3:

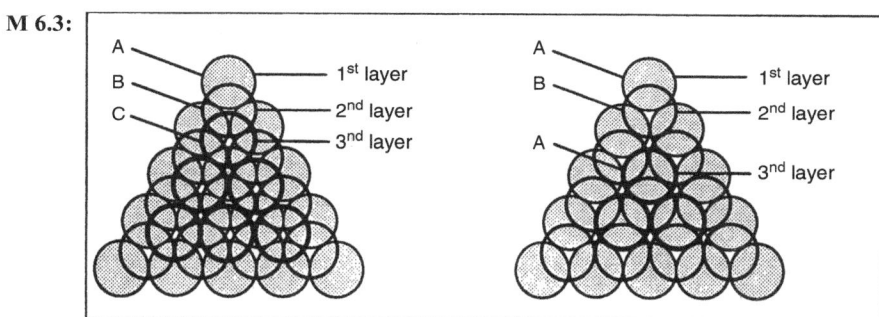

M 6.4:

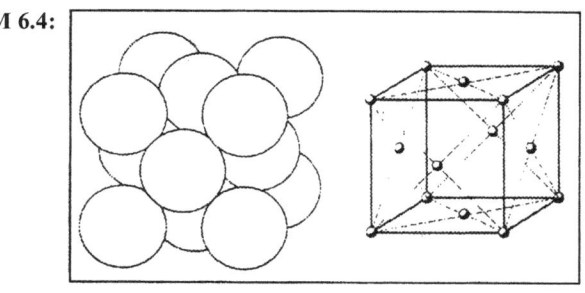

M 6.5:

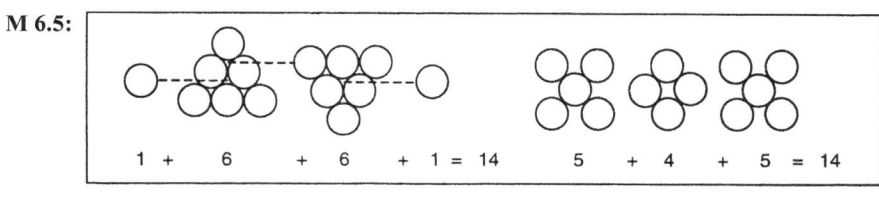

M 6.6:

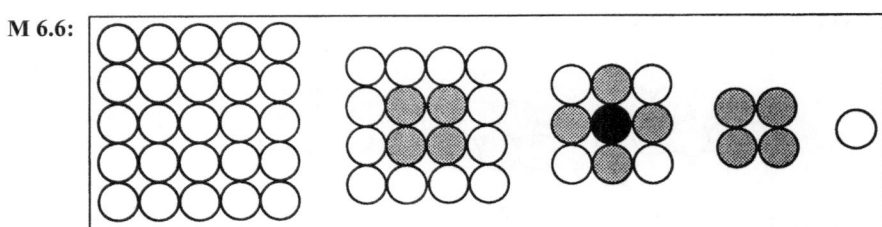

M 6.7:

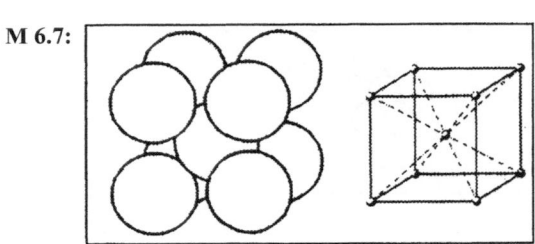

M 6.8:

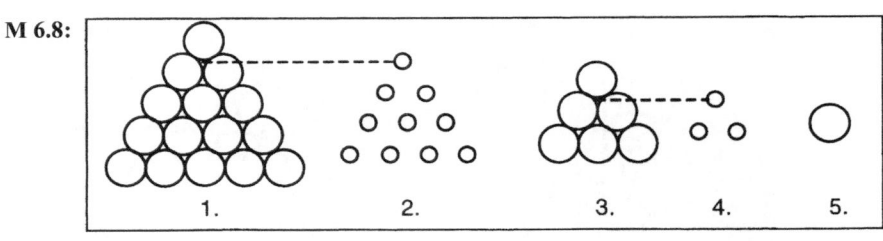

M 6.9:

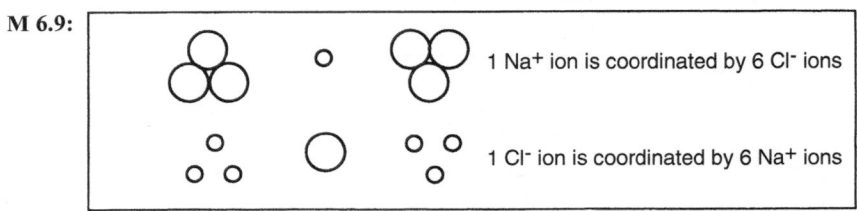

1 Na$^+$ ion is coordinated by 6 Cl$^-$ ions

1 Cl$^-$ ion is coordinated by 6 Na$^+$ ions

M 6.10:

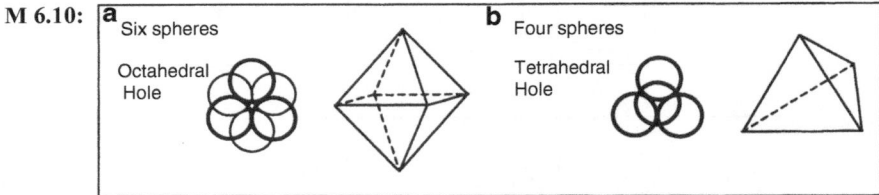

a   Six spheres

Octahedral Hole

b   Four spheres

Tetrahedral Hole

M 6.11:

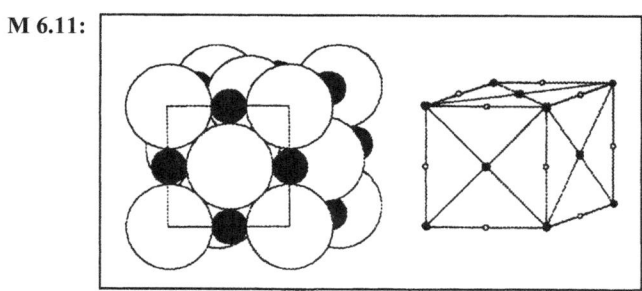

M 6.12:

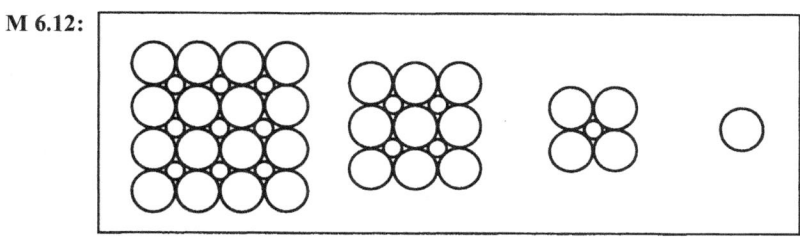

M 6.13:

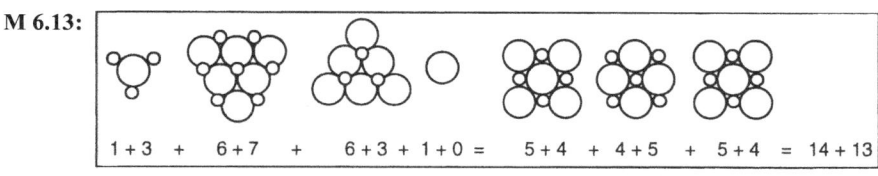

1 + 3 + 6 + 7 + 6 + 3 + 1 + 0 = 5 + 4 + 4 + 5 + 5 + 4 = 14 + 13

References

1. Haeusler K (1998) Highlights in der Chemie. Aulis, Koeln
2. Stachowiak H (1965) Gedanken zu einer allgemeinen Theorie der Modelle. Studium Generale 18:432
3. Steinbuch K (1977) Denken in Modellen. In: Schaefer, G., u. a.: Denken in Modellen. Westermann, Braunschweig
4. Kircher E (1977) Einige erkenntnistheoretische und wissenschaftstheoretische Auffassungen zur Fachdidaktik. Chim.did. 3:61
5. Harsch G (1981) Kristallgeometrie. Packungen und Symmetrie in Stereodarstellungen. Diesterweg, Frankfurt
6. Geomix, Ratec: Koerberstr. 15, 60433 Frankfurt
7. Leybold Didactic: Postfach 1365, 50330 Huerth
8. Institute for Chemical Education: University of Madison, 1101 University Av., Madison WI 53707, USA
9. Harsch G (1985) Vom Wuerfelspiel zum Naturgesetz. Simulation und Modelldenken in der Physikalischen Chemie. VCH, Weinheim
10. Harsch G (1982) Statistische Spiele für den naturwissenschaftlichen Unterricht. Dr. Flad, Stuttgart
11. Harsch G (1984) Kinetics and mechanism – a games approach. J Chem Educ 61:1039

12. Eigen M, Winkler R (1975) Das Spiel – Naturgesetze steuern den Zufall. Piper, München
13. Buck P (1994) Die Teilchenvorstellung – ein "Unmodell". Chem Sch 41:412
14. Buck P (1994) Wie kann man die "Andersartigkeit der Atome" lehren? Chem Sch 41
15. Asselborn W, u.a.: Chemie heute. SI. Braunschweig 2010 (Schroedel)
16. Sauermann D, Barke H-D (1998) Chemie für Quereinsteiger. Schueling, Muenster. Band 1: Strukturchemie und Teilchensystematik, also http://www.wikichemie.de
17. Sauermann D, Barke H-D (1998) Chemie für Quereinsteiger. Schueling, Muenster. Band 2: Struktur der Metalle und Legierungen, also http://www.wikichemie.de
18. Sauermann D, Barke H-D (1998) Chemie für Quereinsteiger. Schueling, Muenster. Band 4: Ionenkristalle mit einfachen Gitterbausteinen, also http://www.wikichemie.de
19. Sauermann D, Barke H-D (1998) Chemie für Quereinsteiger. Schueling, Muenster. Band 3: Moleküle und Molekuelstukturen, also http://www.wikichemie.de
20. Barke H-D (1994) Chemical education and spatial ability. J Chem Educ 70:968
21. Barke H-D (1980) Raumvorstellung im naturwissenschaftlichen Unterricht. MNU 33:129
22. Barke H-D (1983) Das Training des Raumvorstellungsvermoegens durch die Arbeit mit Strukturmodellen im Chemieunterricht. MNU 36:352
23. Barke H-D, Wirbs H (2002) Structural units and chemical formulae. CERAPIE 3:185
24. Dickerson RE, Geis I (1981) Chemie – eine lebendige und anschauliche Einfuehrung. Verlag Chemie, Weinheim
25. Watson JD (1969) Die Doppel-Helix. Rowohlt, Hamburg
26. Faita: Postfach 1146, 83402 Mitterfelden, Germany

Further Reading

Ben-Zvi R, Eylon B, Silberstein J (1987) Students' visualization of a chemical reaction. Educ Chem 24:117–120
Car M (1984) Model confusion in chemistry. Res Sci Educ 14:97–103
De Vos W, Verdonk AH (1996) The particulate nature of matter in science education and in science. J Res Sci Teach 33(6):657–664
Ehrlén K (2007) Children's understanding of globes as a model of the Earth: a problem of contextualizing. Int J Sci Educ 30:221–238
Gabel DL (1993) Use of the particle nature of matter in developing conceptual understanding. J Chem Educ 70(3):193–194
Gabel DL, Samuel KV, Hunn D (1987) Understanding the particulate nature of matter. J Chem Educ 64:695–697
Griffiths AK, Preston KR (1992) Students' misconceptions relating to fundamental characteristics of atoms and molecules. J Res Sci Teach 29:611–628
Grosslight L, Unger C, Jay E, Smith CL (1991) Understanding models and their use in science: conceptions of middle and high school students and experts. J Res Sci Teach 28(9):799–822
Haidar HA, Abraham MR (1991) A comparison of applied and theoretical knowledge of concepts' based on the particulate nature of matter. J Res Sci Teach 28(10):919–938
Harrison AG, Treagust DF (1996) Secondary students mental models of atoms and molecules: implications for teaching science. Sci Educ 80:509–534
Harrison AG, Treagust DF (2002) The particulate nature of matter: challenges in understanding the submicroscopic world. In: Gilbert JK, Jong OD, Justi RV, Driel JH (eds) Chemical education: towards research based practice. Kluwer Academic, Amsterdam
Ingaham AM, Gilbert JK (1991) The use of analogue models by students of chemistry at higher education level. Int J Sci Educ 13:193–202
Johnson P (1998) Progression in children's understanding of a 'basic' particle theory: a longitudinal study. Int J Sci Educ 20(4):393–412

Krishnan SR, Howe AC (1994) The mole concept: developing on Instrument to assess conceptual understanding. J Chem Educ 71:653–655

Larkin JH (1979) Information processing models and science instruction. In: Lochhead J, Clement J (eds) Cognitive process instruction: research in teaching thinking skills. The Franklin Institute Press, Philadelphia, PA, pp 109–119

Nahum TL, Mamlok-Naaman R, Hofstein A, Krajcik J (2007) Developing a new teaching approach for the chemical bonding concept aligned with current scientific and pedagogical content knowledge. Sci Educ 91:579–603

Niaz M, Aguilera D, Maza A, Liendo G (2002) Arguments, contradictions, resistances and conceptual change in students' understanding of atomic structure. Sci Educ 86:505–525

Nussbaum J (1985) The particulate nature of matter in the gaseous phase. In: Driver R, Guesne E, Tiberghien A (eds) Children's ideas in science. Open University Press, Milton Keynes, UK, pp 124–144

Paivio A (1986) Mental representations: a dual coding approach. Oxford University Press, New York

Schnotz W, Preuβ A (1999) Task-dependent construction of mental models as a basis for conceptual change. In: Schnotz W, Vosniadou S, Carretero M (eds) New perspectives on conceptual change. Pergamon, Amsterdam, pp 193–222

Srere M (1985) The gaseous state. In: Driver R, Guesne E, Tiberghien A (eds) Children's ideas in science. Open University Press, Philadelphia, pp 105–123

Vekiri E (2002) What is the value of graphical displays in learning? Educ Psychol Rev 14:261–312

Vosniadou S, Brewer WF (1992) Mental models of the earth: a study of conceptual change in childhood. Cogn Psychol 24:535–585

Chapter 7
Scientific Terminology and Symbols

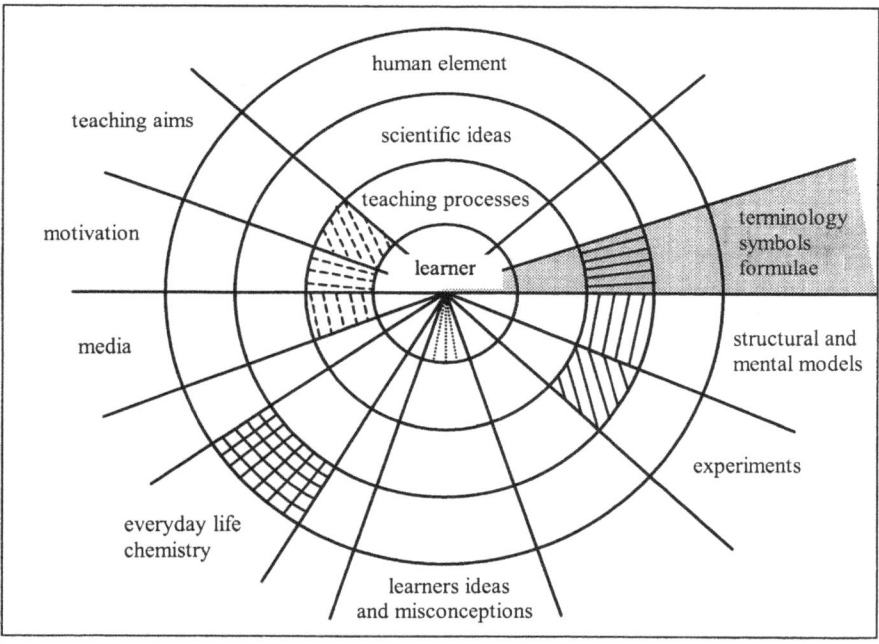

"*Chemical formulae and equations are the language of experts in chemistry, communication across borders is not possible without them. The description of chemical processes would be as complicated as if we had to write our curriculum vitae in cuneiform characters*" [1].

With this statement Peter von Zahn [1] emphasizes the importance of chemical symbols, which he calls a unique communication tool for chemists – whether they work in Europe or America, in China, or Japan. With this in mind it is also an important aim of chemical education to teach chemistry terminology and symbols to students. They have to be able to read about and in newspapers and magazines.

H.-D. Barke et al., *Essentials of Chemical Education*,   189
DOI 10.1007/978-3-642-21756-2_7, © Springer-Verlag Berlin Heidelberg 2012

At the same time, scientists are required to make their methods and conclusions understandable even for nonprofessionals: "Scientists and inventors cannot live isolated in an ivory tower, they require a sounding board. They need the approval of those who will benefit from their scientific work in the end." [1]

Many technical terms are used in everyday language, but chemical terms are also part of it: substance, metal, solution, acid, lye, gas, combustion, are examples. Often the meaning of such terms in the professional language is different or more advanced, but it certainly is always more precise then the one in everyday language. Combustion in our everyday language is an expression for candles or spirits burning, which afterwards are irretrievable destroyed. At the same time, the expert thinks about the reaction with oxygen to form the invisible and colorless gases carbon dioxide and water vapor.

To be able to talk with students about chemical issues in an appropriate way, it is important to lead them from their everyday language to the chemical terminology and symbolic language. They should learn to phrase their observations and discuss explanations in a scientific way.

7.1 Scientific Ideas: Parameters, Units, Terms, Symbols

Much of terminology and nomenclature like parameters and units are historically grown. For example, 1 mile was defined in ancient Rome as "milia passuums," i.e. the distance of 1,000 double steps of a short man (today: 1,475 m). The "statute mile" in England and the United States was changed to 1,000 double steps of a huge man (today: 1,609.30 m), and finally in 1875, the unit meter (m) and kilometer (km) were defined (see the next paragraph).

Those and other units have been redefined a number of times by standard committees like IUPAC (International Union of Pure and Applied Chemistry) to achieve more precise and standardized definitions. These problems of nomenclature and historical changes in the meaning of terms, symbols, parameters, and units are to be considered in the following.

7.1.1 International System and Derived Units

Three examples of parameters and units shall demonstrate historical changes:

Length. The meter as a unit of length has been defined in 1875 as the 10,000,000th part of one earth quadrant (1/4 circumference of earth, today's mean value: 10,000.09 km). It has been set by the length between two marks on a platinum–iridium bar, located in Sèvre near Paris since 1889. A more precise definition of the meter could be obtained with the help of spectroscopic methods: the meter was redefined in 1960 as equal to 1,650,763.73 wavelengths of the orange–red emission line in the electromagnetic spectrum of the krypton-86 isotope in vacuum.

Time. A mean solar second equals the well-known part of a solar day respectively a full rotation of earth around its axis: $1\ s = 1/24 \times 1/60 \times 1/60 = 0.0000115$ days. This was valid until 1964. Due to the beginning of space travel, a more precise definition was needed, and the second was redefined on the basis of oscillation frequencies of the caesium-133 isotope: it is the duration of 9,192,631,770 oscillations.

Thermal energy. The original unit for thermal energy was the calorie; it describes the amount of energy required to raise the temperature of 1 g of water by 1 K; more precisly, from $14.5^{\circ}C$ to $15.5^{\circ}C$. The unit Joule (J) was introduced to make this amount of energy comparable with other kinds of energy and to be able to convert it easily: $1\ J = 1\ Nm = 1\ kg\ m^2/s^2$. The old unit for thermal energy can be converted into the new unit according to the above definition on the basis of the SI units: $1\ cal = 4.18\ J$.

SI units. The last example shows a particularity: the unit J is defined by a term of the basic units kg, m and s. Since 1968, these units are subject to the *Système Internationale d'unités* (*International System of units*), therefore they are called SI units. There are a lot of other parameters that can be derived from SI units or that have been redefined on their basis: area, volume, density, pressure, speed, acceleration, concentration, force, energy, enthalpy, entropy, power, electrical charge, voltage, electric resistance, electric capacity, and others. Table 7.1 contains the seven basic parameters and corresponding SI units.

7.1.2   Parameters and Units at School

Mass and density. If one would consequently take the SI units as a basis, then all masses have to be stated in kg and densities in kg/m^3 (some new tables are doing this). But it seems to be reasonable to use the units g and mg for smaller masses of

Table 7.1 Basic parameters and SI units, their symbols and conversion factors

Parameter	Symbol	SI-unit	Symbol	Definition
Length	l	Meter	m	Wavelength krypton-86
Mass	m	Kilogram	kg	$1\ dm^3$ of water ($4^{\circ}C$)
Time	t	Second	s	Oscillations of caesium-133
Temperature	T	Kelvin	K	$0\ K = -273.15^{\circ}C$
Current	I	Ampere	A	$0.19\ cm^3$ oxyhydrogen per 1 s
Amount of substance	n	Mole	mol	Avogadro constant N_A
Light intensity	I	Candela	cd	Special amyl acetate flame

The following prefixes are common for multiples or parts of the SI-units:

10^9	10^6	10^3	10^2	10^{-1}	10^{-2}	10^{-3}	10^{-6}	10^{-9}	10^{-12}
G	M	k	h	d	c	m	μ	n	p
giga	mega	kilo	hecto	deci	centi	milli	micro	nano	pico

substances, and it is more common for teaching the density of substances to use g/ml or g/cm^3 for solids or liquids, but g/L for gases.

In the literature, one can find different symbols for liter and milliliter: ml and l respectively mL and L. Since the capital L in the latter notation better contrasts with the number 1, the notation mL and L is used in this book.

Atomic mass. In chemistry, there also exists another mass unit than the kg: the unit u for the atomic mass. The former "atomic weight" was dimensionless, it was first defined by comparison with the mass of a hydrogen atom, then by comparison with the mass of an oxygen atom, the values were entered in first atomic mass tables. Later, it was redefined to have an exact value and to make the atomic mass comparable with other parameters: 1 u = 1/12 of the mass of the isotope C-12. It can also be connected to the unit g with the help of Avogadro's constant: 1 g = N_A 1 u.

The molecular mass as well as the molar mass can be obtained from the addition of atomic masses of participating atoms. Examples:

$$m_{1\,C\,atom} = 12\,u, \quad m_{1\,mol\,C\,atoms} = 12\,g, \quad M(C) = 12\,g/mol,$$

$$m_{1\,C_2H_5OH\,molecule} = 46\,u, \quad m_{1\,mol\,C_2H_5OH\,molecules} = 46\,g,$$
$$M(C_2H_5OH) = 46\,g/mol.$$

Mole. The unit for the amount of substance n is the mole with the symbol mol. The meaning of this unit has changed drastically. Some decades ago, the definition was: the mole equals the molecular weight in g, therefore molecular masses in u were easily changed into the unit g. Today, the following IUPAC-definition is used: "one mole is the amount of a substance that has as many particles as atoms in 12 g of carbon, which only consists of C-12 isotopes. The particles of that amount of a substance have to be further characterized: it can be atoms, molecules, ions, electrons, and other particles" [2]. The advantage of this definition is that the imprecise information "1 mol oxygen" is replaced by "1 mol O atoms," or "1 mol O_2 molecules," or even "1 mol O_3 molecules":

$$m_{1\,mol\,O\,atoms} = 16\,g, \quad m_{1\,mol\,O_2\,molecules} = 32\,g, \quad m_{1\,mol\,O_3\,molecules} = 48\,g,$$

the sulfur portion m = 32 g either consists of 1 mol S atoms or 1/8 mol S_8 molecules.

The molar mass M (g/mol), the molar volume V_m (L/mol), the molar quantity N_A (1/mol), the molar charge F (C/mol), molar concentration molarity (mol/L), or molality (mol/kg) can be derived from the mole. The common use of the molarity unit mol/L for concentrations allows precise values according to the amount of particles. For example, the imprecise use of the term "1 molar sulfuric acid" should be replaced by the precise use:

$$c_{H_2SO_4} = 0.1\,mol/L \text{ means } c_{H^+(aq)} \approx 0.2\,mol/L \text{ resp. } c_{SO_4^{2-}(aq)} \approx 0.1\,mol/L.$$

Temperature. The unit used in everyday life and also in science lessons is °C. In 1777, Celsius defined the temperature of freezing water as 0° and of boiling water as 100°. In the United States, the Fahrenheit unit (°F) is also used: in 1777, Fahrenheit defined the lowest temperature of an ice–salt mixture as 0° (-18°C), and the temperature of the blood in our body as 100° (37°C). The formula for conversion is: °F $= 1.8 \times$ °C $+ 32$.

The unit K (not °K) can be used for differences in temperatures and definitely has to be used for terms of the gas laws. The absolute minimum of temperatures 0 K was defined by Lord Kelvin in 1888: 0 K $= -273$°C (exactly: -273.15°C).

7.1.3 Terminology at School

Many terms exist that underwent a change of meaning. Therefore, these terms shall be discussed here: formula, equation, particle, valence, energy, acid and base, isomerism, etc.

Formula and equation. It became established to call symbols like CH_3COOH a formula. In mathematics and physics, a formula is a term in which one can enter numbers and where new numbers result. One should stick to this definition in math and physics and should call the chemical formula a symbol: atom symbol, ion symbol, molecule symbol, etc.

Equations also have a different meaning in mathematics and physics. Chemical "equation" can only mean that the sum of the masses on the left side of the arrow "equals" the total mass on the right side of the arrow in the equation. Substances, properties, or energies are definitely different on both sides of the arrow. The term "equation" might therefore be misleading for beginners and should be replaced by reaction scheme or reaction symbol.

Chemical Symbols. Using the words CH_3COOH *symbol* or molecule *symbol* helps to avoid the term formula and supports a fundamental idea of modern philosophy: "Symbols are not representatives of their object, but a vehicle for the imagination of objects. The denotation of the common use of the term symbol contains four elements: subject, symbol, image, and object." [3] The chemist (subject) notices the term CH_3COOH (symbol) and imagines the structure of the acetic acid molecule (image). They know that the structure of this molecule (object) could always be confirmed by methods of instrumental analysis (see details in [3] and [4]).

Particles. The term "particle" is introduced as a preliminary model of the structure of matter for young students at school. It is used, for example, to illustrate a sugar crystal consisting of sugar particles or an ethanol solution consisting of ethanol particles and water particles. The scientific meaning of the term "model" has to be explained when the first model is introduced, like the "particle model of matter" in grade 7 or 8 (see Chap. 6) (Fig. 7.1).

Two educational problems arise. As long as the term "particle" is only used for metal atoms or gas molecules, no problems occur: a sphere or ball is a good model for

Crystal structure symbols		Molecular symbols	
$\{Ca^{2+} [8c + 12c]$ $F^-_2[4t + 6o]\}$ $\{Ca^{2+} F^-_2\ 8/4\ \}G$	PARTHÉ symbol and NIGGLI symbol for the ionic structure	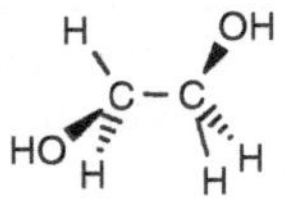	Stereo symbol for the molecular structure
↑	↑	↑	↑
$\{(Ca^{2+})_4(F^-)_8\}$	Symbol for the unit cell	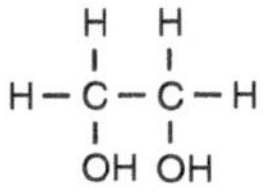	Constitutional symbol
↑	↑	↑	↑
$\{(Ca^{2+})_1(F^-)_2\}$	Symbol for the ion ratio	$HO-CH_2-CH_2-OH$ $C_2H_4(OH)_2$	Half structural symbol
↑	↑	↑	↑
$Ca^{2+}(F^-)_2$ CaF_2	Empirical formula	CH_2OH CH_3O	Empirical formula

Fig. 7.1 Different types of chemical symbols (examples: calcium fluoride and ethane diol)

both particles of matter. But as soon as the term "particle" is used for salts or salt solutions, one gets to the limits of this model. According to the convention, "one ball in the particle model of matter is used for one particle of a pure substance," there are problems with salts: in the case of sodium chloride, one ball would have to be used for one ion pair Na^+Cl^- or even for the unit cell $\{(Na^+)_4(Cl^-)_4\}$. Because salts and salt solutions are built of at least two kinds of ions, any correct model should use two different balls. Due to this problem, the term "particle" in the sense of the particle model of matter is normally not used for solid salt crystals or for salt solutions.

The second problem is the observation of powder mixtures, suspensions or aerosols. Students often describe a mixture of iron powder and sulfur powder with "iron particles and sulfur particles." They use the term "particles" for visible crystals, although they know that one crystal consists of billions of particles. If one attaches importance to the consequent use of the term "smallest particle" for the submicroscopic level of atoms, ions, and molecules, small amounts of substances should be called "crystal, grain, droplet" or similar. Properties should only be assigned to those terms, not to particles: the sulfur particle is not yellow.

Valence. We still use the term "valence" in a way that was historically helpful for hypotheses of the composition of substances. In the nineteenth century, chemists believed that all compounds are built of molecules. "Aluminium is trivalent" was understood as "Al atoms have three valences according to Kekulé's valance theory from 1865." Corresponding to these ideas the symbols AlF_3 or Al_2O_3 were written like for molecular compounds – in 1865, the ions were unknown.

This general use of the term "valence" is useless today and therefore expendable. The term "ion" and the *charge number* of the ion can be used for ionic compounds: starting from Al^{3+} ions, F^- ions, or O^{2-} ions we can describe corresponding symbols like $\{(Al^{3+})_1(F^-)_3\}$ or $\{(Al^{3+})_2(O^{2-})_3\}$ and shorten these symbols – if necessary – to AlF_3 or Al_2O_3.

Nonmetal atoms can be described in their molecules with a special *bonding number*: C atoms have the bonding number 4, O atoms have the bonding number 2 and H atoms 1. Molecular building sets are based on this idea; molecular structures can even be predicted using bonding numbers. The terms "charge number" and "bonding number" should be used until the nucleus-shell model will show further details of ionic and molecular structures.

Energy. Three terms have to be distinguished: energy, enthalpy, and free enthalpy. The scientific term "energy" is used for constant volumes, the term "enthalpy" for constant pressure conditions. So "enthalpy" should be introduced at school since measurements of volumes with a syringe for example normally run under constant pressure. Usually the thermodynamical distinction is not that strict in lessons at school, the terms "reaction heat," "thermal energy," and "amount of energy" are used parallelly. The free reaction enthalpy ΔG is being discussed in the context of "spontaneous" chemical reactions. It depends on both – the reaction enthalpy, ΔH, as well as reaction entropy, ΔS, and the Gibbs–Helmholtz equation shows the relation: $\Delta G = \Delta H - T\Delta S$.

School chemistry often tries to set chemical processes apart from physical processes. But since enthalpies are being measured for melting, evaporation, or dissolution processes, which cannot be distinguished from reaction enthalpies, it does not make sense to make the differentiation between chemical and physical processes: every change of substance accompanied by energy transfer has to be called a "chemical reaction."

Different symbols are common for the qualitative specification of the energy transfer. Terms like exothermic and endothermic or $\Delta H < 0$ are possible:

$$\text{hydrogen (g)} + \text{oxygen (g)} \rightarrow \text{water (l)} + \text{thermal energy} \qquad (7.1)$$

$$\text{hydrogen (g)} + \text{oxygen (g)} \rightarrow \text{water (l)}; \quad \text{exothermic (or } \Delta H < 0) \qquad (7.2)$$

If the thermal energy is put into one line with the substances connected by a "+" [see (7.1)], students might think that the thermal energy is involved in the reaction as some kind of "heat substance," as it was believed centuries ago. Therefore, it is advantageous to separate the substances from the energy transfer with a semicolon [see (7.2)].

If quantitative measured energies are discussed, the values may be different. The formation enthalpy of 1 mol of particles should be written as kJ/mol [see (7.3)], but generally reaction enthalpy is indicated by the unit kJ only because the enthalpy is connected to the number of moles as given by the reaction equation [see (7.4)]:

$$H_2\,(g) + \tfrac{1}{2}\,O_2\,(g) \rightarrow H_2O\,(l); \quad \Delta H = -285\,kJ/mol \tag{7.3}$$

$$2H_2\,(g) + O_2\,(g) \rightarrow 2H_2O\,(l); \quad \Delta H = -570\,kJ \tag{7.4}$$

Acids and bases. There are at least three different levels for the definition of the term "acid" (and also three levels for the term "base"), according to the historical development:

1. The oldest definition was *structure-related*: "An acid is a substance that colors the indicator litmus red or that dissolves limestone by the generation of a gas" (Boyle, 1663).
2. The next definitions were *structure related* after the composition of different acid molecules were analyzed: "acids are hydrogen compounds (i.e. CH_3COOH), the H atom can be replaced by a metal atom" (i.e. CH_3COONa, Liebig, 1838), or "acids dissociate in water, in an aqueous solution there are $H^+(aq)$ ions and acid-rest ions." (Arrhenius, 1884)
3. At last there is a *function-related* definition: "particles that can donate protons are called acids" (Brønsted, 1923): particles like the H_2O molecule or the HSO_4^- ion can be an acid or a base particle depending on the reaction partner, the acid particle in hydrochloric acid is not the HCl molecule but the $H_3O^+(aq)$ ion.

The Arrhenius definition was a big improvement to all earlier definitions: it stated that $H^+(aq)$ ions in acidic solutions are responsible for the acidic properties; in alkaline solutions the $OH^-(aq)$ ions are responsible for the alkaline properties. Therefore, this definition still appears in many of today's curricula in school.

But there are open questions. It was entirely unclear why some acid molecules were ionized completely, whereas some acids only ionized to a certain degree and had to be given a "degree of ionization." The substance ammonia and its alkaline solution in water could not be explained with this definition: OH^- ions do not split off from NH_3 molecules. Therefore, ammonium hydroxide NH_4OH was invented and the aqueous solution was explained with the corresponding ions. However, this substance does not exist as a solid and cannot donate hydroxide ions by dissociation in water, like other solid hydroxides do. Even today one can find old containers with the label "NH_4OH" or "ammonium hydroxide" for a fictitious substance.

The Arrhenius definition was limited to the solvent water. A number of substances were familiar that could indicate acids and bases with an indicator, but did not exist in aqueous solution. The neutralization of ammonium salts and metal amides was shown in liquid ammonia, using common indicators. It can be expressed in analogy with neutralization in aqueous solution:

$$NH_4^+ + NH_2^- \rightarrow 2NH_3$$

This problem could be resolved with the Brønsted definition: an excellent example for the extension of mental models on one side and for the coexistence of two theories on the other side. The Arrhenius definition is still valid today for

Table 7.2 Acid particles in several substances

Substance	Acid-particles	Spectator ions
Hydrochloric acid (aq)	H_3O^+ (aq) ions	Cl^- (aq) ions
Sulfuric acid (aq)	H_3O^+ (aq) ions	SO_4^{2-}(aq) ions, HSO_4^-(aq) ions
Pure sulfuric acid (l)	H_2SO_4 molecules	–
Sodium hydrogen sulfate (s)	HSO_4^- ions	Na^+ ions
Sodium dihydrogen phosphate(s)	$H_2PO_4^-$ ions	Na^+ ions

aqueous solutions and should not be called "false." However, that problem needs to be discussed with the students in chemistry lessons.

According to the Brønsted definition, it has to be emphasized that the term "acid" does not mean a substance, but an acid particle; some examples show this (see Table 7.2). The H_3O^+ (aq) ions are written with the intention to show particles that can donate one proton and leave H_2O molecules; sometimes the symbol H^+ (aq) ion is taken because it is shorter to write.

For alkaline solutions, these two levels can be distinguished by using different terms. One can speak of sodium hydroxide solution or sodium hydroxide (solid) with the meaning of a substance. The term "base" instead should be used for particles that can accept protons: OH^- ions or NH_2^- ions or NH_3 molecules. For acids, one has to deduce the meaning of the term "acid" as a substance or as a particle from its context. Particles cannot generally be categorized into acids and bases. Depending on the reaction partner, some particles react as acids as well as bases: H_2O or NH_3 molecules, OH^- or HSO_4^- ions. These are also called amphiprotic. For acid–base reactions, it is useful to symbolize protolysis and acid–base pairs.

Isomerism. Substances with the same molecular formula, but different molecular structures, are called isomers. There are two main forms of isomerism: constitutional and stereoisomerism. The first one can be subdivided into positional isomerism, tautomerism, valence, as well as functional isomerism (=constitutional isomerism) and geometric isomerism (*cis–trans* isomerism), optical as well as conformational (rotational) isomerism (=stereo isomerism) (see Fig. 7.2). Stereoisomers, whose molecules contain several asymmetrical carbon atoms and that are no optical isomers are called diastereoisomers (for example, α-D-glucose and β-D-glucose).

7.2   Teaching Processes: From Everyday to Expert and Symbol Language

"Every lesson has to start with the child's experience." (Dewey) "New experiences that children gain during lessons are being organized with the help of pre-existing concepts." (Ausubel)

These famous statements are nowadays being confirmed empirically by educational psychologists who support constructivism. Individuals build their own

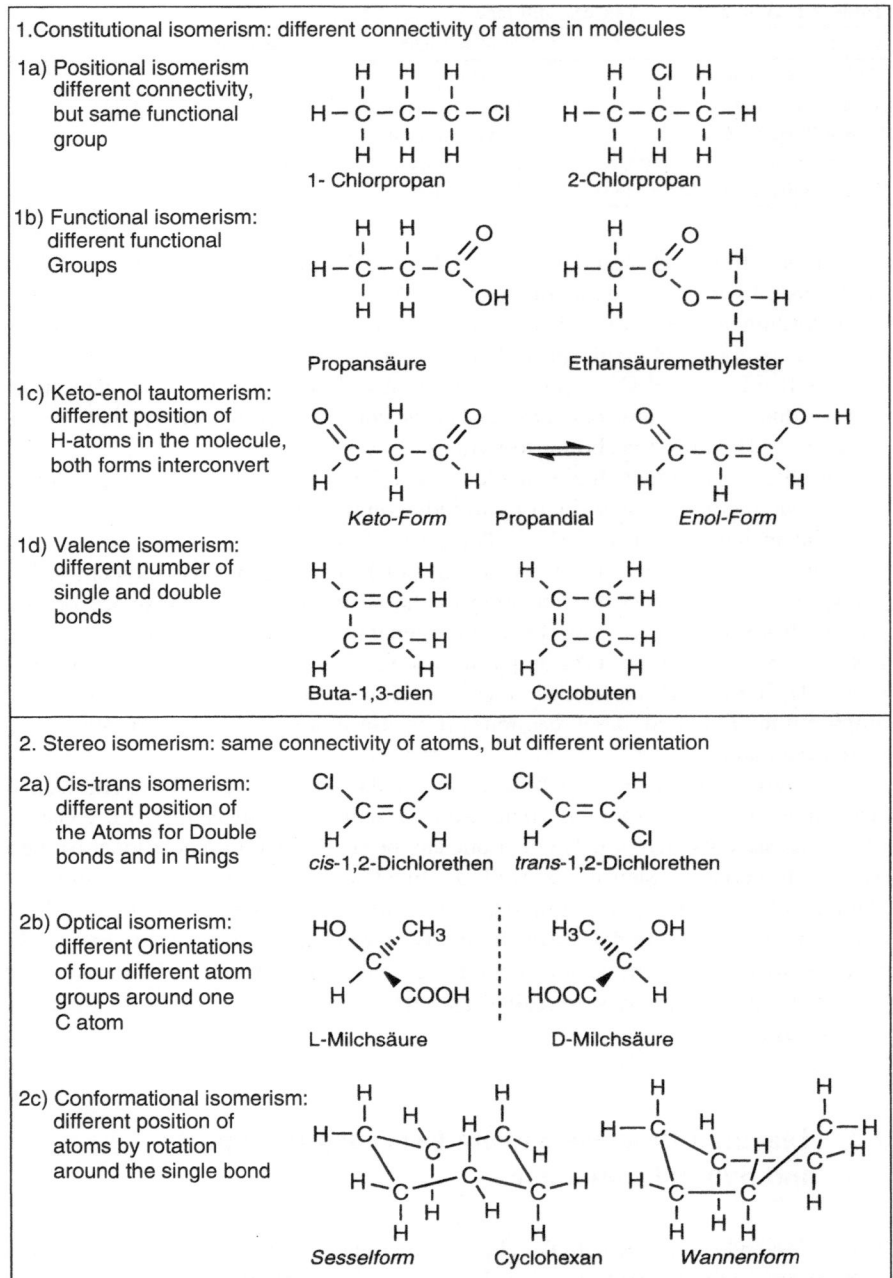

Fig. 7.2 Different forms of isomerism and corresponding terminology [5]

Table 7.3 Description of chemical processes with everyday, expert and symbol language

Everyday language	Scientific language	Symbol language
Lime dissolves in water until something lies on the bottom undissolved	Saturated solution of Calcium hydroxide is at equilibrium with the excess solute; solubility at 20°C: 0.96 g/L water	$\{Ca^{2+}(OH^-)_2\} \leftrightarrows Ca^{2+}(aq) + 2\,OH^-(aq)$ $L(Ca(OH)_2) = 8 \times 10^{-6}\,(mol^3/L^3)$ $c_{(Ca^+)} = 0.013\,mol/L$ $c_{(OH^-)} = 0.026\,mol/L$ (Normbed.)
Lime water tastes soapy	Calcium hydroxide solution colors universal indicator paper blue, pH-value is higher than 7	For 0.005 molar calcium hydroxide solution is valid: $c_{(OH^-)}$=0.01 mol/L, pH = 12.0
Limestone	Calcium carbonate	$\{Ca^{2+}CO_3{}^{2-}\}$, $CaCO_3$
Burnt lime	Calcium oxide	$\{Ca^{2+}O^{2-}\}$, CaO
Production of slaked lime	calcium oxide (s) + water → calcium hydroxide (s); exothermic	$CaO + H_2O \rightarrow Ca(OH)_2$; $\Delta H < 0$ $O^{2-} + H_2O \rightarrow 2\,OH^-$; $\Delta H < 0$ (acid-base reaction)

cognitive structure on the basis of previous experience and existing knowledge. These educational facts apply especially to the extension of everyday language to scientific terminology and ultimately to chemical symbols. An example of "lime reactions" should show related differences (see Table 7.3).

7.2.1   Connecting Everyday Language and Scientific Terminology

Initially phenomena should be described in everyday language and reaction symbols should be put into words [see (7.1) and (7.2)]. Molecular and structural formulae can follow later (see Table 7.3). If a reasonable amount of terminology has been learned successfully, corresponding terms should be used when describing new phenomena. Existing knowledge can thereby be connected to these new phenomena and the student's existing cognitive structures can be extended. When the students are familiar with the terms "acid" and "base," "acidic" and "alkaline," or "hydronium ion and hydroxide ion" from the Arrhenius concept, these terms can be taken up again for the introduction of the Brønsted concept. The terms will be connected to the smallest particles that donate or accept protons. When the students are familiar with the terms "donor" and "acceptor" from acid–base reactions, they can also use them for redox reactions and connect both types of reactions with "donor–acceptor reactions."

Concept maps. Sumfleth [6] described the networking of fundamental terms of chemistry lessons, built such networks, and analyzed them. Figure 7.3 shows two of these concept maps for systematizations regarding "acid–base reactions, redox reactions, and donor–acceptor reactions." These schemata can only be helpful when the students realize and understand the different meanings of the words and the arrows between them. Students can be asked in a quiz to build a concept map. Students have to organize given terms and put arrows between interrelated terms

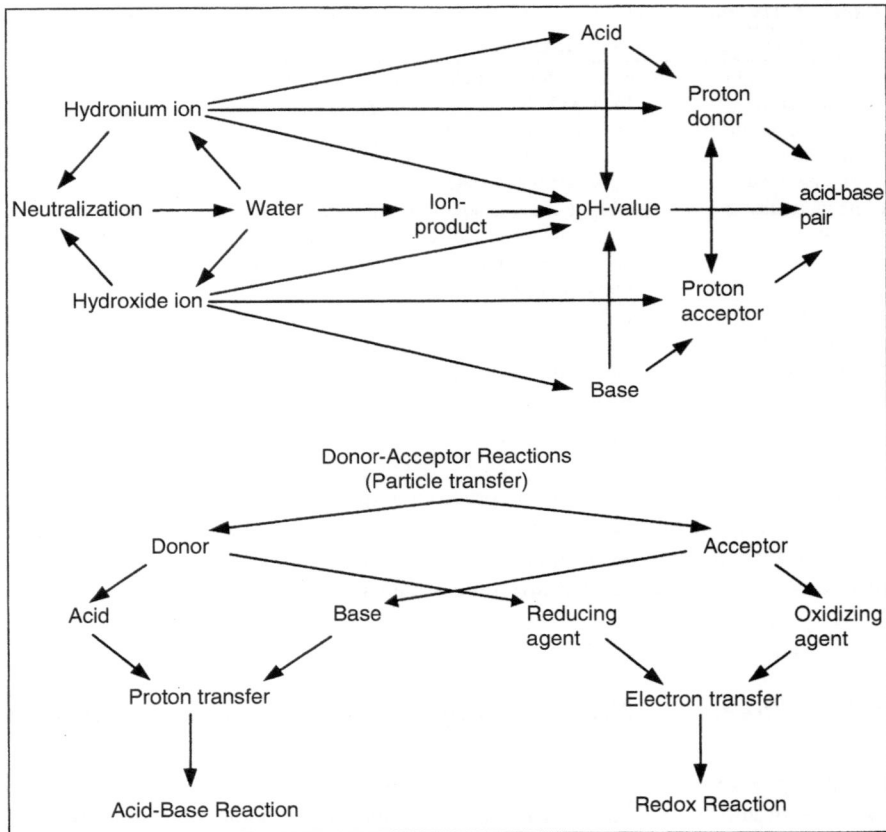

Fig. 7.3 Concept maps according to donor–acceptor reactions [6]

and label them. Figure 7.3 shows a demanding network of "acid–base reactions." Behrendt [7] gives further examples for useful concept maps.

Spiral curriculum. The extension of the meaning of certain terms, models, and symbols as well as the progressive abstraction is often called the spiral curriculum approach: the beginning of a spiral should always include the use of everyday language; the progress on the spiral should ensure the connection of familiar terms with new terms. Figure 7.4 shows examples for "solubility" and "acids": in grade 8, one starts with everyday expressions like "sour taste," vinegar, and citric acid, goes on in grade 10 to indicators and pH, to terms like acidic, alkaline, and neutral, points out the neutralization by the reaction of H^+(aq) ions and OH^-(aq) ions, finishes in grade 12 with Brønsted acids and bases, ionic product of water and pH, strong and weak acids, buffer, etc. For a better coordination of topics and teaching units, Schmidkunz and Buettner described "chemistry lessons in a spiral curriculum." [8]

Levels of terms. Working with mental models (molecular structure, ionic structure, etc) and concrete models (packing of spheres, crystal structure models)

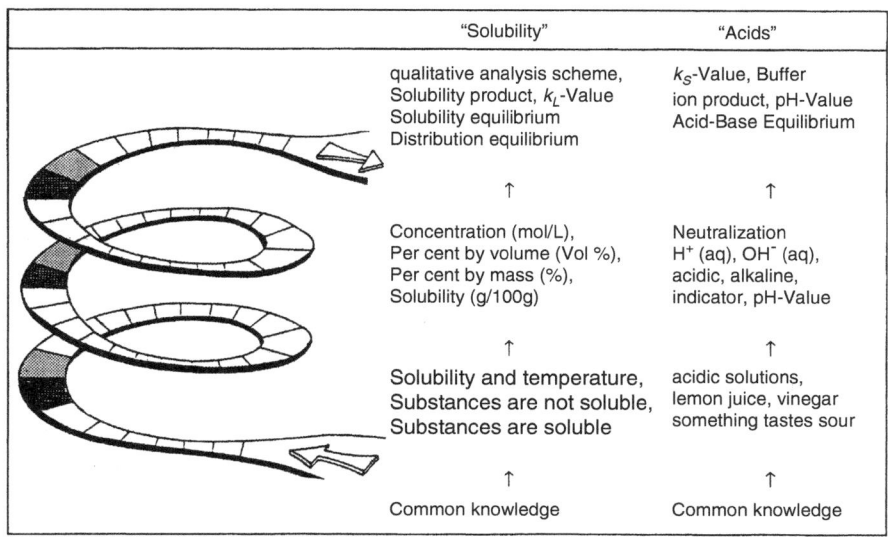

	"Solubility"	"Acids"
	qualitative analysis scheme, Solubility product, k_L-Value Solubility equilibrium Distribution equilibrium	k_S-Value, Buffer ion product, pH-Value Acid-Base Equilibrium
	↑	↑
	Concentration (mol/L), Per cent by volume (Vol %), Per cent by mass (%), Solubility (g/100g)	Neutralization H^+ (aq), OH^- (aq), acidic, alkaline, indicator, pH-Value
	↑	↑
	Solubility and temperature, Substances are not soluble, Substances are soluble	acidic solutions, lemon juice, vinegar something tastes sour
	↑	↑
	Common knowledge	Common knowledge

Fig. 7.4 Terms "solubility" and "acids" in the spiral curriculum from grade 8 to 12

Reality (Original)	Conscious (Mental model)	Reality (Concrete model)
Solution of Sugar in Water	only sugar particles mixed with water particles rapidly moving	
substance, sugar, water, sugar solution, rock sugar crystals, beet sugar, powdered sugar, sweet taste, sirup, melted sugar	particles, particle size movement of particles, sugar particles, water particles, arrangement of particles, crystal structure, C-, H- and O-atoms in the sugar molecule	sphere, ball, disk, cross small and big balls, moving balls, red and blue color, structural model, packing of spheres, molecular model

Fig. 7.5 Terms on the level of substances, mental models and concrete models

has been discussed in Chap. 6, Steinbuch's scheme [9] was illustrated there (see Fig. 6.2). According to the three different levels related terms should be differentiated: terms of the substance level, of the mental models level, and the concrete models level. Figure 7.5 shows one example for the solution of sugar in water and the interpretation on the basis of a particle model of matter. The terms of one level should be connected consequently during a class discussion: sugar crystals, their dissolution in water and the sugar solution are discussed on the substance level. Rapidly moving sugar particles and water particles can work as mental models. Red balls and blue crosses can be used for a model drawing on the concrete models level. Absurd combinations and wrong mental models like "sugar solution consists of red balls and blue crosses" or "sugar particles are red balls and taste sweet," would result if terms from different levels are being mixed.

7.2.2 Chemical Symbols

"When we ask adults about what they still remember from their chemistry lessons at school, we see a problem that our students today also have. Usually we get the following answer: 'oh, there were formulas'. And people are proud when they still know what H_2SO_4 is – not what it means: more chemical knowledge did usually not survive. A senior administrator on a board of education, who had studied science, once said to me: 'Do not tell me about the educational value of chemistry – chemistry means only formulas!'" [10]. Scheible [10] states elsewhere: "The chemical formula has discredited us." These comments show that it seems to be difficult to raise sufficient understanding of chemical symbols or formulae during chemistry lessons.

Chemistry lessons without formulae. The problems of symbol language can be avoided in beginning lessons by some simple methods: using descriptions from everyday life, names for substances instead of formulas, verbal reaction equations with the symbols (s), (l), (g) or (aq):

hydrogen chloride (g) + water (l) → hydrogen chloride (aq);   exothermic

hydrochloric acid + magnesium (s) → magnesium chloride (aq) + hydrogen (g); exothermic

The difference between the symbols (l) and (aq) has to be explained by pointing out the difference between pure liquid hydrogen chloride (l) with a boiling temperature of $-85°C$ and the aqueous solution of hydrogen chloride, which is commonly called hydrochloric acid.

Symbols on the level of the particle model of matter. In addition to verbal reaction symbols one can introduce concrete models on the basis of the particle model for a better understanding of chemical processes (see Fig. 7.5). It is necessary to add the names of the particles in the drawing: sugar particles, water particles ("H_2O particles" are a contradiction because the composition "H_2O" only should be explained on the level of Dalton's atomic model).

Close-packed arrangements of spheres can be built during chemistry lessons to demonstrate the spatial arrangement of metal particles like silver particles or copper particles in crystals (described in Chaps. 6 and 10). Also sugar particles can be taken and their arrangement in crystals can be discussed by comparing the sugar crystals of the same shape ("rock sugar").

Symbols on the level of Dalton's atomic model. Dalton's philosophy, published in 1808, includes the connection of the term "element" with the idea of the "atom": there exist as many different kinds of atoms as there are elements, the atoms of different elements differ in their masses. First atomic mass tables were developed and first atomic symbols used (see Fig. 5.1). Dalton led the search of the Greek philosophers for the basic units of matter to first results with his ideas. He suggested new compounds through a combination of atoms; unfortunately he assigned the molecular symbol HO to the compound water resulting in the mass of $1 + 8$ in first atomic mass tables (instead of $2 + 16$). Avogadro's hypothesis together with the experimental results of Gay–Lussac led to the symbol H_2O for a water molecule (see Fig. 7.6).

In the easiest case combinations of metal atoms of one kind form a closest packing: balls can be piled up as packing of spheres with the coordination number 12 (see Chaps. 6 and 10). It is also possible to substitute metal atoms in a metal structure by other metal atoms: crystals of alloys can be described in this way. They can be illustrated with packing of similar-sized balls and different colors (see details on models of metal structures and alloys in Chap. 10).

Combinations of nonmetal atoms mostly lead to molecules: H_2, O_2, Cl_2, P_4, S_8, for elements, and H_2O, HCl, H_2SO_4, P_4O_{10}, or H_3PO_4 in compounds. In molecular model kits the bonding number of nonmetal atoms is usually represented by push buttons or similar on the balls. They can be used to build molecular models in the form of ball-and-stick or calotte models or to demonstrate the regrouping of atoms in chemical reactions (see Fig. 7.6). Many reactions of organic school chemistry, similar to the previous example, can be illustrated with molecular models, visualized with drawings and symbolized with structural symbols.

Ions would have to be added to the atoms to complete the "kit of the basic units of matter" according to current knowledge. The students get an illustration of important atoms and ions in the form of a periodic system as shown in Fig. 7.7

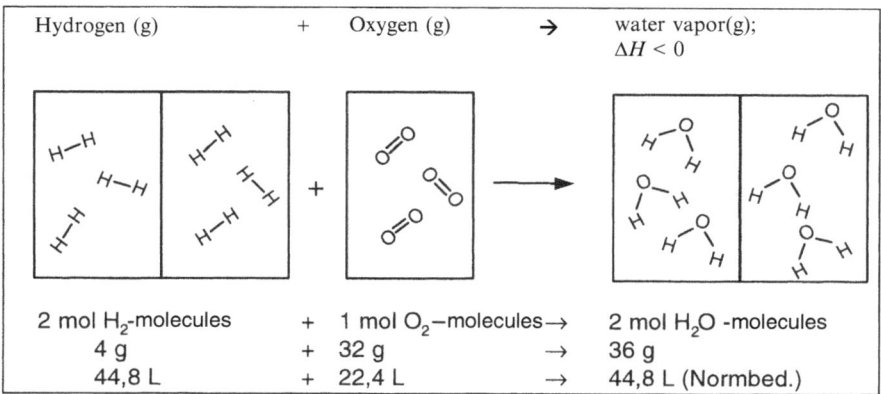

Hydrogen (g)	+	Oxygen (g)	→	water vapor(g); $\Delta H < 0$

2 mol H_2-molecules	+	1 mol O_2–molecules→		2 mol H_2O -molecules
4 g	+	32 g	→	36 g
44,8 L	+	22,4 L	→	44,8 L (Normbed.)

Fig. 7.6 Mental model for regrouping atoms in molecules

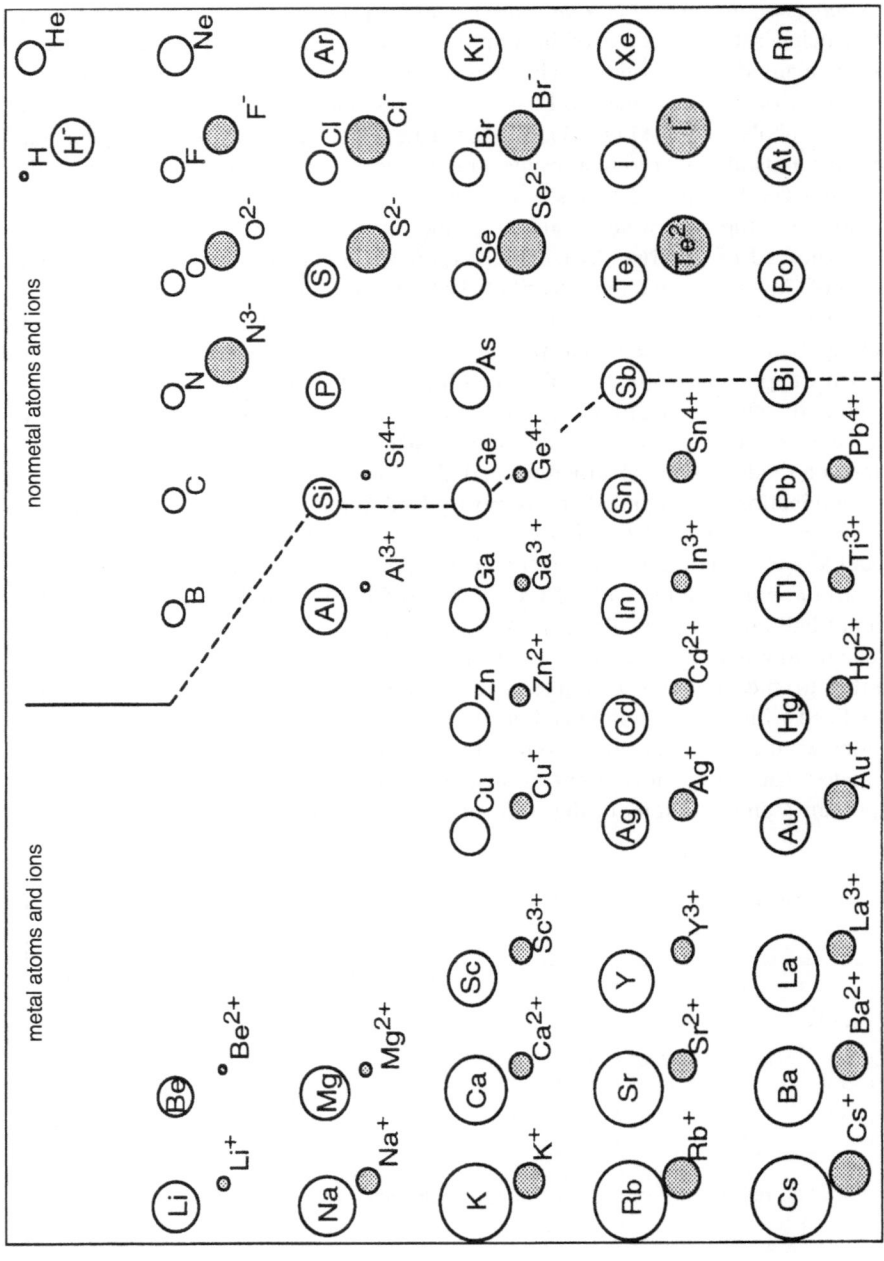

Fig. 7.7 Atoms and ions as fundamental units of matter [11]

(see Fig. 10.4, see also colored Periodic table in www.thinking-chemistry.org). Ions that were left out for reasons of clarity can be added if needed (for example Cu^{2+}, Pb^{2+}). Salt structures can be developed by combination of the ions on the left and right side of the periodic table; they can also be built as concrete models like sphere packing (see Chaps. 6 and 10). Chemical symbols for salt crystals can be derived from the charge number of the ions: this way the students theoretically combine Al^{3+} ions and O^{2-} ions to an ion lattice with an ion ratio of 2:3, they may find familiar symbols like $\{(Al^{3+})_2(O^{2-})_3\}$ or Al_2O_3 (Table 7.4).

In general it is possible to combine atoms and ions systematically with a few specific guidelines:

– Metal atoms "left and left in PSE" can be combined with undirected bonding to giant structures: models of metal and alloy crystals, visualized by packing of spheres (see Chap. 6).
– Nonmetal atoms "right and right in PSE" bond with directed bonding with particular bonding numbers, resulting in molecules or atomic structures (see Chap. 6).
– Ions "left and right in PSE" bond undirected to giant structures of ions with specific charge numbers, so called ionic structures visualized by special sphere packing (see Chap. 6).

On the basis of Dalton's model chemical reactions can only be described in those cases where the nature of particles does not change. Metal–metal reactions to alloys, nonmetal–nonmetal reactions to volatile substances, or reactions between salt solutions with precipitation: for example the precipitation of insoluble green iron hydroxide from corresponding salt solutions (see Fig. 7.8). In these cases the quality of particles does not change during the reaction: atoms remain atoms, ions

Table 7.4 Rules for combining metal atoms, nonmetal atoms and ions (see Fig. 7.7)

Places in the periodic table	Particle quality	Bonding type	Chemical structure
Left + left in PSE	Metal atoms	Spatially undirected	Metal lattice
Right + right in PSE	Nonmetal atoms	Spatially directed	Molecule
Left + right in PSE	Ions	Spatially undirected	Ionic lattice

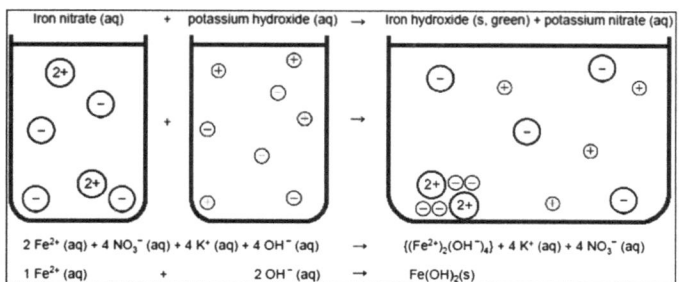

Fig. 7.8 Models and symbols of the precipitation of iron hydroxide from salt solutions

remain ions. Redox reactions of metals with nonmetals or electrolyses of salt solutions cannot be described with this model, because atoms transform into ions or vice versa. These changes can only be explained after the introduction of the nucleus-shell model, including the knowledge about electrons in subshells.

Symbols on the level of the nucleus-shell model of the atom. As long as the main emphasis is on chemical structures involving atoms, ions and molecules, they can be illustrated by concrete models on the level of Dalton's model – a sophisticated atomic model is not necessary. Once we are not only dealing with spatially directed and undirected bonds, but also with differentiated chemical bonding, Rutherford's atomic model including the shell model or the energy level model cannot be avoided. On the one hand terms like nucleus, proton, neutron and isotope have to be introduced, on the other hand also the idea of electrons, electron clouds, electron pair and the electron pair repulsion. The discussion of chemical bonds leads to the concept of ionic and covalent bond, may be connected to the term electronegativity. Also polarized bonds, hydrogen bonds or van-der-Waals bonds can be illustrated, and dipolar molecules should be introduced.

On the basis of the nucleus-shell model it is possible to describe structure and bonding of simple chemical structures derived from the first three periods of the periodic table, as well as chemical reactions in which particle changes can be explained:

Acid–base reactions $\rightarrow$ transition of protons
Redox reactions $\rightarrow$ transition of electrons
Complex reactions $\rightarrow$ transition of ligands

Terms on the level of the atomic orbital model. Simple nucleus-shell models are sufficient as long as the structure of substances with elements from the first three periods is being discussed: successive insertion of electrons from shell to shell can be explained until the element calcium. The differentiation of the main shells into subshells becomes necessary as soon as the build-up of atoms from the subgroups or transitional metals should be explained: s-, p-, d- and f-levels and the distribution of electrons to these subshells.

The theory of wave mechanics needs to be discussed by the wave-particle duality of electrons and by understanding electrons as standing waves. In this sense a certain energy state can be assigned to a special standing wave and described with quantum numbers: principal quantum number n, subsidiary quantum number l, magnetic quantum number m, spin quantum number s.

Starting from the quantum number special energy states can be calculated with the help of wave functions. These calculations lead to specific probability densities or probabilities of finding electrons: s-, p-, d-, and f-electron clouds or orbitals, atomic and molecular orbitals as well as bonding or antibonding orbitals need to be differentiated. It needs to be pointed out that these orbitals are only obtained by mathematical combinations of wave functions and compared to measurements of the laboratory: densities of single electron clouds cannot be measured; only the overall electron density of a molecule or of an atom. On the other hand structure and bonding of electron systems can be predicted by mathematical combinations of wave functions: molecular modeling and molecular design are successful with current computing power.

It should be emphasized that it is difficult to make the orbital model understandable for students, because this model is based on mathematical assumptions and formal combinations of wave functions: electrons have to be regarded as mathematical terms in this context. It can neither be understood as a wave nor as a particle – depending on the chosen experimental instruction; one can use either one or the other description. The wave-particle duality thus reveals that the electron neither solely reacts as a wave nor solely as a particle. Due to these problems the teacher has to consider choosing phenomena, which can be explained with simpler models – the quantum mechanical model of the atom should only be taken into account when experimental results cannot be explained otherwise.

7.2.3 Deduction of Chemical Symbols at School

The introduction and usage of chemical symbols at school plays an essential role due to the importance of chemical symbols as a unique communication tool for chemists. Methods for the introduction have been discussed extensively in all chemistry educational journals over time. The historical–empirical derivation of formulae from the comparison of mass ratios and atomic masses on one hand and the structure-oriented approach on the other hand (see Chap. 10) are those most discussed for chemical education today.

Historical–empirical derivation of chemical symbols. It is possible to deduce empirical formulae from mass ratios of reacting substances and comparison of atomic masses on the level of Dalton's atomic model. One takes the principle "counting by weighing" and finds in the laboratory, for example, 1.00 g of titanium reacts to form 1.67 g of titanium oxide. One can deduce:

$$M\,(\text{titanium}) = 47.67\,\text{g/mol} \Rightarrow 47.67\,\text{g titanium contain } N_A = 6.023 \times 10^{23}$$

$$\text{Ti atoms}$$

$$\Rightarrow 1.00\,\text{g titanium contains } 0.126 \times 10^{23}\,\text{Ti atoms}$$

$$M\,(\text{oxygen}) = 16\,\text{g/mol} \Rightarrow 16\,\text{g oxygen contain } N_A = 6.023 \times 10^{23}\,\text{O atoms}$$

$$\Rightarrow 0.67\,\text{g oxygen contain } 0.252 \times 10^{23}\,\text{O atoms}$$

For the composition of titanium oxide follows: $N_{\text{Ti atoms}}{:}N_{\text{O atoms}} = 0.126{:} 0.252 = 1{:}2$

The ratio of atoms and thereby the composition can be stated in an empirical formula after the conversion of masses (in g) into the absolute number of atoms (see example) or amounts of substance (in mol). This way of analysis is abstract, because the calculation of empirically determined mass ratios with theoretical atomic masses leads to the atomic ratio: most young students are overchallenged by this method.

Kaminski and Jansen [12] therefore propose the number of atoms in 1 mg of substance to calculate formulae: for example 1 mg of carbon contains of 502×10^{17} C atoms, 1 mg of oxygen contains of 376×10^{17} O atoms. After having the numbers of reacting atoms the calculation of formulae is simplified: if laboratory measurements show that 10 mg of carbon react with 26.7 mg of oxygen the numbers of $5,020 \times 10^{17}$ C atoms and $10,040 \times 10^{17}$ O atoms can be calculated very easy and the CO_2 symbol derived. But one problem remains: naturally the numbers are of the order of 10^{17} and young students are not used to work with this kind of big numbers.

Historically analytical balances were used to deduce chemical symbols. Corresponding analysis methods have been developed by Liebig and Berzelius in the mid-nineteenth century and were in use for a long time. Methods of instrumental analysis became available in many chemical institutes step by step from the 1960s: spectral analysis, gas chromatography, X-ray structure analysis, atom absorption spectroscopy, UV-, IR-, NMR-spectroscopy and mass spectrometry. Today analyses of substances are carried out with these instrumental analysis methods.

Deduction of symbols from structural models. Experts are able to identify the structure of crystalline substances with the help of X-ray structure analysis. Lattice constants as well as bond angles and bond lengths can be determined using dedicated software; even spatial illustrations of the chemical structure can be printed out.

The concept of X-ray structure analysis should be interesting for students if some exemplary Laue diagrams are shown (see Fig. 5.3) with the corresponding diffraction lattices being illustrated with laser beam experiments [13]. If students are given either a structural model of a molecule or a unit cell of a crystalline structure, they are able to deduce the empirical formulae from the models by counting the ratio of atoms or ions in these (see Chaps. 6 and 10). This instructional method fits the educational demand to add the level of structural models to facilitate the understanding of chemical symbols [14]. This is schematically described in Fig. 3.8: phenomena (substances and reactions) are introduced first, then corresponding structural models, and finally derived formulae and chemical equations.

This sequence is in tune with Johnstone's "Chemical Triangle" [15], concerning the connections of substances, structural models and chemical symbols (see Fig. 7.9): the Macro level shows substances and reactions and all that "can be seen, touched and smelled," the submicro level contains all considerations of involved atoms, ions, molecules and chemical structures, while the representational level expresses symbols, formulae, equations, calculations, tables and graphs. In this sequence the introduction of chemical formulae should be possible.

Johnstone also warns that mixing these levels has to be done with care: "It is psychological folly to introduce learners to ideas at all three levels simultaneously. Herein lies the origin of many misconceptions. Trained chemists can keep these three levels in balance – but not the learner" [15]. Gabel points out that a big mistake in chemical instruction seems to be the jump from the macro level direct to

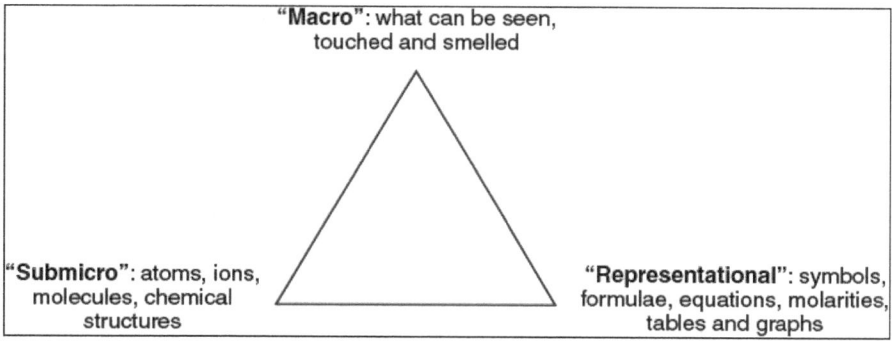

Fig. 7.9 Johnstone's triangle for chemical education [15]

Fig. 7.10 Beaker diagram
for the neutralization of an
acidic solution

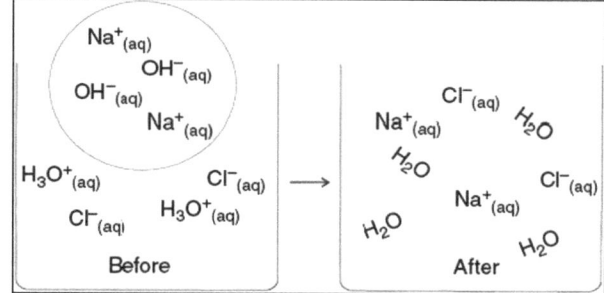

the representational level when she writes: "The primary barrier to understanding chemistry, however, is not the existence of the three levels of representing matter. It is that chemistry introduction occurs predominately on the most abstract level, the symbolic level" [16].

One example for these difficulties is the way the neutralization of acids and bases is often introduced: the teacher shows that the indicator color changes when adding drops of sodium hydroxide solution to a sample of hydrochloric acid (macro level) and writes the equation $HCl + NaOH \rightarrow NaCl + H_2O$ (symbolic level) on the blackboard. Students asked about the particles which are involved in the neutralization reaction mostly state: "HCl molecules or NaOH molecules" – the direct way from the phenomena to chemical symbols develops misconceptions. Instead a "beaker model" (see Fig. 7.10) with hydrated ions in both solutions can be drawn after observing the phenomena. Students will discover that only hydronium ions react with hydroxide ions to form water molecules (submicro level) and would develop modern chemistry understanding without misconceptions [14]. Empirical research also shows that (aq)-symbols are very useful: students understand that ions are not attracted to each other and bond to molecules because the hydration shells of the ions compensate the ionic charges and the hydrated ions move like "billiard spheres on a table."

Instruction with structural models equals modern structure analysis: in all cases results lead to smallest structural units of substances. For molecular substances the molecule represents such a unit and the chemical formula is stated for the molecule. The symbols for acetic acid or the benzene molecule are CH_3COOH and C_6H_6, respectively. These symbols would not be shortened to CH_2O and C_1H_1 or CH, respectively. This agreement should be transferred to symbols for ionic solids. The smallest structural unit should also be symbolized there – the unit cell. Symbols for the unit cell of sodium chloride should be $\{(Na^+)_4(Cl^-)_4\}$ or Na_4Cl_4 (see Fig. 6.3). The teaching process may start with demonstrating models of unit cells and counting of ions (see Chaps. 6 and 10). Symbols like Na_4Cl_4, Li_8O_4 or Zn_4S_4 should be derived for the smallest structural units – similar to the symbols CH_3COOH or C_6H_6 for molecules as smallest structural units of corresponding substances.

In the second step the symbols can be abbreviated to the most common symbols – but these lack information: NaCl, Li_2O or ZnS. In this way students learn about the importance of formulae and their connection to chemical structure and gain a modern concept of chemistry. Tests on spatial ability have shown that the majority of middle school students is able to spatially identify unit cells drawn in two dimensions [17] and to count the number of ions successfully [18].

7.3   Learner–Student Concepts of Structures and Symbols

Every student knows abbreviations like PC or MP3. Therefore students might think that chemical symbols are similar formal abbreviations of substance names: examples like NaCl, CaO or MgO support this concept. If students do not have a structural concept, they are not able to understand the subscripts in symbols like H_2O or Al_2O_3 and just learn them by heart. They keep their incorrect or contradictory concepts and see the chemical symbols as some kind of secret language.

Concepts of combustion. In Chap. 1 we introduced a student's mental model for the combustion of magnesium as an example. The student formulated the correct reaction symbol "2 Mg + O_2 → 2MgO," but his real concept was: "Magnesium consists of two kinds of particles. One vaporizes during combustion, the other remains as magnesium oxide." He also made a drawing accordingly (see Fig. 1.6). Empirical research with about 300 grade 9 and 10 students found that almost all of them wrote down the correct reaction equation, but 70% of them had inadequate mental models and produced wrong drawings [19]. It is evident that mental models from everyday life, which students construct by observation over many years, cannot be transferred to scientific concepts just by the formulation of reaction symbols. Only models of the structure of substances before and after combustion make the reaction equation understandable and can support the understanding of combustion processes [14].

Concepts of the term "ion." The following test was conducted with a group of high school students in grade 10 that already knew the terms "ion" and "electron transfer through redox reaction." In addition this group was able to arrive at chemical equations for demonstrated precipitation reactions. For the questionnaire the following experiments were carried out:

1. The reaction of nickel oxide with aluminium and the corresponding observable bright flash of light was shown to the students. The students were asked to explain their observations, to draw a model of the structure of nickel oxide and aluminium crystals and to write down chemical equations in words, structures and formulae.
2. Concentrated solutions of calcium chloride and sodium sulfate were mixed and a white precipitate could be observed. The students were asked to make draw models of both solutions and write the reaction equations in words, structures and formulae. The students knew from previous examples that "structures" require labeling with ionic symbols in the case of ions, while in case of molecules it means labeling molecular structures.

Only a few students were able to completely solve these problems. Statistics show that nearly 100% of all students wrote down right reaction symbols in words, but only 20% formulated correct structures or formulae [4]. Up to 80% have mental models as illustrated in Fig. 7.11. The main result of this study is that students often do not write down ions of familiar metal oxides, but switch to molecules or mix ions and molecules (marked with a dashed box). Concepts of the formation of ions from corresponding atoms appear in cases, where ions already exist in salt solutions and do not have to be formed anymore. These misconceptions are "school made" and caused by deficits in the teaching processes [14]. It seems to be desirable to hand out a list of "atoms and ions as basic units of matter" to the students, as shown in Fig. 7.7.

Concepts of stoichiometry. With an empirical study on stoichiometric calculations Schmidt [20] could show that only a small part of the students acquires this ability. He noticed the following misconceptions: "No differentiation between equation coefficient and formula indices; for example between 2 O and O_2. No differentiation between amount-of-substance ratio and mass ratio, equal amounts of substance of reagent and product in gas reactions, equal volumes of reagent and product, and others" [20]. Schmidt found these misconceptions through construction and analysis of specific multiple choice questions with suitable distractors: "The distractors were built in a way that the students had to deal with numbers that fit to the correct as well as to the wrong answers, to solve the problems. For example: 2 g of a compound contain 1 g of copper and the rest is sulfur. Which chemical formula fits to this information – CuS, CuS_2, Cu_2S or Cu_2S_2?" [20].

Schmidt tried to find strategies, which led the students to make typical mistakes and get to the wrong answers: "You cannot avoid misconceptions in chemistry lessons. They should not be suppressed, but students should be aware of them: the mistakes in their strategies should be discussed together. If chemistry lessons are not successful in solving stoichiometric exercises, students might turn their back to chemistry: stoichiometry might be the crossroads, where the student's decision is

1a) Draw your model of the structure of a nickel oxide crystal and an aluminium crystal

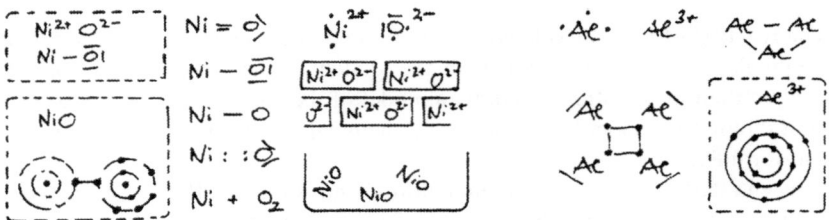

1b) Formulate three reaction equations (words, structures, formulae)

Nickel oxid (s) + Aluminium (s) ⟶ Nickel (s) + Aluminiumoxid (s)

S.o.   S.o.

2a) Draw your model of a calcium chloride solution and a sodium sulfate solution

2b) Formulate three reaction equations (words, structures, formulae)

Calciumchlorid (aq) + Natrium sulfat (aq) ⟶ Calcium sulfat (s) + Natrium chlorid (aq)

S.o.   S.o.

Fig. 7.11 Selected flawed answers from grade 10 high school students (dashed framed answers are from one single student) [4]

taken for or against chemistry – therefore it is very important to know where the difficulties come from" [20].

7.4 Human Element: Laymen's Understanding of Scientific Terminology

Scientific terminology and chemical symbols serve as an important tool for easy and rational communication, as a communication tool with high information content: it can be understood worldwide regardless of culture, language, or social system [1]. At the same time every terminology is a boundary for all those who are not familiar with it. Severe problems for communication even exist between the sciences because of their highly specialized terminology: only those people who are proficient in science's terminology are able to talk about specific issues.

Of course scientific terminology complicates communication with the public and their understanding of science [21]. In a public dialog mistrust towards the experts is usually created, when they are not able to translate scientific terminology into everyday language: this situation led to the development of popular scientific publications that try to make science understandable.

Teachers should not only see themselves as rivals to this popularization in magazines, radio, TV and on the internet, instead they can be mediators who address students' issues. They can also integrate everyday dialogs into their

Name of the salt	involved ions	ion ratio	empirical formula
Calcium fluoride	Ca^{2+}, F^-	$\{(Ca^{2+}), (F^-)_2\}$	CaF_2
Calcium nitride			
Barium chloride			
Aluminium fluoride			
Lithium oxide			
Sodium hydroxide			
Calcium hydroxide			
Magnesium nitrate			
Sodium carbonate			
Calcium sulfate			
Aluminium sulfate			
Potassium aluminium sulfate			

Fig. 7.12 Worksheet on the formulation of symbols for ionic structures

1. Cu atoms form a cubic primitive structure, Zn atoms fill the cubic holes:

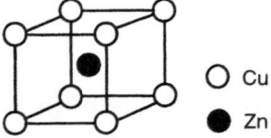

○ Cu

● Zn

2. Cr atoms form a cubic face-centered structure, N atoms fill the octahedral holes

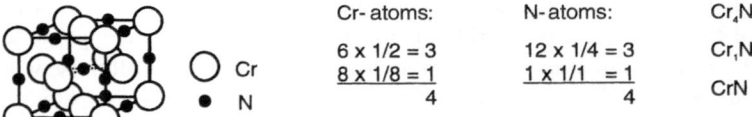

○ Cr

• N

Cr- atoms:

$6 \times 1/2 = 3$
$\underline{8 \times 1/8 = 1}$
                4

N- atoms:

$12 \times 1/4 = 3$
$\underline{1 \times 1/1\ \ = 1}$
                4

Cr_4N_4

Cr_1N_1

CrN

3. Al atoms form a cubic face-centered structure, Sb atoms fill half of the tetrahedral holes

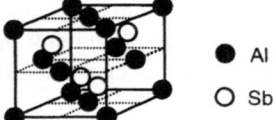

● Al

○ Sb

4. Pb atoms form a cubic face-centered structure, Mg atoms fill all tetrahedral holes

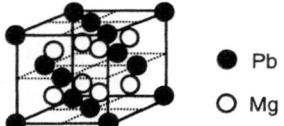

● Pb

○ Mg

5. Superstructures of copper-gold alloys

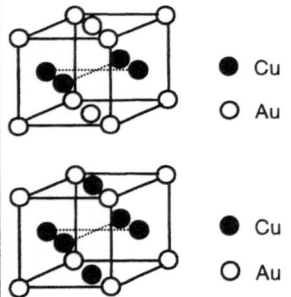

● Cu

○ Au

● Cu

○ Au

Fig. 7.13 Worksheet for developing formulae from mental models of unit cells

chemistry lessons to assist the students in understanding and critically reflecting on scientific issues, which are presented in the media [22]: a demanding goal for teaching chemical terminology.

Problems and Exercises

P7.1 Different types of symbols or formulae exist in chemistry. State different types (a) for three crystalline solids, (b) for three volatile substances. Which information hides behind the different symbols or formulae?

P7.2 For the introduction of chemical symbols it is reasonable to teach the students in everyday language first, then switch to scientific terminology and later to chemical symbols. Describe this way using three different examples.

P7.3 Chemical symbols can look very different on different levels of the spiral curriculum: there may be words, empirical formulae or structural formulae. Choose three different reactions and formulate the reaction equations on these three levels.

P7.4 Theoretically combine "ions left and right in PSE" (see Fig. 7.7). Write down the formulae of three different salt crystals and draw corresponding models of ionic structures. Determine chemical symbols on the basis of ionic charge numbers: use the worksheet in Fig. 7.12.

P7.5 Theoretically combine "metal atoms left and left in PSE" (see Fig. 7.7). Give three examples of alloys and draw your mental models. Determine chemical symbols for the metal structures from unit cells, using the examples of the worksheet in Fig. 7.13.

References

1. Zahn Pv (1981) Freund und Helfer oder heimlicher Feind ? Chemie im Kreuzfeuer der öffentlichen Meinung. CU 12:1
2. Dörrenbächer A (1995) IUPAC-Regeln und DIN-Normen im Chemieunterricht. Aulis: Köln
3. Langer S (1979) Philosophie auf neuen Wegen. Mittenwald
4. Barke H-D (1988) Chemiedidaktik zwischen Philosophie und Geschichte der Chemie. Frankfurt
5. Jäckel M et al (1998) Chemie heute Sekundarstufe II. Schroedel, Hannover
6. Sumfleth E et al (1989) Stoffe: Eigenschaften und Reaktionen. Modelle: Teilchenanordnungen und -umordnungen. Eine mit Lernhilfen gestützte Einführung in die Chemie. MNU 42:411
7. Behrendt H (1997) Concept mapping. Schülerinnen und Schüler konstruieren eigene Begriffsnetze. NiU-Physik 8:18
8. Schmidkunz H, Büttner D (1985) Chemieunterricht im Spiralcurriculum. NiU-P/C 33:19
9. Steinbuch K (1977) Denken in Modellen. In: Schäfer, G., u.a.: Denken in Modellen. Westermann, Braunschweig
10. Scheible A (1969) Ist unser Chemieunterricht noch zeitgemäß ? MNU 22:449
11. Sauermann D, Barke H-D (1998) Chemie für Quereinsteiger. Schüling, Münster

12. Kaminski M, Jansen W (1994) Die Ermittlung der chemischen Formel im Anfangsunterricht. NiU-Chemie 25:12
13. Barke H-D, Rölleke R (1999) Max von Laue: ein einziger Gedanke – zwei große Theorien. PdN-Ch 48:16
14. Barke H-D, Hazari A, Sileshi Y (2009) Misconceptions in chemistry – addressing perceptions in chemical education. Springer, Heidelberg
15. Johnstone AH (2000) Teaching of chemistry – logical or psychological? CERAPIE 1:9
16. Gabel D (1999) Improving teaching and learning through chemistry education research: a look to the future. J Chem Educ 76:548
17. Barke H-D (1993) Chemical education and spatial ability. J Chem Educ 70:968
18. Wirbs H, Barke H-D (2002) Structural units and chemical formulae. CERAPIE 3:185
19. Barke H-D (1982) Probleme bei der Verwendung von Symbolen im Chemieunterricht. NiU – P/C 131
20. Schmidt HJ (1990) Stolpersteine im Chemieunterricht. Diesterweg, Frankfurt
21. Becker HJ (1988) Verbraucherfragenim RIAS-Telefonstudio: Gegenstand fachdidaktischer Forschung? Chim Did 14:69
22. Becker HJ (1995) Ein Alltagsdialog über Jughurt – Chance für fächeraufweitenden Chemieunterricht. PdN-Ch 44:17

Further Reading

Gilbert JK, Treagust AF (2008) Reforming the teaching and learning of the macro/submicro/symbolic representational relationship in chemical education. In: Paper presented at the 19th symposium on chemical and science education, University of Dortmund, Germany
Herron JD (1996) The chemistry classroom: formulas for successful teaching. American Chemical Society, Washington, DC
Johnstone AH (2000) Teaching of chemistry-logical or psychological? Chem Educ Res Pract 1(9)
Kauffman GB (1979) Principal net equations for expressing chemical reactions. J Coll Sci Teach 9:83–85
Marais P, Jordan F (2000) Are we taking symbolic language for granted? J Chem Educ 77:1355–1357
Yarroch WL (1985) Student understanding of chemical equation balancing. J Res Sci Teach 22(5):449–459

Chapter 8
Everyday Life and Chemistry

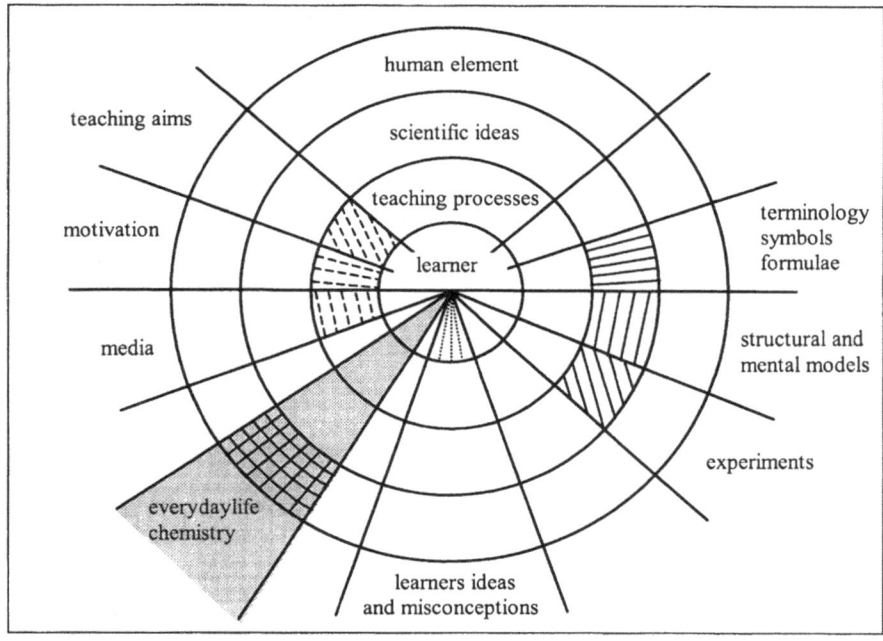

"In her chemistry lessons Sarah hears a lot about sulfur dioxide and about catalytic oxidation to sulfur trioxide, about the contact process to synthesis and the importance of sulfuric acid for the chemical industry. One evening her father reads something about the increase of acid rain in the news paper and asks Sarah: You've now had chemistry for a long time in school. Tell me – what is acid rain? How do specialists explain this phenomenon? Sarah answers: I don't know, my teacher didn't say anything about that" [1].

David Waddingon [1] caricatured chemical education from his point of view. Students of classes 9–11 of a German secondary high school also expressed similar views with their comments: "Chemistry class does not seem to be so senseless if you can use the content in everyday life; there should be more references to practice

H.-D. Barke et al., *Essentials of Chemical Education*,   217
DOI 10.1007/978-3-642-21756-2_8, © Springer-Verlag Berlin Heidelberg 2012

then chemistry class would not be an abstract complex of formulae; in particular references to everyday life are good for the back-ground and they benefit those who won't choose a profession in the chemical industry" [2].

The Society of German Chemists (GDCh) explains their point of view this way: "The mission of chemical education is to make the central significance of chemical knowledge comprehensible and tangible. The connection between chemistry and the living area of the learners has to be established and used in order to dispose them to handle the environment in a responsible way. It is important to enable learners to incorporate their chemistry knowledge into their life." [3].

The great significance of establishing or using references to everyday life is well known by most teachers, textbook authors, or guideline experts – only the amount of everyday life chemistry in the curriculum and the importance of systematic chemistry seem rather controversial. Today chemical education is much influenced by the movement of "chemistry in context" (see later in this chapter) and one generally starts with everyday life examples before the systematics of chemistry is taught. These discussions and how to teach these contexts from everyday life will be presented in this chapter.

8.1   Learner: Curiosity and Interest

Young adolescents have an intrinsic interest in learning something about them-selves as well as objects and procedures from their immediate living environment. Chemistry lessons are able to tie in with wise questions from everyday life to this curiosity and deal with it age-appropriately – questions referring to this problem will be raised and discussed.

1. *In which areas do learners have everyday life experience?* In the first instance, they certainly are experiences of the childhood home, from the kitchen, the bathroom, the garage, or the garden. Pfeifer, Häusler, and Lutz [4] combine further areas in a diagram (see Fig. 8.1) from which possible experiences of everyday life can be derived.
2. *How do specific surroundings of the learners affect their images?* On the one hand, if there is a large industrial complex, and many mothers and fathers work there, certain images of this industrial plant are developed. If students live in the countryside, they have a different attitude to questions about agriculture, fertilizers, or pesticides than those living in the city.
3. *What kind of chemical phenomena do students experience every day?* They often experience the destruction concept when they observe burning paper, wood, or coal, when they talk about ink eraser or fuel consumption by cars (see Chap. 1). They also get much information about environmental problems regarding food, water, air, and soil and develop corresponding attitudes against "chemistry." The big importance of chemistry in our lives and also the problems with our environment should be differentiated and discussed. Daily newspapers, magazines, television, and radio publish a partial view of substances and

Fig. 8.1 Aspects of everyday
life and sources of students'
experiences [4]

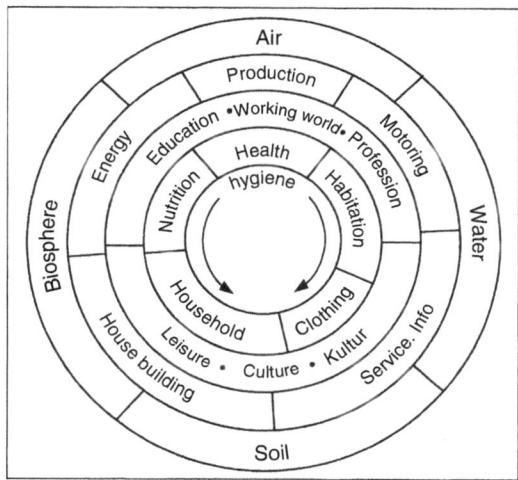

materials through their advertisement, which impress pupils in a dedicated way.
Such connections should also be identified and discussed.

4. *How can objective criticism and attitude changes be encouraged?* Chemistry
   classes, which adequately consider everyday life and environmental issues,
   would account – next to similar education by the parents – for critical abilities
   and attitude change. Well-prepared excursions to industrial complexes or purification plants and recycling companies are also qualified to revise preconceived
   opinions and to develop new images.

Students' interests. Another issue stems from questions before: What kind of
interest do pupils have in dedicated themes of everyday life and environmental
chemistry? The answer to this question can be interesting for teachers because one
intended project, often the only project in the school year, should appeal to
students' interests – otherwise the teacher perhaps teaches against the interests of
all students, or of the boys or of the girls.

To answer this question, an empirical study was planned and a questionnaire was
developed (see Fig. 8.2). It was distributed to 200 students of age 15–18 years
(classes 9–11) in 1986 at a secondary school near Hannover in Germany, and in
1995 at a secondary school in Jena, also Germany [5]. The evaluation took gender
into account to elicit special interests of boys and girls (see Table 8.1).

Some of the proposed topics are interesting only for boys or only for girls, and
many topics suggest very little interest by boys as well as girls (see Table 8.1).
Every teacher, if possible, should arrange their own specific survey on location
because the range of interests is greatly controlled by the region around the school;
and they should be able to evaluate which topic is qualified for the class or which
excursion is favored by the students.

Household chemicals and interest. A further pilot study shows how far the
interest of students can be increased if household chemicals instead of laboratory

1. In the area of **everyday life** I would like to hear more about the following topics
(mark your choices from 1 - 4; 1 for the most wanted topic, 2 for the next, etc.):

O food	O bathroom cleaner	O fuel
O conservation additives	O washing powder	O cement, mortar
O alcohol, drinks	O cosmetics	O fertilizer

2. In the area of **nature and environment** I would like to hear more about the following topics
(mark your choices from 1 - 4; 1 for the most wanted topic, 2 for the next, etc.):

O water pollution	O acid rain	O treatment of waste
O air pollution	O exhaust gas	O recycling of paper, glass
O soil pollution	O over fertilization	O waste oil and treatment

3. In the area of **chemical processes** I would like to hear more about the following topics
(mark your choices from 1 - 4; 1 for the most wanted topic, 2 for the next, etc.):

O photo production	O glue	O batteries
O galvanization	O explosives	O fuel cells
O coloring textiles	O metal alloys	O rocket engines

4. In the area of **chemical industries** I would like to hear more about the following topics
(mark your choices from 1 - 4; 1 for the most wanted topic, 2 for the next, etc.):

O steel and metals	O sugar from sugar beets	O plastics
O petrol and diesel	O salt from salt mines	O paints
O sulfuric acid	O paper from wood	O medicines

Fig. 8.2 Part of a questionnaire for finding students' interests in everyday life chemistry [5]

Table 8.1 Results of the questionnaire (see Fig. 8.2)

Great interest of boys and girls	Food, alcohol, photo production, explosives, paper of wood
Great interest of girls	Cosmetics, treatment of waste, coloring textiles, medicines
Great interest of boys	Petrol and diesel, exhaust gas, rocket engines
Mixed interest of boys and girls	Conservation, water pollution, air pollution, acid rain, recycling of paper and glass, batteries, fuel cells, steel and metals, plastics
Very low interest of boys and girls	Bathroom cleaner, washing powder, cement and mortar, fertilizers, pollution of soil, over fertilization, waste oil and treatment, galvanization, metal alloys, sulfuric acid, sugar from sugar beets, salt from salt mines.

chemicals are used in chemistry lessons. Wanjek [6] scheduled the unit "acids and bases" for several 9th classes of a comprehensive school in Münster, Germany. In student-run experiments, household chemicals were tested with universal indicator, and the usage of acidic and basic cleaners was analyzed and neutralizations of special samples of these cleaner solutions were performed. The lessons were given by different chemistry teachers of the school; they needed about 6 school hours in 3 weeks. The students in five classes were asked about their interests before and after the lectures – and also about the experiments they had carried out themselves.

The results of the questionnaire show that students' interest increases after those lectures concerning "acids and bases." Especially the girls expressed a lack of interest before the lessons, but afterward the interest increased noticeably. The effect of the experiments students carried out was even bigger: girls as well as boys expressed an interest in these hands-on experiments. Regarding the combination of household chemicals and experiments, the results show a higher interest by the girls in comparison to the boys: for the girls, the household chemicals combined with hands-on experiments are more interesting than for boys.

Attitudes to chemistry and chemistry classes. In a study with a large number of participants, Müller-Harbich, Wenck und Bader [7] discovered that students barely make differences between attitude to chemistry and chemical education: "Students who possess a positive attitude to their chemical lessons at school also have an open-minded attitude to chemistry. But especially students with an affective positive attitude to environmental problems have a negative attitude to chemistry and vice versa. The observed attitude of the girls expresses the general opinion: who is involved in ecological activities, refuses the chemistry" [7].

Heilbronner and Wyss [8] assigned students from Switzerland to paint their images of chemistry and wanted to detect their attitudes to chemistry. The frequent disasters of chemical industries in the late 1970s were reflected in the results: two-thirds of the images offer negative motives of environmental damages (see Fig. 8.3a), of threatening mankind and of animal experiments. The conclusion of the authors has been: "The chemistry teacher is probably the only one who starts his instruction in front of a class which has formed a bad opinion about the worthiness of the new school subject chemistry" [8].

In the late 1990s, Hilbing invited students in the area of Münster, Germany, to paint "their image of chemistry," in order to investigate to what extent this problem still exists [9]. These pictures feature positive motives mixed with negative ones (see Fig. 8.3b). Hilbing [9] could assert that only 35% of the boys and 16% of the girls have painted motives, which predominantly reflect negative attitudes. Compared to the results from Switzerland, the percentage of negative motives has nearly divided in half.

In addition to the paintings, a questionnaire was given to the students. The results referring to the attitude of *chemistry* [9], assert that 65% of the boys are giving a positive opinion but girls only to the extent of 33%. After asking about the attitude toward *chemistry lessons* only a smaller number gave a positive answer, namely 31% of the boys and 18% of the girls. So the positive attitudes toward chemistry in general are unexpectedly higher than those toward chemical education.

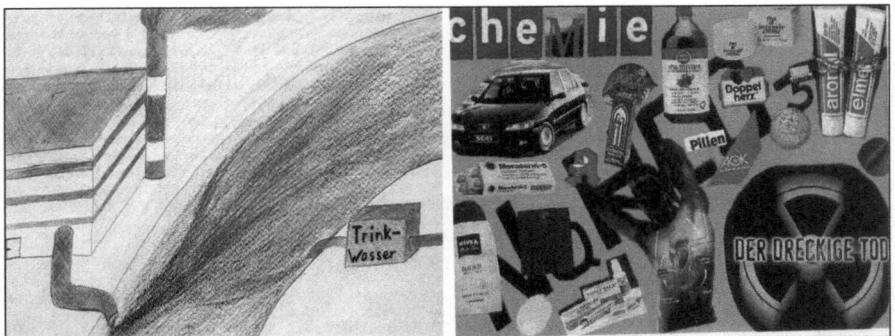

Fig. 8.3 Negative (**a**, [8]) and mixed motive (**b**, [9]) from: "paint your image of chemistry"

This result correlates with the motives of the pictures, which were painted by the same students: here, positive attitudes toward chemistry were found as well. A successful chemical education should be built upon these positive attitudes. Negative attitudes toward chemistry lessons may result from the way the school chemistry is taught: chemistry teachers and researchers of chemistry didactics have to find ways and means to improve chemical education.

8.2 Scientific Ideas: Chemistry Systematic Versus Everyday Life Chemistry

Concerning the controversy of systematic versus everyday life chemistry Just [10] stated: "*Everyday life chemistry* works with substances or processes of everyday life. *Everyday life oriented chemistry lessons* mean in contrast more, namely to focus chemistry lessons – as far as it is possible – consistently on everyday life."

Christen [11] replied: "Everyday life or environmental chemistry is no alternative to the scientific systematic way. Chemistry lessons may be oriented on everyday terms, but my opinion: scientific chemistry should serve as the only guideline for chemical education."

This dispute runs through the entire literature of the didactics of chemistry (see also Chap. 8.3). To get better acquainted with the scientific view of this topic, some scientific aspects of everyday life chemistry are presented.

Everyday phenomena and chemistry. Chemistry lessons can help students to acquire the scientific view for explanations of everyday phenomena. In particular, it is possible to "undress the packaging" of the everyday phenomena and to "translate" them into chemical processes. Some examples:

To brew coffee:	Extraction
To remove a stain:	Solubility
To wash clothes:	Emulsify and disperse
To use ink eraser:	Redox reaction
To produce photos:	Redox reaction, complex reaction
To polish silver:	Redox reaction
To dissolve mineral tablets:	Acid–base reaction
To bake with backing powder:	Acid–base reaction
To get rid of lime:	Acid–base reaction
To work with mortar:	Acid–base reaction
To use oven spray:	Solubility, saponification

In chemistry lessons, the mutual connections between chemical knowledge, technological advance, and individual living habits can also be presented and discussed in the historically development: soaps, detergents and cosmetics, preservation of food, fertilization and pest control, drugs and pharmaceuticals.

Taking the diagram of Pfeifer, Häusler, and Lutz (see Fig. 8.1) as a basis of everyday life experiences, there are a multitude of experiments with everyday chemicals in accordance with this diagram and related scientific facts to be considered. Beginning with "hygiene" in the middle of the diagram and continuing with "health" and other aspects, these relations are introduced with examples of special experiments (see E8.1–E8.16).

The reactions of the substances are sketched with reaction equations. For those equations, it shall be indicated whether it is a redox reaction (RR), an acid–base reaction (ABR), or a complex reaction (CR). In this sense, it is helpful to first teach those three basic concepts systematically before everyday chemicals and their reactions are shown by experiments and explained with the ideas of the basic concepts.

8.2.1 Hygiene: Example "Bathroom Chemicals"

Tab cleaner "NaOH type" (E8.1): This cleaner contains solid sodium hydroxide and decomposes organic material by highly concentrated alkaline solutions – thus blockages in the washbowl or toilet can be eliminated. If aluminum is added to the cleaner, there is a reaction with water to form hydrogen and produce a vortex that increases the cleaning effect:

$$Al\,(s) + 3H_2O\,(1) + OH^-\,(aq) \rightarrow [Al(OH)_4]^-\,(aq) + 3/2H_2\,(g) \qquad \textbf{RR, CR}$$

Hydrogen gas is detectable, but first "nascent" H atoms also react with nitrate ions of sodium nitrate, which are added for safety:

$$8\,\{\,H\,\} + NO_3^-\,(aq) \rightarrow NH_3\,(aq) + OH^-\,(aq) + 2H_2O\,(1) \qquad \textbf{RR}$$

Toilet cleaner "HSO_4^- type" (E8.2): This cleaner contains solid sodium hydrogen sulfate, which reacts with water and is highly acidic. The acidic solution dissolves lime spots:

$$NaHSO_4\,(s) + H_2O \rightarrow Na^+\,(aq) + H_3O^+\,(aq) + SO_4^{2-}\,(aq) \quad \textbf{ABR}$$

$$CaCO_3\,(s) + 2H_3O^+\,(aq) \rightarrow Ca^{2+}\,(aq) + 3H_2O + CO_2\,(aq,\,g) \quad \textbf{ABR}$$

Sanitary cleaner "HOCl-type" (E8.3): This solution forms "nascent" O atoms: they can destroy dangerous germs through oxidation:

$$HOCl\,(aq) + H_2O \rightarrow H_3O^+\,(aq) + Cl^-\,(aq) + \{O\} \quad \textbf{ABR and RR}$$

If this solution is mixed with strong acids, for instance with solid sodium hydrogen sulfate, gaseous chlorine develops; because of the toxicity of chlorine, there is always a warning message on the label "don't mix the cleaner with acidic substances!":

$$HOCl\,(aq) + Cl^-\,(aq) + H_3O^+\,(aq) \rightarrow Cl_2\,(aq,\,g) + 2H_2O \quad \textbf{ABR and RR}$$

8.2.2 Personal Hygiene: Example "Deodorants"

Deodorant "Al^{3+} type" (E8.4): Some deodorant substances – such as "Hydrofugal" – are produced on the basis of aluminum chloride hexahydrate. Both the acidic reaction and the presence of aluminum ions have germicidal effects:

$$\{[Al(H_2O)_6]^{3+}(Cl^-)_3\}\,(s) + H_2O \rightarrow H_3O^+\,(aq) + [Al(H_2O)_5OH]^{2+}\,(aq) + 3Cl^-\,(aq)$$
$$\textbf{ABR and CR}$$

8.2.3 Health: Example "Mineral Tablets"

Mineral tablets "Ca^{2+}/Mg^{2+} type" (E8.5): Substances containing calcium ions or magnesium ions are available as chewable tablets and as fizzy solutions if water is added. They contain mixtures of carbonates and citric acid (shortened: HCit molecules). When dissolved in water, bubbling gaseous carbon dioxide escapes, Ca^{2+}(aq) or Mg^{2+}(aq) ions are set free:

$$MgCO_3\,(s) + \ 2HCit\,(s) \rightarrow aq \rightarrow Mg^{2+}\,(aq) + 2Cit^-\,(aq) + H_2O + CO_2\,(aq,\,g)$$
$$\textbf{ABR}$$

8.2.4   Nutrition: Example "Table Salt"

Table salt "iodate type" (E8.6): In today's table salts, in addition to the usual sodium chloride, often minerals in small concentrations are mixed, such as calcium carbonate, sodium phosphate, or sodium iodate ("iodine salts"). They serve not only as supplementary nutrients (promotion and preservation of the teeth) but also as a technical medium to improve the flow ability. Mixing sodium iodate solution first with an acid and afterward with colorless potassium iodide solution, the mixture produces brownish-colored iodine precipitation:

$$IO_3^- \ (aq) + 5I^- \ (aq) + 6H^+ \ (aq) \rightarrow 3I_2(aq, brown) + 3H_2O \quad \textbf{RR}$$

If the concentration of iodine is too small after using table salt and no brown color appears, starch solution is able to indicate smallest iodine concentrations by the specific blue color.

8.2.5   Household: Example "Baking Process"

Baking soda "sodium bicarbonate type" (E8.7): For the production of bread and cake, baking soda is mostly used. It has the task to develop the gases carbon dioxide and water vapor by heating the mixture: it provides the dough with cavities and produces loose structure of bread. In most cases, sodium bicarbonate is mixed with solid citric acid – the gas carbon dioxide forms:

$$NaHCO_3 \ (s) + HCit \ (s) \rightarrow aq \rightarrow Na^+ \ (aq) + Cit^- \ (aq) + H_2O \ (l) + CO_2 \ (aq, g)$$
$$\textbf{ABR}$$

Baking powder "ammonium carbonate type" (E8.8): Using ammonium carbonate as baking powder, in addition to carbon dioxide and water vapor also ammonia is produced. To get rid of this gas after the baking process, cake or cookies must be flat and ammonia can escape:

$$(NH_4)_2CO_3 \ (s) \rightarrow 2NH_3 \ (g) + H_2O \ (g) + CO_2 \ (g) \quad \textbf{ABR}$$

8.2.6   Habitation: Example "Fuels for Heating"

Almost all homes are heated by fossil fuels: one either burns wood or coal in a fireplace or stove, or natural gas from the gas pipeline, or propane or heating oil from storage tanks near the house. In all cases, the chimney sweep periodically checks levels of soot, concentrations of carbon monoxide and carbon dioxide, exhaust temperatures to ensure an optimal combustion:

$$C\,(s) + O_2\,(g) \rightarrow CO_2\,(g); \quad \Delta H = -393\,\text{kJ/mol} \quad \textbf{RR}$$
$$CH_4\,(g) + 2O_2\,(g) \rightarrow CO_2\,(g) + 2H_2O\,(g); \quad \Delta H = -800\,\text{kJ/mol} \quad \textbf{RR}$$

8.2.7 Clothing: Example "Textile Dye Remover"

Dye remover "dithionite type" (E8.9): For bleaching of textiles or to remove stains, the substance sodium dithionite is often used as a "reduction bleach." The alkaline solution creates "nascent" H atoms, which destroy oxygen compounds (such as dyes or inks):

$$Na_2S_2O_4\,(s) + 2OH^-\,(aq) + H_2O \rightarrow 2Na^+\,(aq) + SO_4^{2-}\,(aq) + SO_3^{2-}\,(aq)$$
$$+ 4\{H\} \quad \textbf{RR}$$

8.2.8 Leisure: Example "Black and White Photography"

Developer "hydroquinone type" (E8.10): While color photography is difficult to describe, black and white photography can be realized relatively easy just by reactions on the surface of silver bromide photographic paper. The exposure of the photographic paper leads to invisible silver crystals, the development with the help of alkaline hydroquinone solution to visible amounts of separated silver and black areas on the exposed parts of the paper:

$$Ag^+Br^-\,(s) \rightarrow \text{light} \rightarrow Ag(s, \text{silver crystals}) + Br\,(\text{dissolved in AgBr}) \quad \textbf{RR}$$

$$2AgBr\,(s) + (C_6H_4)(OH)_2\,(aq) + 2OH^-\,(aq) \rightarrow 2Ag + (C_6H_4)O_2\,(aq)$$
$$+2H_2O + 2Br^-\,(aq) \quad \textbf{RR}$$

Fixer "thiosulfate type" (E8.11): The fixation is necessary because after developing, the unexposed silver bromide will cling to the photographic paper and would blacken further. It is extracted by complex reactions with the help of sodium thiosulfate solution:

$$AgBr\,(s, \text{ unexposed}) + 2S_2O_3^{2-}\,(aq) \rightarrow [Ag(S_2O_3)_2]^{3-}\,(aq) + Br^-\,(aq) \quad \textbf{CR}$$

8.2.9   Working World: Example "Metal Processing"

Etching chemical "Fe^{3+} type" (E8.12): For the production of special electronic components copper-coated plastic plates are used. To produce specific conductor lines for electric current, the plate is prepared with wax-coated corresponding lines. The rest of the copper layer is dissolved in special solutions, e.g. an iron(III) chloride solution:

$$Cu\,(s) + 2Fe^{3+}\,(aq) \rightarrow Cu^{2+}\,(aq) + 2Fe^{2+}\,(aq) \quad \textbf{RR}$$

8.2.10   Electric Power Supply: Example "Accumulators"

Accumulator "Pb/PbO_2 type" (E8.13): Accumulators are able to supply electricity and to be charged again. The best known is the starter battery in the car; it provides the electrical energy available to start the engine of the car. In the charged condition the electrodes are in the form of metallic lead and of red–brown lead oxide, during electron transfer these electrode materials are changing into lead sulfate:

$$\text{Minus pole}: \quad Pb\,(s) \rightarrow Pb^{2+}(PbSO_4) + 2e^- \quad \textbf{RR}$$

$$\text{Plus pole}: \quad PbO_2\,(s) + 4H^+\,(aq) + 2e^- \rightarrow Pb^{2+}(PbSO_4) \quad \textbf{RR}$$

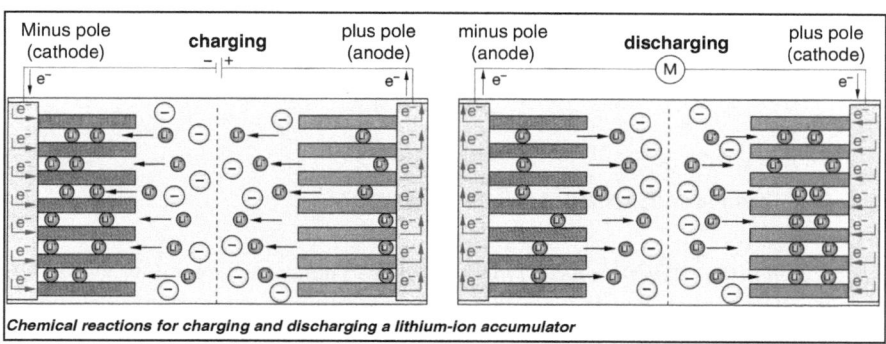

Chemical reactions for charging and discharging a lithium-ion accumulator

Accumulator "Li^+ ion type": For energy supply in computers or mobile phones, the lithium-ion battery is already common; and also in future electric cars this type of accumulator should run. The following equations and both pictures for charging (left) and discharging (right) explain the electron transfer for delivering electric power: [16]

$$\text{Minus pole}: \quad LiC_6 \rightarrow C_6(\text{graphite}) + Li^+ + e^- \quad \textbf{RR}$$
$$\text{Plus pole}: \quad Li^+ + LiCo_2O_4 + e^- \rightarrow 2LiCoO_2 + 2H_2O \quad \textbf{RR}$$

Martin Winter [17] at University of Muenster will develop a new lithium-ion-battery for cars in the next years, Marco Oetken and Martin Hasselmann [18] presented school experiments according to the lithium-ion-battery:with graphite electrodes they electrolyzed a solution of lithium perchlorate in propylen carbonate and got a voltage of 3.5 V [18].

8.2.11 House Building: Example "Hardening of Mortar"

Mortar "Ca(OH)$_2$ type" (E8.14): The lime in the mortar mixture of sand, lime, and water means chemically calcium hydroxide, it can be produced of solid calcium oxide by the reaction with water. In the setting of mortar, the reaction of calcium hydroxide with carbon dioxide out of the air occurs and forms solid calcium carbonate and water. Depending on the thickness of the wall and the temperature, it takes months or years until the setting is complete:

$$CaO\,(s) + H_2O\,(l) \rightarrow Ca(OH)_2\,(s) \quad \textbf{ABR}$$

$$Ca(OH)_2\,(s,\ aq) + CO_2\,(aq,\ g) \rightarrow CaCO_3\,(s) + H_2O\,(g,l) \quad \textbf{ABR}$$

Destruction of lime stone by acidic rain: Calcium carbonate (such as marble in the natural rock of buildings) can be attacked and destroyed by acidic rain and converted to crystal–water-containing calcium sulfate. Because it increases the volume, lime stone or natural marble can be weathered very strong on its surface:

$$CaCO_3\,(s) + 2H_3O^+(aq) + SO_4{}^{2-}\,(aq) \rightarrow CaSO_4 \cdot 2H_2O\,(s) + H_2O + CO_2\,(g)$$
$$\textbf{ABR}$$

8.2.12 Service: Example "Fire Fighting"

Fire extinguisher "bicarbonate type" (E8.15): There are several types of fire extinguishers, most operate on the basis of compressed carbon dioxide. If firefighters like to set a foam carpet on the runway of an airport, the reaction can be described by solid aluminum bicarbonate and sodium bicarbonate with water for firefighting, which develops extinguishing foam with special soaps:

$$Al^{3+}\,(s) + 6H_2O \rightarrow [Al(H_2O)_5OH]^{2+}(aq) + H^+(aq) \quad \textbf{ABR and CR}$$

$$HCO_3{}^-\,(s) + H^+\,(aq) \rightarrow H_2O + CO_2\,(aq, g) \quad \textbf{ABR}$$

8.2.13 Motoring: Example "Alcohol Test"

Alcohol test "chromate type" (E8.16): To check the blood alcohol content of drivers in road traffic, the police use either detection tubes or test equipment on the basis of infrared spectroscopy. The detection tubes are used to first estimate the alcohol content in breathing air. These tubes contain yellow crystals of potassium chromate mixed with sodium hydrogen sulfate. In the presence of a wet steam of alcohol, a reduction to green-colored chromium(III)-compounds occurs, so the color changes from yellow to green:

$$2CrO_4^{2-} \text{ (aq, yellow)} + 2H^+ \text{ (aq)} \rightarrow Cr_2O_7^{2-} \text{ (aq)} + H_2O \quad \textbf{ABR}$$

$$3\ CH_3CH_2OH \text{ (g)} + Cr_2O_7^{2-} \text{ (aq)} + 8H^+ \text{ (aq)} \rightarrow 3CH_3CHO \text{ (aq)}$$
$$+ 2Cr^{3+} \text{(green)} + 7H_2O \quad \textbf{RR}$$

8.2.14 Manufacturing: Example "Fertilizer"

In addition to natural fertilizers (dung, slurry), we have mineral fertilizers. On the one hand, they are extracted from salt stocks under the Earth's surface: potassium-, calcium- and magnesium salts, nitrates, phosphates, etc. On the other hand, nitrate and ammonium salts are produced artificially from nitrogen in the air by the Haber process:

$$N_2(g) + 3H_2 \text{ (g)} \rightarrow 2NH_3 \text{ (g)} \quad \textbf{RR}$$

$$2NH_3 \text{ (g)} + 3^1/_2\,O_2 \text{ (g)} \rightarrow 2NO_2 \text{ (g)} + 3H_2O \text{ (g)} \quad \textbf{RR}$$

$$4NO_2 \text{ (g)} + 2H_2O \text{ (l)} + O_2 \text{ (g)} \rightarrow 4HNO_3 \text{ (aq)} \quad \textbf{RR}$$

$$NH_3 \text{ (aq)} + HNO_3 \text{ (aq)} \rightarrow NH_4NO_3\text{(aq)} \quad \textbf{ABR}$$

Insoluble calcium phosphate – from deposits in North Africa – is converted with pure sulfuric acid to soluble dihydrogen phosphate, which is ready to be used in fertilizing mixtures:

$$Ca_3(PO_4)_2 \text{ (s)} + 2H_2SO_4 \text{ (l)} \rightarrow 2CaSO_4 \text{ (s)} + Ca(H_2PO_4)_2 \text{ (s)} \quad \textbf{ABR}$$

8.2.15   Air: Example "Smog"

In inverted atmospheric conditions, a warm air layer is lying on the cold air above ground like a cover: these air layers do not allow the diffusion of gases like sulfur dioxide, nitrogen oxides, and carbon monoxide into higher regions of the atmosphere. There is also solid dust that cannot escape and creates "smog" – a word combining smoke and fog. Gases and dust are considerably polluting the air and consequently the breathing. Nitrogen oxides are mainly produced by the reaction of air in hot automotive engines operating at high speed:

$$N_2 \,(g) + O_2 \,(g) \rightarrow 2NO \,(g); \quad 2NO \,(g) + O_2 \,(g) \rightarrow 2NO_2 \,(g, \text{ brown}) \quad \mathbf{RR}$$

In the catalytic convertor of a car, finely dispersed platinum crystals on the ceramic body function as catalyst material, reducing nitrogen oxides and oxidizing carbon monoxide:

$$NO \,(g) + CO \,(g) \rightarrow 1/2 \, N_2 \,(g) + CO_2 \,(g) \quad \mathbf{RR}$$

8.2.16   Water: Example "Sterilizing Drinking Water"

To sterilize drinking water, one uses chlorine or ozone – in both cases "nascent" O atoms $\{O\}$ are able to oxidize the organic impurities:

$$Cl_2 \,(g) + H_2O \rightarrow 2H^+ \,(aq) + 2Cl^- \,(aq) + \{O\} \quad \mathbf{RR}$$

$$O_3 (aq) \rightarrow O_2 (aq) + \{O\} \quad \mathbf{RR}$$

8.2.17   Soil: Example "Soil Acidification"

Acid deposition is caused by industry and by car emissions into the air: droplets of hydrochloric acid, sulfuric acid, or nitric acid solution occur causing acidification of the soil after rain. On the one hand, this causes fine roots to be damaged and carbonates from mineral nutrients to be dissolved and washed out. On the other hand, Al^{3+} ions that are fixed and harmless in solid aluminum salts are set free and able to damage the roots of trees and plants:

$$Al(OH)_3 \,(s) + 3H^+ \,(aq) \rightarrow Al^{3+} \,(aq) + 3H_2O \quad \mathbf{ABR}$$

8.3 Teaching Processes: Systematic Chemistry Plus Everyday Life Chemistry

"The term everyday chemistry includes all chemical processes and related substances that play a role in our daily lives. However, it would result in an immense quantity of actual themes and other contents that must be differentiated still according to individual interests. It is therefore clear that a life-related chemical lesson in the unreflective sense is no alternative to a clearly structured teaching. Major efforts are needed to bridge the gap between chemistry teaching and the everyday world, that is, to develop strategies to combine everyday life and meaningful learning" [12]. Lutz and Pfeifer [12] have expressed these statements and offered their reflections in many publications. Possible strategies for combining chemistry and everyday life teaching should be discussed; also methods by which the teaching processes can be designed to include everyday chemistry.

Methods of teaching processes: The mediation role of chemical education between everyday life and scientific chemistry can be achieved in various ways and thereby contribute to the diversity of methods:

Learning through active hands-on experimentation: Take river water or soil samples and show some ways how to analyze them (for example, by use of Aquamerck sets for different ions), or test some fertilizer mixtures and show the presence of important ions.

Field trips to places of interest outside school: Visit the regional waste water treatment plant or a plant for drinking water treatment, look to the waste processing and recycling station, prepare and conduct interviews, create photo reports or an exhibition of posters. Visit also specific industries that are located near your school.

Presentations and discussions with special experts: Invite, for example, a firefighter of the nearest fire department, a technician of the car industry, a CD production engineer, a food inspector, a paint manufacturer. Let them present their work and information about any chemistry related issues.

Audio-visual media or multimedia: Show and discuss CDs about special ways of the production of important substances that cannot be seen near the school: an iron smelter, aluminum by electrolysis of molten aluminum oxide, copper refining by electrolysis, production of batteries and accumulators, crude oil distillation, etc. Produce your own self-designed material, make a CD, and show it to other classes or parents.

Subject-related role plays: Use or create role plays for themes that cannot be taught through experiments, like the question "steak or grains" [13]: the roles of different meat consumers, a butcher, an agricultural scientist, and a food chemist are assumed by students, who perform a role play. The other students of the class can agree or disagree with the arguments and putting their own arguments forward until the discussion has revealed everyone's point of view.

Project-based teaching or project-oriented lessons [14]: "Water and Environment" [15] can be a useful project for 9th or 10th graders; done after a visit to a waste water purification plant near the school. The students may arrange their themes themselves; they design and present their own posters and organize an exhibition in school for their friends and parents.

Textbooks and everyday chemistry: Textbook authors have always tried to enrich chemistry topics for all different school types with references to everyday life and environment [16]. If you class topics according to their level of references to daily life from those without any reference to those with a high level of everyday life implications, one obtains the following list (based on the German textbook):

Topics without references to everyday life: fundamental laws of chemistry; gas laws; Dalton's atomic model; structure of atoms, ions, and molecules; chemical bonding; and mental models concerning quantum mechanics.

Topics with references to everyday life as initial motivation: for the topic "chemical reactions" you may start with a burning match, with dissolving mineral tablets, before the standard experiments are performed and interpreted. The topic "water and the H_2O molecule" may be started with the use of water in everyday life, e.g. with purification of waste water and production of drinking water, before the decomposition of water into the two elements hydrogen and oxygen is shown and before the synthesis of water from those elements will follow.

Topics ending with references to everyday life for reinforcement: after the scientific explanation of electron transfers concerning standard reactions of metals and salt solutions for the topic "redox reactions and electron transfer," students may encounter batteries of everyday life or iron rust and corrosion and their interpretation.

Topics with full references to everyday life: after teaching most basic concepts, for example, acid–base reactions, redox reactions, equilibrium and energy, many everyday life phenomena can be described scientifically, and many projects concerning everyday life can be carried out. Such projects may deal with environmental or food chemistry, with energy production and alternatives, with industrial processes, and the production of important chemicals.

Curricula with full references to everyday life: a number of curricula, which are based fully on everyday life or on environmental chemistry, have been developed. One example is ChemCom, short for "Chemistry in the Community" [17] (see Fig. 8.4) that offers as the first topic "the quality of our water" and provides many applications of water in our life. Under the "molecular view of water," ball-and-stick models of water molecules appear, and also colored balls as models for H atoms and O atoms appear. The formula H_2O is introduced as the first formula in the curriculum.

Later "electrical nature of matter" follows and dipole molecules are discussed. The dipole molecule of water and particular ions are introduced, with certain ions used as a basis for specific water tests. All this information is completely without the common scientific structure of chemistry – all scientific information is taken only as required for the topic "water."

Another curriculum is named *Salters Advanced Chemistry*. It is divided into three parts: "Chemical Storylines" [18] delivers the everyday life topics and environmental themes (see Fig. 8.5), "Chemical Ideas" [19] contains the scientific ideas, the systematic of chemistry technically prepared for looking at all the

Preface

To You, the Student

Safety in the Laboratory

Supplying Our Water Needs
I The Quality of Our Water
II A Look at Water and Its Contaminants
III Investigating the Cause of the Fish Kill
IV Water Purification and Treatment
V Putting It All Together: Fish Kill in Riverwood—Who Pays?

Conserving Chemical Resources
I Use of Resources
II Why We Use What We Do
III Can We Continue to Use Things Up?
IV How Much Do We Have and for How Long?
V The Promise of New Materials

Petroleum: To Build or to Burn?
I Petroleum in Our Lives
II Petroleum: What Is It? What Do We Do with It?
III Petroleum as a Source of Energy
IV Making Useful Materials from Petroleum
V Alternatives to Petroleum
VI Putting It All Together: Choosing Petroleum Futures

Understanding Foods
I Foods: To Build or to Burn?
II Food as Energy
III Foods: The Builder Molecules
IV Essential Materials Needed in Small Amounts
V Food Additives
VI Putting It All Together: Nutrition in Many Cultures

Acknowledgments

Preface

To You, the Student

Safety in the Laboratory

Nuclear Chemistry in Our World
I Nuclear Phenomena in Everyday Life
II Discovering Atomic Energy
III Radioactive Decay
IV Nuclear Energy: Power of the Universe
V Living with Benefits and Risks
VI Putting It All Together: Separating Fact from Fiction

Chemicals, Air, and Climate
I Life in a Sea of Air
II Investigating the Atmosphere
III Atmosphere and Climate
IV Human Alteration of the Atmosphere
V Putting It All Together: Is Air a Free Resource?

Chemistry and Health
I Maintenance of Health
II Your Body as a Chemical Factory
III The Chemistry of Exercise
IV The Chemistry of Personal Health Care
V Chemical Control: Drugs and Toxins in the Human Body
VI Putting It All Together: Assessing Personal and Public Health Risks

The Chemical Industry: Promise and Challenge
I A New Industry for Riverwood
II An Overview of the Chemical Industry
III Nitrogen Products and Their Chemistry
IV Chemical Energy ↔ Electrical Energy
V Pharmaceuticals
VI Putting It All Together: The Role of Chemical Industry Past, Present, and Future

Acknowledgments

Fig. 8.4 Contents of ChemCom: "Chemistry in the Community" [17]

chemistry contents concerning the stories from Chemical Storylines (see Fig. 8.6). Finally "Activities and Assessment" and a CD provide instructions for laboratory experiments and assessment tasks for all described topics.

In the lectures according to "Chemical Storylines," appropriate references to the scientific information and corresponding experiments in the book of Chemical Ideas are given. These references ensure that students will also recognize and learn the structure and ideas of chemistry.

The advanced lessons of the Storylines on the "Advanced Level" may allow this way after students have had a 2-year chemistry introductory on the "ordinary level." All additional information are recognizable more easily and handled better by students than it is possible in the curriculum "ChemCom," which is designed for beginning instruction in chemistry.

Chemistry in Context: Based on Anglo-American curricula as "Chemistry in the Community" [17], Salters Advanced Chemistry [18], and also "Chemistry in Context" [20], a working group around Parchmann, Ralle, and Demuth [21] developed a German written curriculum called "Chemie im Kontext" [22]. This curriculum contains similar contexts and stories as in "Chemical Storylines" [18]. The material is, however, contained in only one book: the first part shows about 20 stories (100 pages), the second part of the textbook shows the scientific chemistry

Introduction for students

The Elements of Life

EL1 What are we made of?
EL2 Take two elements …
EL3 Looking for patterns in elements
EL4 Where do the chemical elements come from?
EL5 The molecules of life
EL6 Summary

Developing Fuels

DF1 Petrol is popular …
DF2 Getting energy from fuels
DF3 Focus on petrol
DF4 Making petrol – getting the right octane rating
DF5 Trouble with emissions
DF6 Methanol – the key to future fuels?
DF7 What other solutions are there?
DF8 Summary

From Minerals to Elements

M1 Chemicals from the sea
M2 Copper from deep in the ground
M3 Mining Cornish tin
M4 Summary

The Atmosphere

A1 What's in the air?
A2 Screening the Sun
A3 Ozone: A vital sunscreen
A4 The CFC story
A5 How bad is the ozone crisis?
A6 Trouble in the troposphere
A7 Keeping the window open
A8 Focus on carbon dioxide
A9 Coping with carbon
A10 Summary

The Polymer Revolution

PR1 Designer polymers
PR2 The polythene story
PR3 Towards high density polymers
PR4 Conducting polymers – breaking the rules
PR5 The invention of nylon
PR6 Kelvar
PR7 Taking temperature into account
PR8 Throwing it away… or not?
PR9 Summary

What's in a Medicine?

WM1 The development of modern ideas about medicines
WM2 Medicines from nature
WM3 Identifying the active chemical in willow bark
WM4 Instrumental analysis
WM5 The synthesis of salicylic acid and aspirin
WM6 Delivering the product
WM7 Development and safety testing of medicines
WM8 Summary

Using Sunlight

US1 The Sun: sustainer of life
US2 Exciting light
US3 Photosynthesis
US4 Fuel for the future
US5 Solar cells
US6 The hydrogen economy
US7 Summary

Engineering Proteins

EP1 Christopher's story
EP2 Protein building
EP3 Genetic engineering
EP4 Proteins in 3-D
EP5 Giving evolution a push
EP6 Enzymes
EP7 Summary

The Steel Story

SS1 What is steel?
SS2 How is steel made?
SS3 A closer look at the elements in steel
SS4 Rusting
SS5 What happens inside a 'tin' can?
SS6 Recycled steel
SS7 Summary

Aspects of Agriculture

AA1 What do we want from agriculture?
AA2 The world at your feet
AA3 How does the soil remain fertile?
AA4 Competition for food
AA5 Summary

Colour by Design

CD1 Ways of making colour
CD2 The Monastral Blue story
CD3 Chrome Yellow
CD4 Chemistry in the art gallery
CD5 At the start of the rainbow
CD6 Chemists design colours
CD7 Colour for cotton
CD8 High-tech colour
CD9 Summary

The Oceans

O1 The edge of the land
O2 Wider still and deeper
O3 Oceans of energy
O4 A safe place to grow
O5 Summary

Medicines by Design

MD1 Alcohol in the body
MD2 The drug action of ethanol
MD3 Medicines that send messages to nerves
MD4 Enzyme inhibitors as medicines
MD5 Targetting bacteria
MD6 Summary

Visiting the Chemical Industry

VCI1 Introduction
VCI2 The operation of a chemical manufacturing process
VCI3 People
VCI4 Raw materials and feedstock preparation
VCI5 The best conditions for the process
VCI6 Safety matters
VCI7 Environmental issues
VCI8 Costs
VCI9 Location

Index

Fig. 8.5 Contents of Salters Chemistry, "Chemical Storylines" [18]

Introduction for Students

Chapter 1 Measuring amounts of substances

1.1 Amount of substance
1.2 Balanced equations
1.3 Using equations to work out reacting masses
1.4 Calculations involving gases
1.5 The Ideal Gas law
1.6 Concentrations of solutions

Chapter 2 Atomic structure

2.1 A simple model of the atom
2.2 Radioactive decay
2.3 Electronic structure: shells
2.4 Electronic structure: sub-shells and orbitals

Chapter 3 Bonding, shapes and sizes

3.1 Chemical bonding
3.2 The size of ions
3.3 The shapes of molecules
3.4 Structural isomerism
3.5 Geometric isomerism
3.6 Optical isomerism

Chapter 4 Energy changes and chemical reactions

4.1 Energy out, energy in
4.2 Where does the energy come from?
4.3 Entropy and the direction of change
4.4 Energy, entropy and equilibrium
4.5 Energy changes in solutions

Chapter 5 Structure and properties

5.1 Ions in solids and solutions
5.2 Forces between molecules: induced dipoles
5.3 Forces between molecules: permanent dipoles and hydrogen bonding
5.4 The structure and properties of polymers
5.5 Water: hydrogen bonding in action
5.6 Bonding, structure and properties: a summary

Chapter 6 Radiation and matter

6.1 Light and electrons
6.2 What happens when radiation interacts with matter?
6.3 Radiation and radicals
6.4 Where does colour come from?
6.5 Ultra-violet and visible spectroscopy
6.6 Chemistry of colour
6.7 Infra-red spectroscopy
6.8 Nuclear magnetic resonance spectroscopy
6.9 Mass spectrometry

Chapter 7 Equilibrium in chemistry

7.1 Chemical equilibrium
7.2 Equilibrium constants
7.3 Partition equilibrium
7.4 What decides the position of equilibrium?
7.5 Ion exchange
7.6 Chromatography
7.7 Solubility equilibria

Chapter 8 Acids and bases

8.1 Acid-base reactions
8.2 Weak acids and pH
8.3 Buffer solutions

Chapter 9 Redox

9.1 Oxidation and reduction
9.2 Redox reactions and electrode potentials
9.3 Predicting the direction of redox reactions
9.4 The effect of complexing on redox reactions

Chapter 10 Rates of reactions

10.1 Factors affecting reaction rates
10.2 The effect of concentration on rate
10.3 The effect of temperature on rate
10.4 Catalysis

Chapter 11 The Periodic Table

11.1 Periodicity
11.2 The *s* block: Groups 1 and 2
11.3 The *p* block: Group 4
11.4 The *p* block: Group 5
11.5 The *p* block: Group 7
11.6 The *d* block: characteristics of transition elements
11.7 The *d* block: complex formation

Chapter 12 Organic chemistry: frameworks

12.1 Alkanes
12.2 Alkenes
12.3 Arenes
12.4 Reactions of arenes

Chapter 13 Organic chemistry: modifiers

13.1 Halogenoalkanes
13.2 Alcohols and ethers
13.3 Carboxylic acids and their derivatives
13.4 The –OH group in alcohols, phenols and acids
13.5 Esters
13.6 Oils and fats
13.7 Amines, amides and amino acids
13.8 Azo compounds

Chapter 14 Organic synthesis

14.1 Planning a synthesis
14.2 A summary of organic reactions

Index

Fig. 8.6 Contents of Salters Chemistry, "Chemical Ideas" [19]

and includes the important five basic chemistry concepts (250 pages); both parts are connected through specific references.

Parchmann, Ralle, and Demuth [21] point out that the structure of the curriculum "Chemie im Kontext" is based on three principles (see Fig. 8.7): context-based approach, development of basic concepts of chemistry, and diversity of teaching methods [21].

Context-based approach means – teaching on the basis of "Situated Learning" – learning that involves concrete learning situations: "Cars of the future" instead of "Electrochemistry," "Ocean, climate and greenhouse effect" instead of "Chemical equilibrium" [23]. The context-oriented approach is accompanied by the widely accepted constructivist teaching principles.

Basic concepts of chemistry should be developed by the students to understand the chemistry behind the stories and to decontextualize the stories in this special approach later. These basic concepts are: Substances and particles, chemical structure and properties, donor–acceptor reactions (acid–base reactions, redox reactions, complex reactions), energy and entropy, and chemical equilibrium. The students should develop the five basic concepts by working with those stories from the first part of the book: the solubility equilibrium of carbon dioxide, for example, may be taught with the story "Ocean, climate and greenhouse effect," but it may also be taught by the stories "Drinks" and "Breathing." [21] In addition, the transfer to other equilibria in other stories is necessary for decontextualization and for general understanding of the basic concept "equilibrium."

Teaching methods should be connected to the different stories and should differ from story to story; self-confidence, personal activity, and hands-on experiments are the most important guide lines. Usually the topics should be separated in four phases: encounter phase, curiosity phase, working phase, in-depth phase, and networking phase [21].

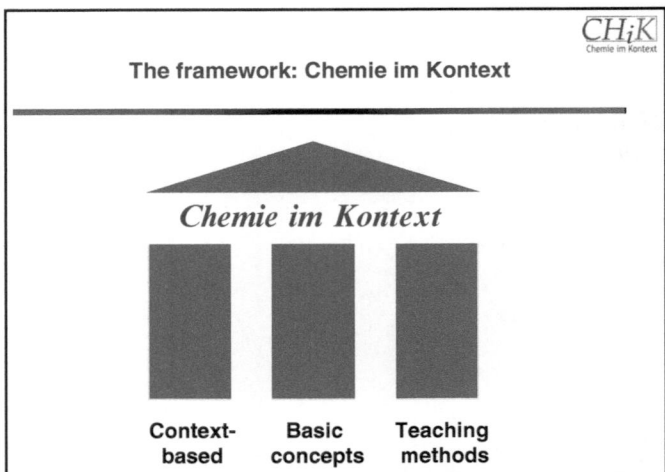

Fig. 8.7 Framework of the German curriculum "Chemie im Kontext" [21]

8.4   Human Element: Role Play and Environmental Education

Social relations have already been mentioned in this chapter. Some details remain, some social relations to chemical and environmental education can be taught through role playing. For Hellweger [13], "the goals of role-plays are to simulate discussion groups, where everyone can be active. Depending on whether some players are designated either as facilitator or moderator, whether as a professional who has to answer to certain issues in more detail, or whether all participants have the same status: the game may have more the character of expert consultation or the form of a free discussion board, it runs more strictly directed or free and spontaneous." Role-playing games have been developed concerning the following issues [13]:

Chemistry lessons – why? The students discuss various rolls about the "what and how" of a meaningful chemistry education. They look for justification, whether to continue to force everybody to learn chemistry, or whether chemistry should be abolished in favor of disciplines that are more helpful for students at school.

The river Elbe is overturning! A river is so polluted that fish have not been seen for a long time. May it be worth to do something for the quality of the water and against its overturning, or are the costs too high and the loss of many jobs from industries near the river too risky?

"Steak or grains?" Are there serious arguments and attitudes of students and adults that meat consumption should be reduced in the highly industrialized countries in favor of more vegetarian products? What are the advantages and disadvantages for the production of much meat for the rich countries in the Western area?

". . . and he has yet drilled a tooth!" More and more toothpaste is consumed by people, but the teeth are getting worse and worse – maybe due to high consumption of sugar. Perhaps a healthy diet is more important than dental hygiene? Should we treat sugar as a dangerous drug?

"All the best with butter?" Do we really improve our health if we restrict the consumption of butter in favor of more margarine? The audience will show in the form of an expert survey that there are two conflicting theories for the development of myocardial infarction, and how these theories can affect our everyday life.

Energy forum 2020. Can Germany by the year 2020 renounce the controversial nuclear energy without compromising living standards and jobs? Students play politicians of different political parties and different philosophies; in the roles of delegates they bring arguments, facts, and pleadings for or against nuclear power.

Otto [24] commented that such role plays "are open to questions about the relationship between scientific knowledge and its consequences for mankind, of science research and social development, of economic interests, environmental pollution and human health." He also points out that "the natural sciences became in the last 20 years those subjects, where students are expected to learn for life and not only for school. Of course this is true only if there is an apparent link between what happens in the classroom and what is going on in TV, in newspapers, in the citizens' initiative. Role-playing games and decision games are based on existing awareness and can expand it, put positions in question, confront with opposite views, differentiate positions concerning specific themes of society. Chemistry teaching requires not only students' motivation but must also motivate further learning – particularly to motivate learning outside of school." [24]

Finally, the social relation of everyday life and chemistry should be targeted for environmental education of young students. Demuth [25] suggested: "For an environmental education not all possible environmental topics should be aspired, but much more important is to work intensive with a few projects." Some project ideas are: "Where to deposit our waste, what is recycling, how to save energy, what have refrigerators to do with the ozone problem?" [25]

In the project "Nitrogen analysis," Demuth [26] undertakes in collaboration with some schools "to investigate the concentration of ammonium and nitrate ions in the soil of small gardens of relatives or friends or of the field of a farmer. These investigations of soil were carried out to understand related environmental problems; and also for farmers and gardeners to help them to fertilize selectively." [26] Projects are a helpful way to realize environment education (see also Chap. 3).

Problems and Exercises

P8.1 Many phenomena of everyday life or nature can be "translated" into chemistry or chemical reactions. Give five examples for those phenomena and their scientific interpretation. Look for useful experiments according to these examples and write chemical equations.

P8.2 In some cases, chemicals of your kitchen or bathroom can replace the chemicals of the laboratory. Give five examples and explain the experiments or reactions. Describe topics or contexts of chemical education where the use of those chemicals may be helpful.

P8.3 For students' motivation, everyday life phenomena should be put at the beginning of a topic in chemistry class. Give five examples, describe the experiments, and their interpretation. How do the lessons continue after successful motivation?

P8.4 The attitudes of students concerning chemistry or chemical education are often negative ("chemistry is bad for our health," "in chemistry classes we have mostly formulae we cannot understand"). What steps or actions could be helpful to turn those attitudes around?

P8.5 There are some days free of classes and you have to plan a chemistry project. What topic, what context you like to choose for those days? What tasks or problems should the students solve in these days? Which questionnaire would you develop to ask about the interests of your students? In which way would you respect different wishes of boys and girls?

Experiments

In Chap. 8.2, many reactions of different household chemicals are described already so that the "problem" and even the "explanation" are given with those interpretations. Therefore, the paragraph "problem" will not be shown in the following short instructions – only the "procedure" is given. Because the "material" consists mostly of some test tubes, this paragraph is also removed.

E8.1 Tab Cleaner "NaOH Type" Put a spoon of the cleaning powder on a watch glass and observe white salt crystals and silver-colored metal splinters. A part of the substance is mixed with a small amount of water in a test tube: an exothermic reaction begins and a smell of ammonia is noticeable. Pieces of wool fabric are added: they decompose slowly.

In a second experiment, pure sodium hydroxide is mixed with aluminum shavings and little water, the produced gas is collected in a second test tube. After igniting the gas, a little bang is observed: the produced gas is hydrogen.

E8.2 Toilet Cleaner "HSO_4^- Type" A small amount of substance is given in a test tube, white salt is observed. It is dissolved in water, the solution is examined with universal indicator paper: acidic reaction. A small amount of calcium carbonate is added to the solution: the sample will be dissolving with gas development. A burning wooden splint is extinguished: the gas is carbon dioxide.

E8.3 Sanitary Cleaner "ClO^- Type" Cleaner liquid is given into a test tube, a strip of indicator paper is dropped, some drops of methyl blue solution are added: indicator paper turns white, blue solution turns colorless.

A second sample is mixed in a test tube with toilet cleaner (E8.2): the smell and color of the developing gas indicate chlorine (dilute the solution in the vent to stop the reaction).

E8.4 Deodorant "Al^{3+} Type" One brand of deodorant is Hydrofugal spray. Indicator paper is moistened: acidic reaction. Little aluminum chloride hexahydrate is dissolved in water (test tube) and examined with indicator paper: acidic reaction.

E8.5 Mineral Tablets "Ca^{2+} and Mg^{2+} Types" One tablet is given into a glass of water: gas bubbles, formation of carbon dioxide. The reaction is repeated with a mixture of calcium carbonate and citric acid, water is added.

In a gas developer, water is dripped on a tablet, the developed gas is collected in a syringe. As soon as the 100 mL mark is reached, the gas is moved out and new developed gas is collected again. The quantity of gas is determined; the gas is examined with a burning splint.

The reaction of one tablet is performed in a pneumatic tub, in such a way that the gas is collected in a water filled cylinder. The volume will be marked. A second tablet will be dissolved in the same way: the observed volume of the gas portion is much larger than previously with the first tablet (compare with E2.3).

E8.6 Table Salt "Iodine Type" Ingredients of the salt will be checked on the label of the container. Iodinated table salt will be mixed with potassium iodide solution and acidified with sulfur acid solution: brown-colored iodine solution appears. If you add starch solution, the specific blue-colored solution shows free iodine, as well. The experiment will be repeated with pure sodium iodate.

E8.7 Baking Powder "Sodium Bicarbonate Type" The baking powder will be floated with a small amount of water and heated: gas development. It will be heated strongly in a dry test tube; the gas that forms will be collected in a syringe and examined with a burning splint: carbon dioxide.

E8.8 Baking Powder "Ammonium Carbonate Type" The experiments of E8.7 will be repeated with this baking powder and pure ammonium carbonate. The mixture of gases formed will be examined with wet indicator paper: alkaline reaction. The smell also indicates ammonia.

E8.9 Textile Decolorizer "Sodium Dithionite Type" Ingredients are checked on the label of the container. In a test tube, a sample of a methyl blue solution will be mixed with a small amount of textile decolorizer powder; in a second test tube with pure sodium dithionite: reaction of the blue solution until decolorization.

E8.10 Developer "Hydrochinone Type" Silver chloride will be precipitated in two test tubes: keep one test tube in the cupboard without light for ten minutes, the other in bright light. The last precipitate turns much darker compared with the first one. Alkaline hydroquinone solution (Xn) is added to both test tubes: both precipitates turn black.

In the photolab under red light, a key ring or something like that is placed on photographic paper and then exposed for a short time. The photographic paper is then put into an already prepared developer solution: the black-and-white picture develops in one minute. The photo has to be rinsed with diluted acetic acid to stop further developing of the photo. The photo is then fixed in special fixer solution before it can be viewed in white light (compare with E8.11).

E8.11 Fixer "Sodium Thiosulfate Type" Silver chloride will be freshly precipitated in a test tube, and the resulting suspension will be diluted. Concentrated sodium thiosulfate solution is given into the dilute suspension and then shaken: the white precipitate of silver chloride dissolves into a clear solution. In the photolab, the developed photo (E8.10) is dropped into the fixer solution and left there for a while. After that, another newly developed photo and the fixed photo are brought into normal white light: the fixed photo stays unchanged, while the entire nonfixed photo turns totally black.

E8.12 Etching Chemical "Fe^{2+} Type" Iron(III) chloride solution (Xn) is prepared in a big test tube. On a copper-coated synthetic strip, a short name is written on it using a wax pen. This strip is dropped into the prepared solution: after a few minutes, only the written name can be seen, the remaining copper is dissolved.

E8.13 Accumulator "Pb/PbO$_2$ Type" A car battery is demonstrated; first the voltage of one cell is measured (2 V) and then the voltage of all six cells together (12 V). A beaker is filled to three quarters with sulfuric acid solution (20%, C), two lead strips are arranged and fixed in such a way that they do not contact each other. The lead strips are connected to a transformer by two cables, a direct current voltage has to adjust in this way so that a gas development can be seen: on one lead strip a layer of a red–brown substance is generated (PbO$_2$), the other lead strip stays as before (Pb). The transformer is removed after a few minutes; the voltage between the two strips is measured: about 2 V. An electric motor is connected, which runs for a while and then stops (compare with E4.10).

E8.14 Flashing Cement "$Ca(OH)_2$ Type" Ingredients are checked on the label of the container. The mixture of the substance with little water is tested with indicator paper: alkaline reaction.

In a beaker, little water is given to fresh calcium oxide (Xi): increase of the volume under hissing noises; strong exothermic reaction. The white product is inspected with a wet strip of indicator paper: strong alkaline reaction.

The product (or calcium hydroxide (Xi) out of a supply bottle) is mixed with water in an Erlenmeyer flask, carbon dioxide is added from a steel bottle and a syringe filled with carbon dioxide is connected to the flask. The flask is moved so that the mixture spreads inside the flask: the piston of the syringe moves; the mixture heats up.

E8.15 Fire Extinguisher Model "Wet and Foam Extinguisher Type" Put on safety glasses. A plastic spray bottle is filled to half with concentrated sodium carbonate solution. A small test tube, which contains a few milliliters of concentrated sulfuric acid, is put into the solution in a way that it swims. The bottle is shut with a capillary glass tube. Close to a sink, the spray bottle is briefly inverted; the spurt of the spray bottle is pointed into the drain: the content of the extinguisher model empties itself with a cutting spurt. In a repetition of this experiment, foam concentrate can be mixed to the sodium carbonate solution: model experiment of a foam extinguisher. Caution in the disposal: remains of concentrated sulfuric acid!

E8.16 Alcohol Test "Chromate Type" One alcotest tube is prepared; a small amount of alcohol is spread in the mouth and ethanol vapor is blown with the breathing air through the tube into a synthetic plastic bag: the color in the indicator zone turns from yellow to green.

A yellow-colored potassium chromate solution (T/N) is acidified, and a small amount of ethanol (F) is added: a color change from yellow to orange occurs first, then the change from orange to green is observed.

References

1. Waddington D (2000) The Salters Chemistry Project: 15 years on. Presentation during the 15th Symposium of Chemical Education in Dortmund/Germany, University of York, 15 June 2000
2. Barke H-D (1987) Chemieunterricht erscheint nicht so sinnlos, wenn man den Stoff auch im Alltag anwenden kann. In: Lindemann H Alltagschemie. NiU P/C 35:Heft 25
3. Gesellschaft Deutscher Chemiker (1992) Denkschrift zur Lehrerausbildung für den Chemieunterricht auf der Sekundarstufe II. Frankfurt
4. Pfeifer P, Haeusler K, Lutz B (1992) Konkrete Fachdidaktik Chemie. Oldenbourg, München
5. Barke H-D (1996) Lebenswelt und Alltag im Chemieunterricht. In: Behrendt H Zur Didaktik der Physik und Chemie. Leuchtturm, Alsbach

6. Wanjek J, Barke H-D (1998) Einfluss eines alltagsorientierten Chemieunterrichts auf die Entwicklung von Interessen und Einstellungen. In: Behrendt H Zur Didaktik der Physik und Chemie. Leuchtturm, Alsbach
7. Mueller-Harbich G, Wenck H, Bader HJ (1990) Die Einstellung von Realschuelern zum Chemieunterricht, zu Umweltproblemen und zur Chemie. Chim. did. 16:151 und 233
8. Heilbronner E, Wyss E (1983) Bild einer Wissenschaft. Chemie CiuZ 17:69
9. Hilbing C, Barke H-D (2000) Male dein Bild von der Chemie. Zum Image von Chemie und Chemieunterricht bei Jugendlichen. CiuZ 34
10. Just E (1998) Missverständnisse zur Aufgabe und zur Wirkung des Faches Chemie in allgemeinbildenden Schulen. CHEMKON 5:96
11. Christen HR (1997) Chemie – faszinierend oder ein Horrorfach? Zur Akzeptanz des Chemieunterrichts. CHEMKON 4:175. Leserbrief (1998) CHEMKON 5:211
12. Lutz B, Pfeifer P (1989) Chemie in Alltag und Chemieunterricht – Gegensatz oder Chance für ein besseres Chemieverstaendnis? MNU 42:281
13. Hellweger S (1981) Chemieunterricht 5–10. Skriptor, Muenchen
14. Frey K (1982) Die Projektmethode. Beltz, Weinheim
15. Barke H-D (1999) Wasser und Umwelt. In: Muenzinger W, Frey K: Chemie in Projekten. Aulis, Koeln
16. Jaeckel M, Risch KT (2010) Chemie heute SII. Schroedel, Braunschweig
17. Winter, M (2009) Nano-porous SiO/carbon composite anode for lithium-ion batteries. J. Applied Electrochemistry
18. Hasselmann M, Oetken M (2011) Elektrische Energie aus dem Kohlenstoffsandwich – Lithium-Ionen-Akkumulatoren auf der Basis redox-amphoterer Graphitintercalationselektroden. CHEMKON 18
19. The University of York Science Education Group (1994) Salters advanced chemistry: chemical ideas. Heinemann, York
20. Parchmann I, Demuth R, Ralle B (2005) Chemie im Kontext. Cornelsen, Berlin
21. Huntemann H, Paschmann A, Parchmann I, Ralle B (1999) Chemie im Kontext – ein neues Konzept für den Chemieunterricht? CHEMKON 6:191
22. Otto G (1981) Nachwort: Zur Problemlage in den naturwissenschaftlichen Didaktiken. In: Hellweger S: Chemieunterricht 5–10. Skriptor, München
23. Demuth R (1992) Umwelterziehung im Chemieunterricht – Ziele, Inhalte, Methoden. NiU-Chemie 3:47
24. Demuth R (1992) Stickstoffanalytik im Chemieunterricht der Sek. I. NiU-Chemie 3:67

Further Reading

American Chemical Society (1993) ChemCom: chemistry in the community, 2nd edn. Kendal/Hunt, Dubuque, IA
Greeno JG, Smith DR, Moore JL (1993) Transfer of situated learning. In: Dettermann DK, Sternberg RJ (eds) Transfer on trial: intelligence, cognition, and instruction. Norwood, Ablex, pp 99–167
Hill G, Holman J (2000) Chemistry in context, 5th edn. Nelson Thornes, Cheltenham, UK
Mahaffy P (2006) Moving chemistry education into 3D: a tetrahedral metaphor for understanding chemistry. J Chem Educ 83(1)
Millar R (2000) Science for public understanding: developing a new course for 16–18-year-old students. In: Cross R, Fensham P (eds) Science and the citizen: for educators and the public. Melbourne Studies in Education, Melbourne, pp 201–214
Sadoski M (2001) Resolving the effects of concreteness on interest, comprehension, and learning important ideas from text. Educ Psychol Rev 13:263–281

Salili F, Chou G, Hong Y (eds) (2001) Student motivation: the culture and context of learning. Kluwer, Amsterdam

Salters Advanced Chemistry Project (1994) Chemical storylines; chemical ideas; activities and assessment pack. Heinemann Education, Oxford, UK

Shamos M (1995) The myth of scientific literacy. Rutgers University Press, New Brunswick, NJ

Stanitski C (2000) Chemistry in context: applying chemistry in society. A project of the American Chemical Society, 3rd edn. McGraw-Hill Higher Education, New York

Stanitski CL, Eubanks LP, Middlecamp CH, Stratton WJ (2000) Chemistry in context: applying chemistry to society, 3rd edn. McGraw-Hill, New York

Van Oers B (1998) From context to contextualizing. Learn Instruct 8(6):473–488

Wiser M, Amin T (2001) "Is heat hot?" Inducing conceptual change by integrating everyday and scientific perspectives on thermal phenomena. Learn Instruct 11:331–355

Chapter 9
Students Discover Organic Chemistry: A Phenomena-Oriented and Inquiry-Based Network Concept (PIN-Concept)

The Phenomena-oriented and Inquiry-based Network-Concept (PIN-Concept) is a curriculum for the training of interconnected thinking in the field of fundamental organic chemistry. It has been developed by Harsch and Heimann [1–21] for the chemical and didactical education of prospective teachers at universities, and for practicing chemistry teachers and their classes at grammar schools. The PIN-Concept turned out to be motivating and effective for teachers' training and for chemistry classes at stage 10–11 (age 16–17). Good experience has also been gained with some simplified components from the PIN-Concept at stages 8–9, but this has not yet been investigated systematically.

Schlösser [22], Wenck [23, 24], and Christen [25] highlighted the advantages of an early introduction to organic chemistry in the 1970s and 1980s already. This good idea has unfortunately not yet been realized in the German syllabus to this very day, despite the fact that the possible benefits to biology and nutrition education are well known. It is necessary to improve this deficiency in the near future and to foster competences across subject borders.

The usability of a curriculum for such a diversified group of students requires a modular concept that leaves a margin for contents and methods. Therefore, the PIN-Concept is not an "all-or-nothing-concept," but a flexible modular system. The aspirational level can be varied for different addresses depending on their preknowledge and also depending on the intended complexity and connectivity of contents and methods.

The PIN-Concept can be characterized with seven didactically reasoned principles ([1], p. 1–29):

- Criterion of concreteness:
  New contents, concepts, and methods should always be introduced in class in a concrete way, i.e., on the basis of actual learners' experiences. This will mostly be experimental experience with substances and their observed properties including their chemical change under specific conditions. The transition to an abstract level of understanding should not be expected by the teacher before the students have become acquainted with a sufficient broad basis of phenomena and

H.-D. Barke et al., *Essentials of Chemical Education*,
DOI 10.1007/978-3-642-21756-2_9, © Springer-Verlag Berlin Heidelberg 2012

concretely ordered perceptions, which require explanation, generalization, and abstraction by the students themselves. In short: first the macroscopic level (substances and their properties under specified conditions), and only then the submicroscopic level (of particles and their properties) and the symbolic level (of abstract formulas and reaction symbols).

– Criterion of connection:
The degree of organization for the contents and methods should be as high as possible in the learning process. Different empirical facts (e.g., analytical, synthetic and spectroscopic data) should be analyzed amongst themselves and compared in relation to each other. New terms should always be introduced as knots in a web of relations – an important aspect that Sumfleth [26–28] and others emphasized and used for concept maps and advanced organizers. Note, however, that for beginners the discovery of experimentally observable relations between concrete substances and their chemical change should always precede the definition and connection of abstract terms and formulae ($\rightarrow$ criterion of concreteness). But this of course does not exclude a quick jump to the abstract level of representations for more advanced students. Indeed, the learning process can and should be accelerated more and more by constructing meaning in an abstract web of relations.

– Criterion of constructivist development:
Terms and reasoning patterns should always be built methodically, step by step, and in a guided inquiry-based learning process. A progressive, methodical structure of terms and arguments is necessary to make their interconnections understandable. According to Aebli [29] the learner has to construe every term in essence for themselves. Therefore, they need enough time and opportunities. The teacher's job is to encourage the construction of a convincing systematization and explanation of empirical facts by the students themselves and to provide alternative steps for such a learning process if they should fail and need help.

– Criterion of limitation:
New terms, methods, and thought patterns should be taught with only a few suitable examples until they are consolidated within this limited cognitive domain. Afterwards they have to be applied to other examples, to guard against functional fixedness. Unnecessary information, needless devices, and dispensable substances and formulae should be avoided. The limitation to relevant aspects of a learning situation prevents misdirecting stimuli ("noise") and encourages problem-solving processes. This has to be considered especially when experiments in the laboratory have to be planned: the learner, who is dealing with new substances, equipment, and complex experimental procedures, easily lapses into the mistake of thoughtless cooking procedures. It is well known that students often perform experiments in the science class with only a rudimentary idea of what they are doing, with virtually no understanding of the purpose of the experiment or of the reasons for the choice of procedure. Too often it seems that they are doing little more than following recipes. Their limited working memory cannot take the wealth of information that has to be coordinated in such a complex learning situation. Information overload

not only causes decreased learning efficiency, but also a loss of motivation in the longer term.

- Criterion of intelligent consolidation:
  Terms, methods, and thought patterns, which were built up certain examples, should be deepened in different variations and contexts in order to make them flexible for further transfer. Verbalization is therefore essential. Only things that one can put into one's own words consistently have truly been understood. Intelligent consolidation also supports "chunking," i.e., the organization of independent schemes to a new unit. According to Miller [30] working memory can keep only a fixed number of "chunks of information" that can be processed at the same time. In particular, grown-ups have only 7 ± 2 simultaneously operating memory spaces available for processing such "chunks of information." For younger students, this number is even smaller because of developmental psychological reasons. This underlines once more the correlation between the two previously mentioned criteria (i.e., the necessity of limitation and of consolidation). Overloading memory causes a rapid decrease of performance, as Johnstone [31–33] and others have convincingly demonstrated with empirical studies and analyses.

- Criterion of supporting cognitive skills:
  Cognitive skills can and should be supported, while respecting the natural limits set by maturation. The exposure to concrete problems in meaningful and everyday contexts [34–36] that require the use of these skills is important. The learner's repertoire of facts, operations, and strategies has a big impact on the cognitive performance and should therefore be extended. Consolidation and application in changing contexts and in the form of different types of exercises are also essential here. Students should become aware of their own thought patterns and should be able to abandon concrete external support step by step. This is necessary for improving their self-concept.
  Self-directed autonomy is essential for the cognitive development, according to Piaget (see Gräber and Stork [37]) and Aebli [29]. The learner has to be confronted with situations where the cognitive skills, which are to be supported, are required and experience can be gained. Cognitive structures in students' brains are differentiated, coordinated, and thereby developed gradually, but also changed if necessary. Conceptual change is based on the interdependency of self-consistent cognitive structures and the awareness of the specific requirements of the problem at hand.

- Criterion of scientific enculturation:
  Students should get to know contents and investigative methods in chemistry lessons that are representative for chemistry as a science, and that help them at the same time to understand their own everyday world better, and to develop their own cognitive skills. They should understand the chemist's language and argumentation to be led to an attitude that is empirical and rationally oriented. They should become aware of the value of scientific knowledge, as well as of its limits (Stork [38], Reiners [39]).

9.1   The Criteria in Relation

The criteria of concreteness and of connection are fundamental to the PIN-Concept right from the beginning. As the knowledge system grows, the other criteria that regulate the learning process become more and more important. However, the more global criteria of supporting cognitive skills and of scientific enculturation require long term and continuous efforts in reasonable contexts. It should be noted, however, that we do not implement these criteria in randomly replaceable contexts, but in the course of problem-solving and conceptualization processes that the growing knowledge system demands. These processes need to be built upon each other without gaps and to proceed step by step. The whole curriculum is laid out for continuous cognitive training and for the establishment of an empirical attitude. This can only be achieved, when all mentioned criteria are being used, but not one to the detriment of another.

In the following, one possible way will be described to explain the methodology of the PIN-Concept by example. The further development of the PIN-Concept, including possible variations and combinations of single components as well as options for the educational reduction and for the integration of interdisciplinary aspects and connections to everyday life, is described in the book "Organic chemistry education according to the PIN-Concept" [1].

9.2   Sorting Unknown Substances with Unknown Reagents

The identification of six unknown pure substances A–F is the focus of this unit. These are all colorless liquids, which will be labeled later. To gather information about them, they will be compared by means of tests under standard conditions with the help of six reagents, which are also unknown at first ([1], S. 295). The results will be displayed in a matrix (Fig. 9.1, see color chart).

It might seem unusual to try to gain information by a combination of unknowns with unknowns. This method was even recommended by the famous chemist Justus von Liebig, who wrote in his popular chemical letters in 1844 ([40], S. 11): "We study the characteristics of bodies, the changes they suffer, when coming into contact with others. All observations together form a language. Every characteristic, every change that we discover on the body, is a word in this language. The bodies show relations to others, they resemble them with certain characteristics or they differ from them. These differences are as manifold as the words of the richest language are. In their meaning, in their relation to our senses they are not less different. But to be able to read this book with its unfamiliar ciphers and to understand it, one has to get to know the alphabet first."

Liebig advises to thoroughly study the "ABC of phenomena" first (see Figs. 9.1 and 9.2), till sufficient similarities and differences between the unknown substances have been found on the phenomenological level. In this way, students discover that

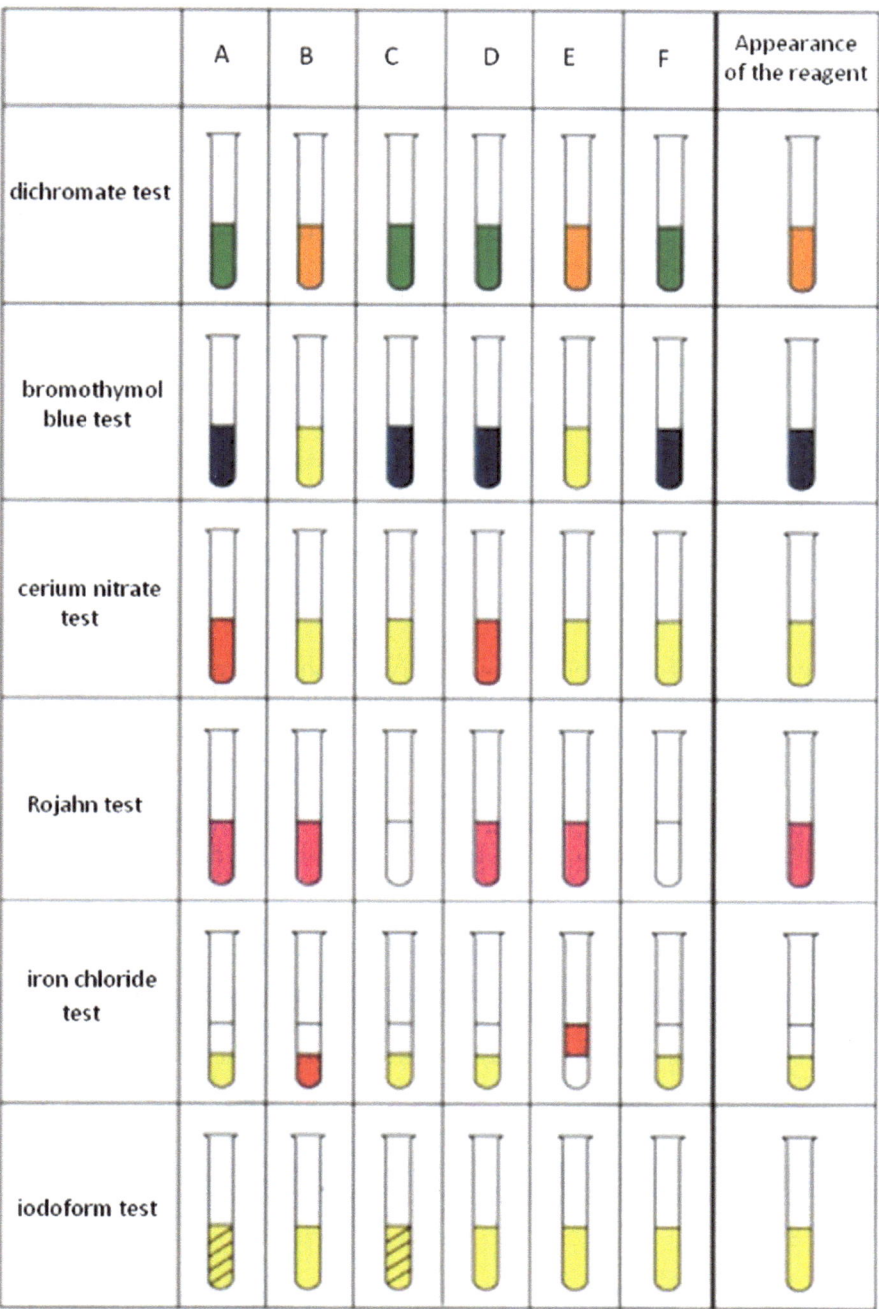

Fig. 9.1 Behavior patterns of the substances A–F when they are tested under standardized conditions with six reagents (dichromate test, bromothymol blue test, cerium nitrate test, phenol-phthalein or Rojahn test, iron chloride test, iodoform test)

the six substances can be categorized into three groups of two according to their test results. They are now labeled per definition (in accordance with the chemical nomenclature which chemists use for these substances) with family names (Fig. 9.2, see color chart):

$$A/D = alcohols \quad B/E = carboxylic\ acids \quad C/F = esters.$$

Therefore (as a consequence of this definition), the cerium nitrate test (reddening) is a test for alcohols, the bromothymol blue test (yellowing) responds to acids, and the phenolphthalein test (decoloration), which is also called Rojahn test, is a test for esters.

Differentiations within the families can be made with the iron chloride or the iodoform tests. The esters are not miscible with water – contrary to the alcohols and carboxylic acids. This miscibility pattern also supports the above classification.

9.3   Interconnection of Substances by Means of Chemical Synthesis

Since both esters and alcohols respond positively to the dichromate test (the orange color of the reagent changes to green) a chemical reaction must have taken place. Which products have been formed? To answer this question, the dichromate test will be performed work-sharing on a preparative scale. Please note safety and disposal instructions for chromium salts! ([1], p. 36–37). A few milliliters of the reaction product are to be distilled off using the standard apparatus (Fig. 9.3). Standard tests are to be run with the distillates. The results are displayed in Table 9.1 ([1], p. 239).

Comparing the behavior patterns of the distillates (Experiments 1–4) with the reference substances (see Table 9.1) indicates the following results:

Experiment 1: alcohol A has been transformed to the carboxylic acid B, i.e., A → B.
Experiment 2: alcohol D has been transformed to the carboxylic acid E, i.e., D → E.
Experiment 3: ester C has been transformed to the carboxylic acid B, i.e., C → B.
Experiment 4: ester F has been transformed to the carboxylic acid E, i.e., F → E.

Substances A–F obviously form a system that is characterized by horizontal synthesis connections and vertical analytical connections (Fig. 9.4).

The discovery of a system allows for the introduction of systematic names for the six substances:

$$A = ethanol \quad B = ethanoic\ acid \quad C = ethanoic\ ethyl\ ester$$
$$D = propanol \quad E = propanoic\ acid \quad F = propanoic\ propyl\ ester$$

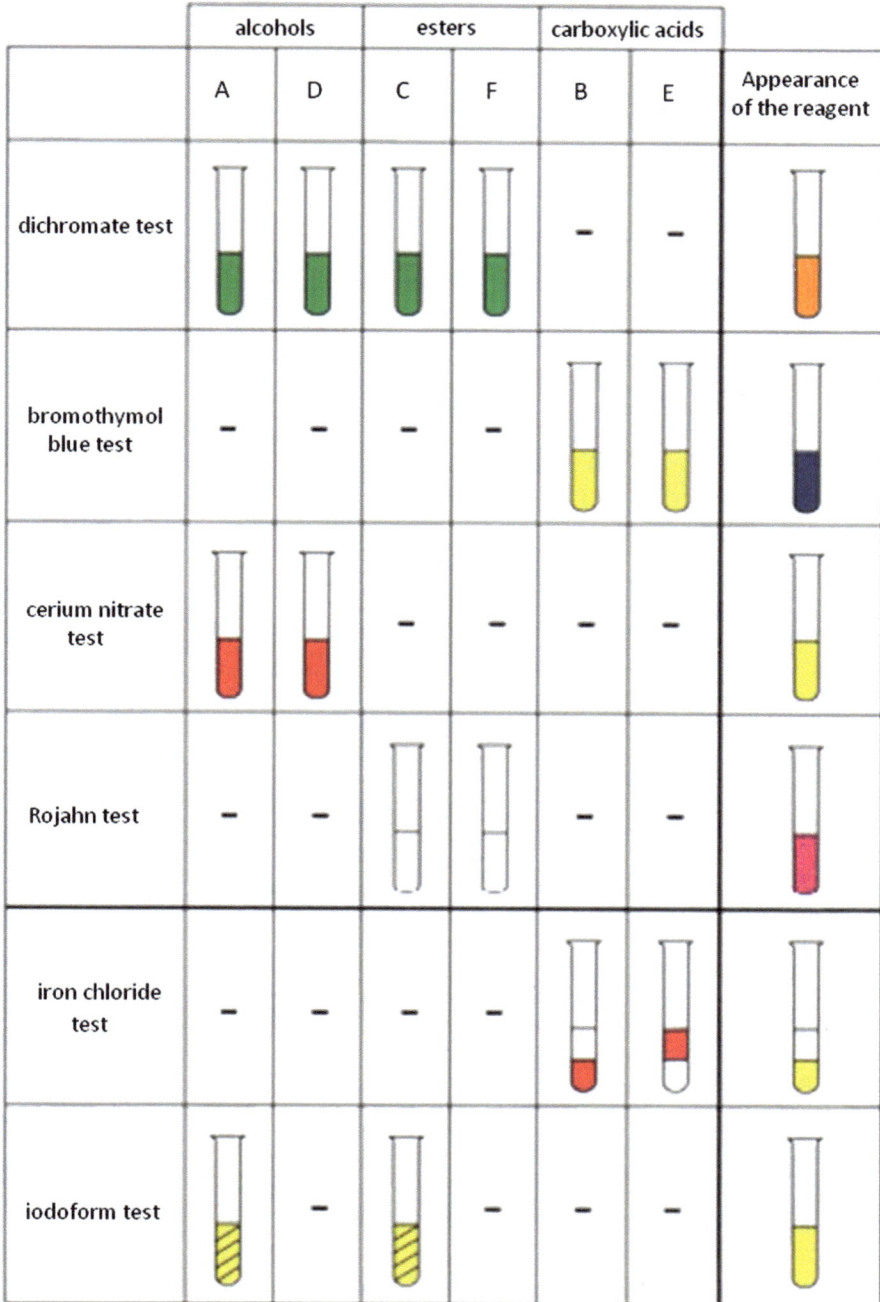

	alcohols		esters		carboxylic acids		Appearance of the reagent
	A	D	C	F	B	E	
dichromate test					–	–	
bromothymol blue test	–	–	–	–			
cerium nitrate test			–	–	–	–	
Rojahn test	–	–			–	–	
iron chloride test	–	–	–	–			
iodoform test		–		–	–	–	

Fig. 9.2 The rearranged table for the behavior patterns of the substances A–F. Note that negative test results are represented by a *dash*

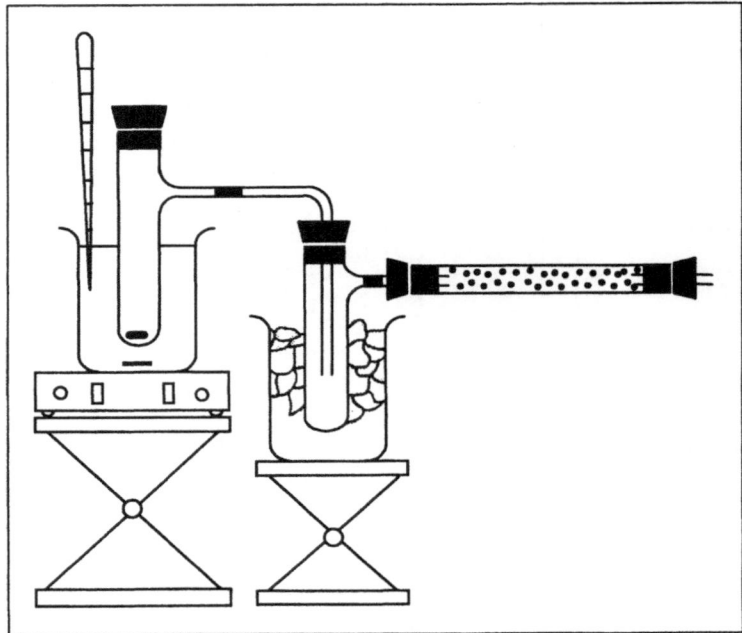

Fig. 9.3 A typical synthesis apparatus ([1], p. 239), which can be used for the investigation of many reactions within the PIN-Concept

Table 9.1 Investigation of the distillates from the reactions of the alcohols A/D and the esters C/F with the dichromate reagent

	Alcohols		Esters		Carboxylic acids		Distillates			
	A	D	C	F	B	E	E1	E2	E3	E4
Dichromate test	+	+	+	+	−	−	−	−	−	−
BTB test	−	−	−	−	+	+	+	+	+	+
Cerium nitrate test	+	+	−	−	−	−	−	−	−	−
Rojahn test	−	−	+	+	−	−	−	−	−	−
Iron chloride test	−	−	−	−	$+_L$	$+_u$	$+_L$	$+_u$	$+_L$	$+_u$
Iodoform test	+	−	+	−	−	−	−	−	−	−

Exp. 1: Reaction of alcohol A. Exp. 2: Reaction of alcohol D. Exp. 3: Reaction of ester C. Exp. 4: Reaction of ester F

9.4   Analysis of Household Substances

Household substances will be tested next. They are to be provided for the students in their original packing. The results are to be presented as described in Table 9.2 ([1], p. 313).

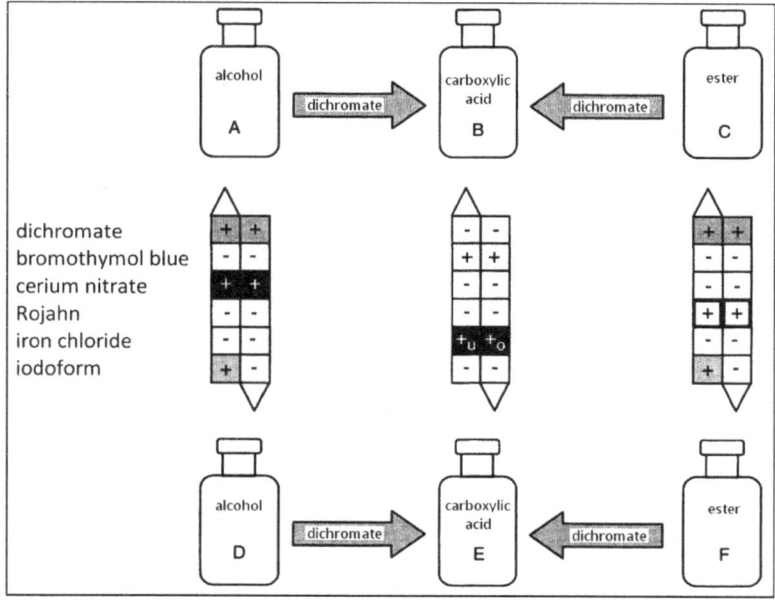

Fig. 9.4 The six substances A–F under investigation build a system, which is characterized by horizontal relationships (dichromate syntheses) and by vertical relationships (test reactions)

Table 9.2 Test results for everyday products

	Alcohols		Esters		Carboxylic acids		Household substances				
	A	D	C	F	B	E	1	2	3	4	5
Dichromate test	+	+	+	+	–	–	–	+	+	+	–
BTB test	–	–	–	–	+	+	+	–	–	–	–
Cerium nitrate test	+	+	–	–	–	–	–	+	+	+	–
Rojahn test	–	–	+	+	–	–	–	–	+	+	–
Iron chloride test	–	–	–	–	$+_L$	$+_u$	$+_L$	–	–	–	–
Iodoform test	+	–	+	–	–	–	–	+	+	+	+

1 = vinegar or vinegar essence
2 = alcoholic drinks
3 = spot remover (e.g., "spot devil")
4 = nail polish cleanser (e.g., Ellocar)
5 = nail polish cleanser (e.g., Margret Astor)

The following conclusions can be drawn (from the reference table, i.e., from the student's point of view):

– Vinegar only contains carboxylic acid B = ethanoic acid = acetic acid
– Alcoholic beverages contain alcohol A = ethanol. Therefore, ethanol is drinking alcohol. But the test results do not show if alcohol A is the only alcohol. The

substances might additionally contain alcohol D = propanol. It has to be made clear that alcoholic beverages only contain ethanol.

– Spot remover does certainly not contain carboxylic acids, but at least one alcohol and one ester. The following combinations are possible:

Combinations of two:	A/C	A/F	D/C	(not: D/F!)
Combinations of three:	A/C/D	A/F/D	D/C/F	A/C/F
Combinations of four:	A/C/D/F			

– Nail polish remover no. 4 shows the same test results as spot remover. Different household products might contain the same substances.
– Contrary to nail polish remover no. 4, nail polish remover no. 5 does not contain any of the six substances A–F. Similar household products might contain different substances.
– Additionally, it can be readily discovered effort that glue contains at least one ester (positive Rojahn test) ([1], p. 315). A smell test with substance C indicates that the glue contains ethanoic ethyl ester. This is confirmed by the positive iodoform test. Note that the ester functions as a solvent, which evaporates when the glue is being applied.

Testing household substances provides good opportunities to connect the system building substances A–F with aspects of everyday life. Synonymous trivial names that have to be learned should be added to the systematic names. The working language should not disconnect itself from communication in everyday life, media, and science:

A = ethanol = drinking alcohol (in alcoholic drinks)
B = ethanoic acid = acetic acid (in vinegar)
C = ethanoic ethyl ester = ethyl acetate (solvent for glue)
D = propanol = propyl alcohol (solvent for cosmetics)
E = propanoic acid = propionic acid (preserving agent for bread)
F = propanoic propylester = propyl propionate (nonpolar solvent)

The microbiological production of acetic acid can be discussed in detail in this context: Ethanol is transformed by *Acetobacter* bacteria with the oxygen in air to acetic acid. One hundred gram of vinegar contain about 5–15 g of acetic acid. Vinegar essence contains 60–80 vol% acetic acid and therefore it is a highly corrosive substance, which can only go on sale under safety measures. It is being diluted with water to produce table vinegar.

Students should be told that for safety reasons only diluted acids are being used in the student experiments in Sect. 9.1.

The function of the dichromate reagent as an oxidizing agent can be understood by comparing the microbiological acetic acid synthesis with the acetic acid synthesis discovered in Sect. 9.2 (from ethanol and dichromate). Since the orange-colored dichromate is being reduced to green-colored chromium salts, the students are now able to understand the green coloration in the positive dichromate test.

The alcohol test used by the police in former times is based on this color change: the detector tube contains the dichromate reagent, which oxidizes the exhaled alcohol vapor from the traffic offender. It should not be talked about the primary oxidation product (acetaldehyde) before the students could gain experience with this substance (see [1], p. 90 ff.).

9.5   Discovery of Other Synthesis Relations

In the next step, we will test how the six standard substances react with sodium hydroxide solution ([1], p. 249).

Ethanoic ethyl ester (substance C) reacts quickly with sodium hydroxide solution. The ester phase and the ester smell have already disappeared after a couple of minutes. The reaction mixture has to be distilled for further testing. Distillate and residue are to be tested with the standard tests.

In contrast to these observations, propanoic propyl ester (substance F) reacts very slowly with sodium hydroxide solution. The volume of the ester phase has reduced only by half after 1 week. Ester phase and aqueous phase are to be tested without distillation.

The results of these two experiments can be found in Table 9.3.

The following can be concluded from Table 9.3:

- Ethanol can be found in the distillate in experiment 5 (it might also contain propanol).
  Acetic acid can be found in the residue (after acidifying with sulfuric acid). Therefore, the following reaction has taken place:
  Ethanoic ethyl ester has been transformed by sodium hydroxide solution to a mixture of acetic acid and ethanol (and possibly propanol).
- Apart from remaining ester, propanol can be found in the upper phase in experiment 6. The presence of ethanol can be excluded due to the negative result of the iodoform test. Propanoic acid can be found in the lower phase (after acidifying with sulfuric acid). The following reaction has taken place:

Table 9.3 Investigation of the reactions of esters with sodium hydroxide solution

	Alcohols		Esters		Carboxylic acids		Exp. 5		Exp. 6	
	A	D	C	F	B	E	Dist.	Res.	UP	LP
Dichromate test	+	+	+	+	−	−	+	−	+	−
BTB test	−	−	−	−	+	+	−	+	−	+
Cerium nitrate test	+	+	−	−	−	−	+	−	+	−
Rojahn test	−	−	+	+	−	−	−	−	+	−
Iron chloride test	−	−	−	−	$+_L$	$+_u$	−	$+_L$	−	$+_u$
Iodoform test	+	−	+	−	−	−	+	−	−	−

Exp. 5: Reaction of ester C (both the distillate and the residue are tested). Exp. 6: Reaction of the ester F (both the upper phase (UP) and the lower phase (LP) are tested)

Propanoic propyl ester has been transformed by sodium hydroxide solution to propanoic acid and propanol.

– For reasons of analogy and simplicity, it can be assumed that the only products of the alkaline hydrolysis of ethanoic ethyl ester are acetic acid and ethanol, not propanol.

To support this hypothesis it will be tried to reproduce the corresponding ester from the cleavage products of the ester hydrolysis in the next step. Sulfuric acid can be used as synthesis reagent (as an antagonist for the sodium hydroxide solution, used in the ester hydrolyses). Note that the antagonistic reason is not compulsory, but it makes the choice of the synthesis reagent reasonable for the student. Mixtures of ethanol/acetic acid and propanol/propionic acid, respectively, are thus to be laced with concentrated sulfuric acid ([1], p. 247). The reaction mixtures can already be poured onto water after a couple of minutes. Ester phases indeed separate out in both cases. They show the behavior pattern of ethanoic ethyl ester and propionic propyl ester, respectively (after washing with sodium carbonate solution). The discovered synthesis relations are displayed and summarized in a net of synthetic pathways (Fig. 9.5).

9.6   Leaping to the Particle Level

The leap from the phenomena to the particle level is abstract by nature and requires formal operational thinking skills for a deep understanding: "Science is by its very nature formal" (Herron [41]). However, it cannot be ignored that the majority of students at high school level (and certainly at middle school level) has substantial problems with reflective, formal operational thinking (Gräber and Stork [37], Shayer and Adey [42], Lawson [43], Häußler, Bünder, Duit, Gräber, Mayer [44]).

For educational psychological reasons, but also due to time limits, it is not possible to simultaneously work on the structural formulae of alcohols, carboxylic acids, and esters with conventional methods of structure analysis. There is no best way that directly leads from the phenomena to the formula. Constructive assumptions are necessary, where eligibility cannot be derived from single phenomena. The explanation for the formula rather results in retrospective from the immediate success of its application in the interpretation of a system of interrelated phenomena, which have been discovered empirically without formulae.

At this point we make use of spectroscopic measurements (simplified ^{13}C-NMR and mass spectra [17] of the substances A–F). Students cannot measure these spectra by themselves, of course, but they are able to work on their concrete operational analysis. The fundamentals of these analytical techniques do not have to be explained in detail before starting with the analysis.

The following assumptions are necessary:

– Molecules of the substances A–F only contain carbon, hydrogen, and oxygen atoms [this can also be determined experimentally, if necessary ([1], p. 317)].
– Covalence and relative masses of the three kinds of atoms C/H/O, as well as the possibility of double bonds, have to be familiar or have to be assumed.

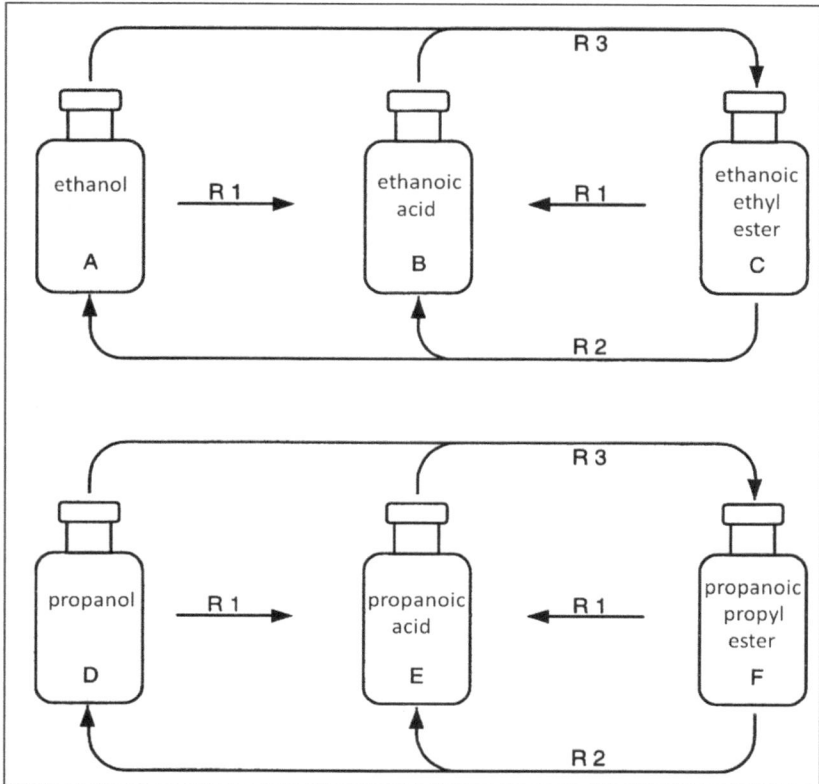

Fig. 9.5 Synthetic pathways of the substances A–F. R1: Reactions with the dichromate reagent. R2: Reactions with sodium hydroxide solution (ester hydrolysis). R3: Reactions with sulfuric acid (ester synthesis)

- It has to be explained that one can find out the number of carbon atoms in a molecule with the help of the ^{13}C-NMR spectrum: Every signal in the spectrum can be matched with a carbon atom of the molecule. Carbon atoms with the same substituents give the same signal (the signals coincide at the same spot of the spectrum).
- Regarding the mass spectrometry students have to know that a mass spectrum gives information about the molecular mass and about the masses of molecular fragments.
- The theoretical background information of the spectroscopic methods can be discussed at this stage or delivered later as needed.

The spectral analysis (Fig. 9.6) will be explained using the example substance A (ethanol):

The molecular mass 46 u can be extracted from the mass spectrum. This leads to three structural hypotheses H1–H3 (taking into account the assumed atomic masses and bonding rules):

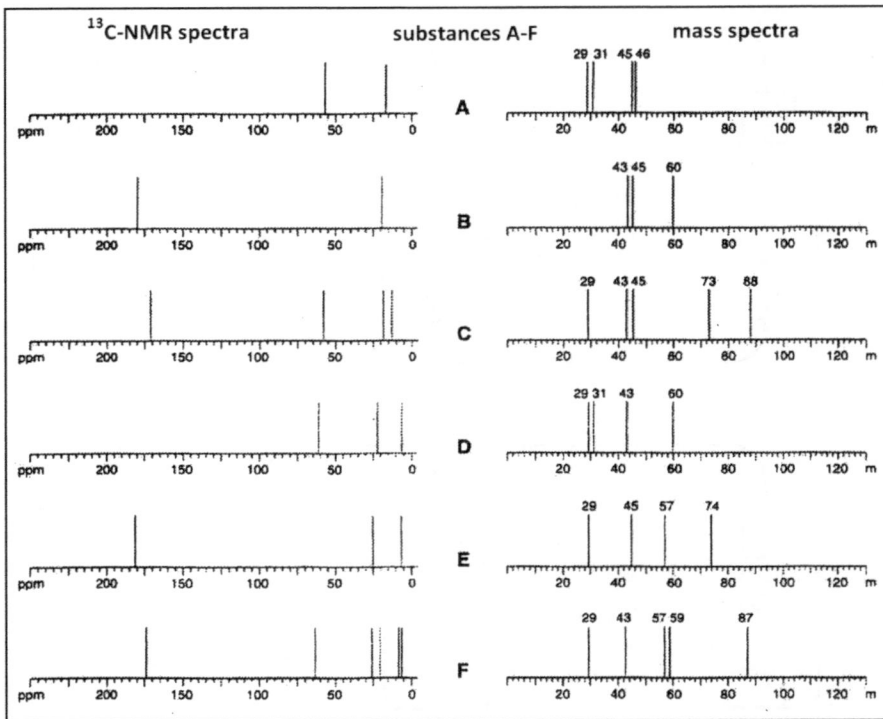

Fig. 9.6 Simplified ^{13}C-NMR spectra and mass spectra of the substances A–F

$$
\begin{array}{ccc}
\overset{\displaystyle O}{\underset{\displaystyle H1}{\overset{\|}{H-C-OH}}} & \underset{\displaystyle H2}{H_3C-CH_2-OH} & \underset{\displaystyle H3}{H_3C-O-CH_3}
\end{array}
$$

- Only one signal (not two) can be expected in the ^{13}C-NMR-spectrum, when starting from hypothesis H1 and H3. Therefore, these hypotheses are not suitable. Only H2 is compatible with the ^{13}C-NMR spectrum.
- Hypothesis H2 can also explain the other signals in the mass spectrum (29 u, 31 u, 45 u), because the corresponding fragments are contained in the formula H2.

$$
\begin{array}{ccc}
H_3C-CH_2- & -CH_2-OH & H_3C-CH_2-O- \\
29\ u & 31\ u & 45\ u
\end{array}
$$

The spectra of the five other substances can be analyzed in the same way [17]. A mass table that shows the most important molecular fragments, arranged according to their masses, should be available for the students as a tool (Fig. 9.7). The results are shown in Fig. 9.8.

mass
possible fragments of the molecules
15 u
29 u
31 u
43 u
45 u

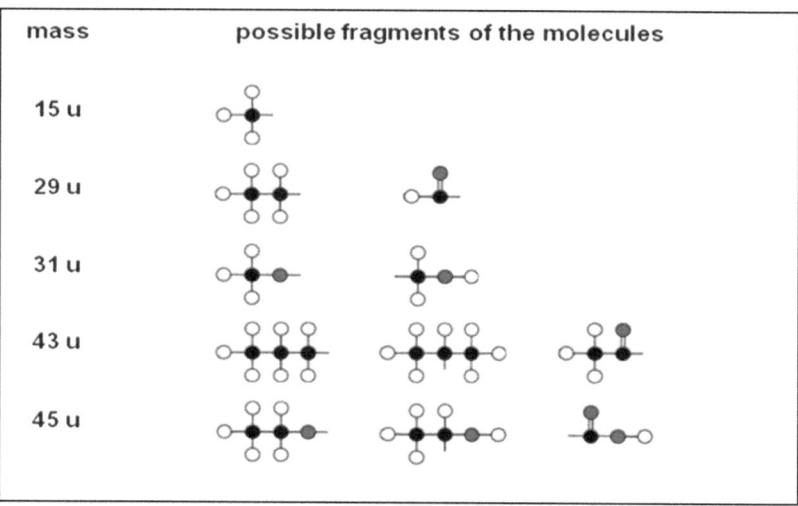

Fig. 9.7 A mass table with possible fragments is helpful for analyzing mass spectra

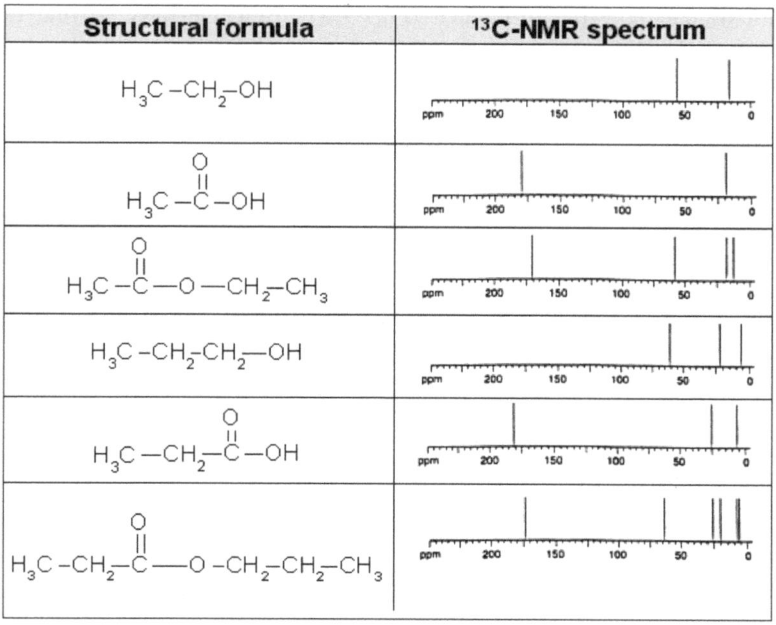

Structural formula
13C-NMR spectrum
H3C-CH2-OH
H3C-C(=O)-OH
H3C-C(=O)-O-CH2-CH3
H3C-CH2-CH2-OH
H3C-CH2-C(=O)-OH
H3C-CH2-C(=O)-O-CH2-CH2-CH3
ppm
200
150
100
50
0

Fig. 9.8 ^{13}C-NMR spectra of the substances A–F and the formulae of the corresponding molecules

To give an impression of the 3D structures, it is advisable to build the corresponding molecular models with the help of a molecule kit. The only thing, however, that is important for the explanation at this level is the constitution of the molecules.

9.7 Use of Formulae ("Combining Structure and Properties")

The advantage of the depicted approach is that students are able to construct formulae rather quickly. What is even more important is that students can immediately assure themselves of the formulae's potential value. They can use these symbols for the interpretation of all the analytical and synthetic relations between the substances A–F, which they have already discovered on the phenomena level:

- The pair wise similar analytical properties (Fig. 9.2, see color chart) of substances A/D (alcohols), B/E (carboxylic acids), and C/F (esters) can apparently be ascribed to the pair wise similar molecular structures. The functional groups that are responsible for the similar properties in pairs can now be identified (Fig. 9.9).
- The pair wise similar synthesis behavior (Fig. 9.5) of substances A/D, B/E, and C/F can also be ascribed to the functional groups. First (very simplified) reaction symbols can now be formulated (Fig. 9.10). Students have to find out the invariant structure characteristics of the educt and product molecules, by comparing their structural formulae. They recognize the partial structure of the educt molecule that survives the reaction unaltered. The synthesis reagent and, if applicable, the solvent (water) have to be responsible for the difference. This leads to simplified chemical equations ([1], p. 60 ff.). Formal operational skills are not necessary for their deduction. Complete redox reaction equations can be discussed here or later.
- The pair wise similar spectroscopic properties (Fig. 9.6) of substances A/D, B/E, and C/F do not only support the functional group concept, but also make the discovery of important rules possible which govern the positions of the signals in the ^{13}C-NMR spectra. Pattern recognition reveals (Fig. 9.8):

  – Carbon atoms that are bonded to *two* oxygen atoms give a signal at a high ppm value (165–185 ppm). Partial structures of this kind are typical for carboxylic acid and ester molecules.
  – Carbon atoms that are bonded to *one* oxygen atom give a signal at an intermediate ppm value (55–65 ppm). Partial structures of this kind are typical for alcohol and ester molecules.
  – Carbon atoms that are bonded to *no* oxygen atom give a signal at a low ppm value (5–30 ppm). This is typical for the hydrocarbon chains of all molecules including the alkanes.

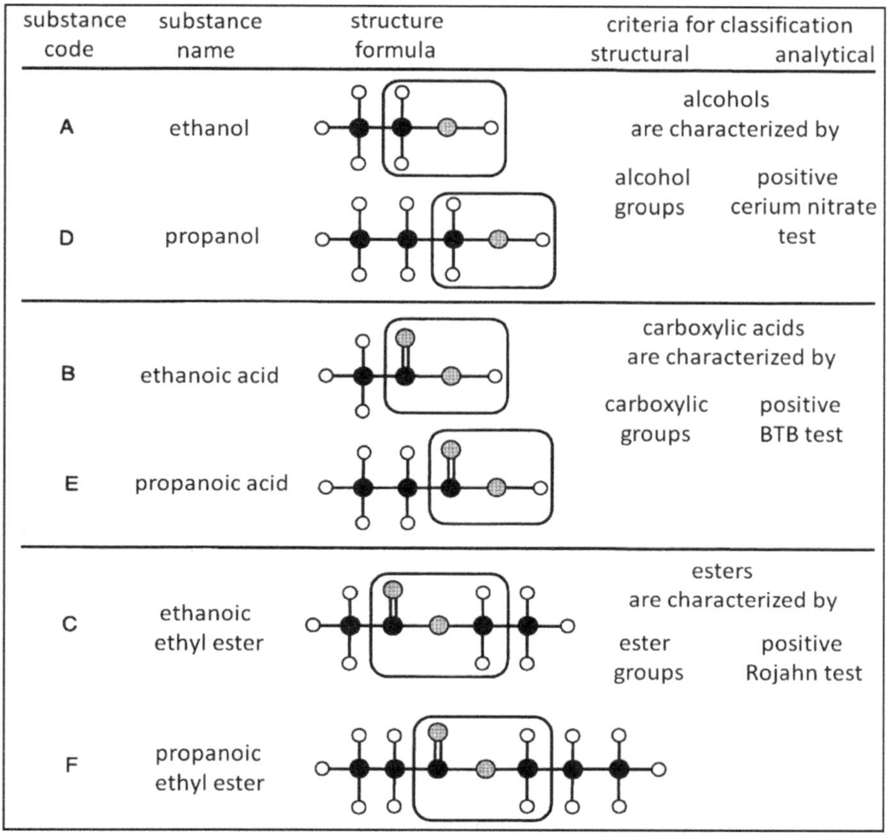

substance code
substance name
structure formula
criteria for classification
structural
analytical
A
ethanol
D
propanol
alcohols
are characterized by
alcohol groups
positive cerium nitrate test
B
ethanoic acid
E
propanoic acid
carboxylic acids
are characterized by
carboxylic groups
positive BTB test
C
ethanoic ethyl ester
F
propanoic ethyl ester
esters
are characterized by
ester groups
positive Rojahn test

Fig. 9.9 The functional groups of the molecules A–F and their relation to the test results

ethyl alcohol
dichromate
ethanoic acid
dichromate
ethanoic ethyl ester

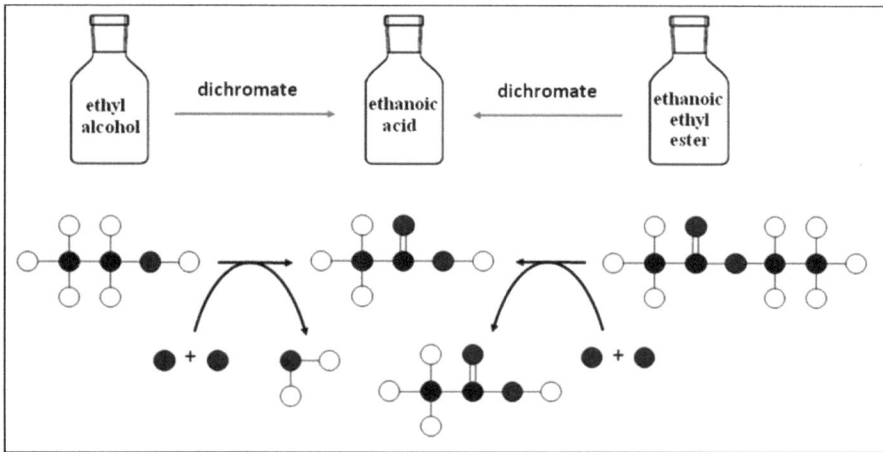

Fig. 9.10 Correlating structural formulae of the molecules A–C with the results of the dichromate syntheses. The oxygen atoms are supplied by the dichromate reagent (oxidizer)

9.8   Integration of Other Substances

With their experience of the exemplary substances A–F, the students now have a good basis for the integration of other substances. On this occasion, they can strengthen, extend, and deepen their skills for the structure–properties-thinking. Three examples will explain this matter.

9.8.1   Analysis of a Natural Substance

The students analyze an unknown natural substance X that can be found in the leaves of wild vines. They find out: X has a positive result in the BTB test (group test for carboxylic acids) and in the cerium nitrate test (group test for alcohols). The dichromate test (for oxidizable substances) is also positive. All other test results are negative. Possible formulae for the molecule of substance X are ([1], p. 70):

$$
\underset{(1)}{\underset{\underset{OH}{|}}{H_2C}-\overset{\overset{O}{\|}}{C}-OH}
\qquad
\underset{(2)}{H_3C-\underset{\underset{OH}{|}}{CH}-\overset{\overset{O}{\|}}{C}-OH}
\qquad
\underset{(3)}{HO-\overset{\overset{O}{\|}}{C}-\underset{\underset{OH}{|}}{CH}-\underset{\underset{OH}{|}}{CH}-\overset{\overset{O}{\|}}{C}-OH}
$$

The ^{13}C-NMR spectrum of X gives two signals (at 176 ppm and at 59 ppm). Therefore, only formulae (1) and (3) are possible. These formulae are also in accordance with the rules for the ppm values which students have discovered before (see Sect. 9.6) by pattern recognition. The unknown substance could therefore be glycolic acid (1) or tartaric acid (3), but not lactic acid (2).

The mass spectrum of X gives three signals (31 u, 45 u, and 76 u). This result goes with the fragments $–CH_2OH$ (31 u) and $–COOH$ (45 u). Formula (3) is therefore excluded, because it cannot explain the fragment with a mass of 31 u, and its molecular mass (150 u) is too large. Formula (1), however, is in line with all results. The natural substance X is glycolic acid, which is contained in the leaves of wild vines.

9.8.2   Development of the Term "Homologous Series"

Students analyze unknown substances, which are only labeled by numbers. The test results are displayed in Table 9.4 (see [1], p. 65 and [14]):

Apparently, the substances are oxidizable alcohols. All four alcohols can be esterified in a test tube experiment with acetic acid (in the presence of sulfuric acid).

Table 9.4 Some properties of the unknown substances 1–4 being investigated by the students

	1	2	3	4
Dichromate test	+	+	+	+
Cerium nitrate test	+	+	+	+
Esterifiable	+	+	+	+
Soluble in hexane	−	+	+	+
Soluble in water	+	+	+	−
Soluble in salt water	+	+	−	−
Molecular mass (± 2 units)	32 u	46 u	61 u	73 u

Due to their solubility, they can be characterized as follows: alcohol 1 (methanol) is the most polar, alcohol 4 (butanol) is the least polar.

The molecular masses of the four alcohols can be determined experimentally according to the method of Heimann and Harsch [16]. Typical results are also listed in Table 9.4.

On this basis, the students can hypothesize on the structural formula of the alcohol molecules. The hypotheses can be tested and verified with the help of the ^{13}C-NMR spectra (Fig. 9.11). The term "homologous series" can now be defined on an experimental and structural basis [14].

– Macroscopic definition of the homologous series:
   Substances being members of a homologous series belong to the same substance class. They have similar chemical properties (e.g., test behavior and syntheses). Their physical properties (e.g., boiling temperatures, solubilities, molar masses) follow a monotonous trend thus forming a rank.
– Submicroscopic definition of the homologous series:
   Molecules being members of the homologous series have the same functional groups, but they differ with respect to the length of their hydrocarbon chains by a methylene group from one member to the next in the rank.

A demonstration experiment (Fig. 9.12) on an overhead projector works well for the visualization of the graded water solubility of homologous alcohols [13]. A graded movement during the solution process can be observed for the short-chained alcohols due to the better water solubility. The long-chained alcohols react rather "sluggishly."

It has to be pointed out that alcohol molecules like methanol and ethanol, which only differ by one CH_2-group have similar chemical properties, but regarding their physiological effects, the difference of one CH_2-group can kill:

– An adult, who drinks 60 ml of ethanol, gets a blood alcohol concentration of about 1 promille. This makes him or her dizzy without serious harms.
– Somebody who drinks 60 ml of methanol will probably be dead after only 20 h of latency.

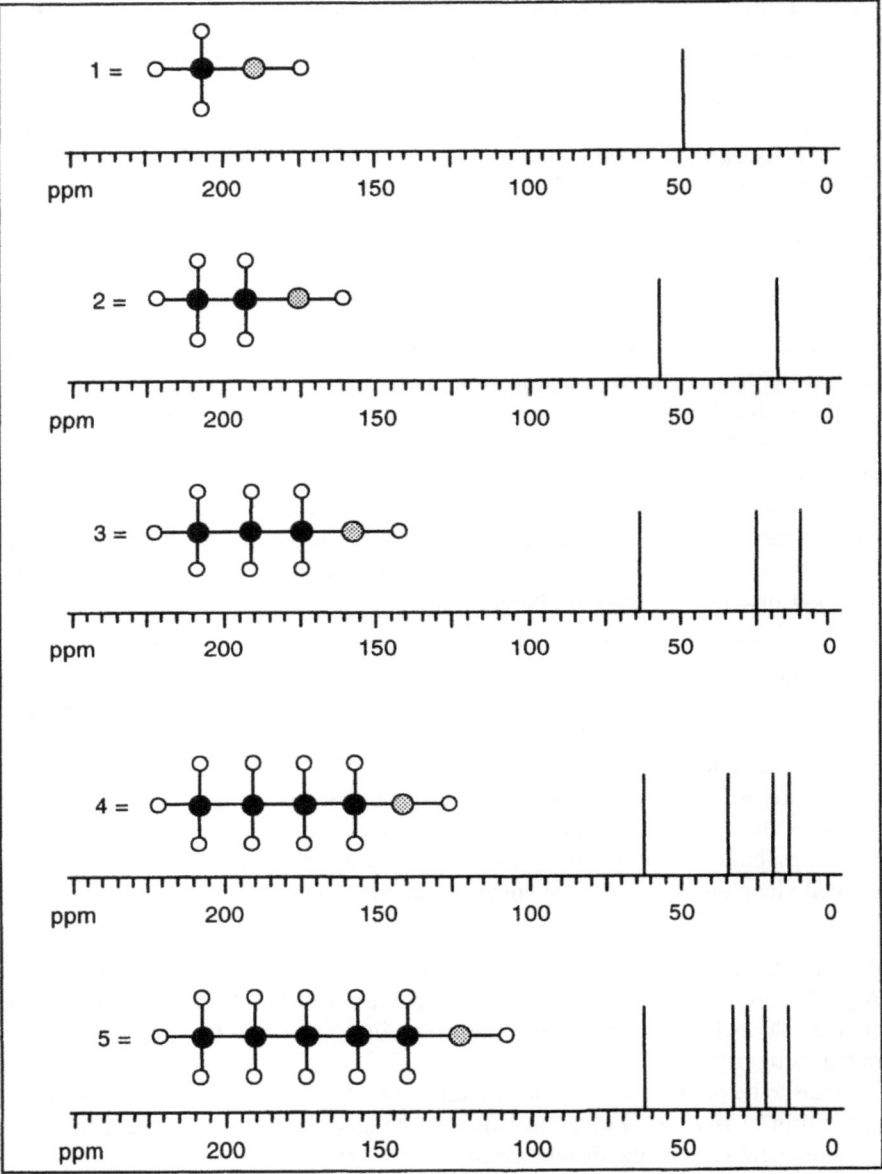

Fig. 9.11 ^{13}C-NMR spectra of the homologous series of alcohols 1–5 (from methanol to 1-pentanol). Note that the signals near 50–60 ppm can be assigned to the carbon atoms being attached to an oxygen atom

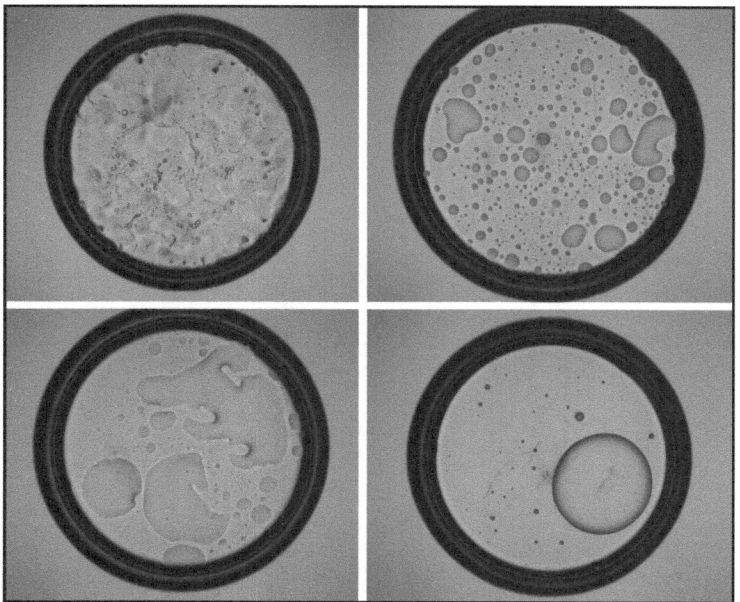

Fig. 9.12 Mixing other alcohols with water in Petri dishes in overhead projection [13]. 1 = butanol, 2 = pentanol, 3 = hexanol, 4 = octanol

9.8.3   Many Ways Lead to Acetic Acid

Synthesis experiments can be effectively used for a further integration of knowledge. A whole range of substances can be converted into acetic acid with the help of the dichromate reagent ([1], p. 239): ethanol, acetaldehyde, acetal, ethanoic ethyl ester, pyruvic acid, lactic acid, acetone, and malonic acid (Fig. 9.13). Acetic acid can be identified in the distillates with the help of the BTB and the iron chloride test. Note that for the oxidation of acetone to ethanoic acid stronger reaction conditions are necessary.

By oxidative decarboxylation of pyruvic acid, lactic acid, malonic acid, and acetone, two substances form: acetic acid and carbon dioxide. The gas can be identified easily.

The discovered reactant–product relations can also be interpreted on the molecular level, when the structural formulae of the reactant molecules are familiar (or when they are being introduced with the help of spectroscopic methods or by other assumptions). Substances, which belong to different families, can thus be connected with the help of the dichromate reagent to a system of synthetic relations that end in one shared center (acetic acid). This unexpected connection of substances can support the coherency of the knowledge and the student's systematic thinking.

The interdisciplinary aspect is also very interesting: in biology lessons students get to know the central role of the activated acetic acid (acetyl-coenzyme A) within

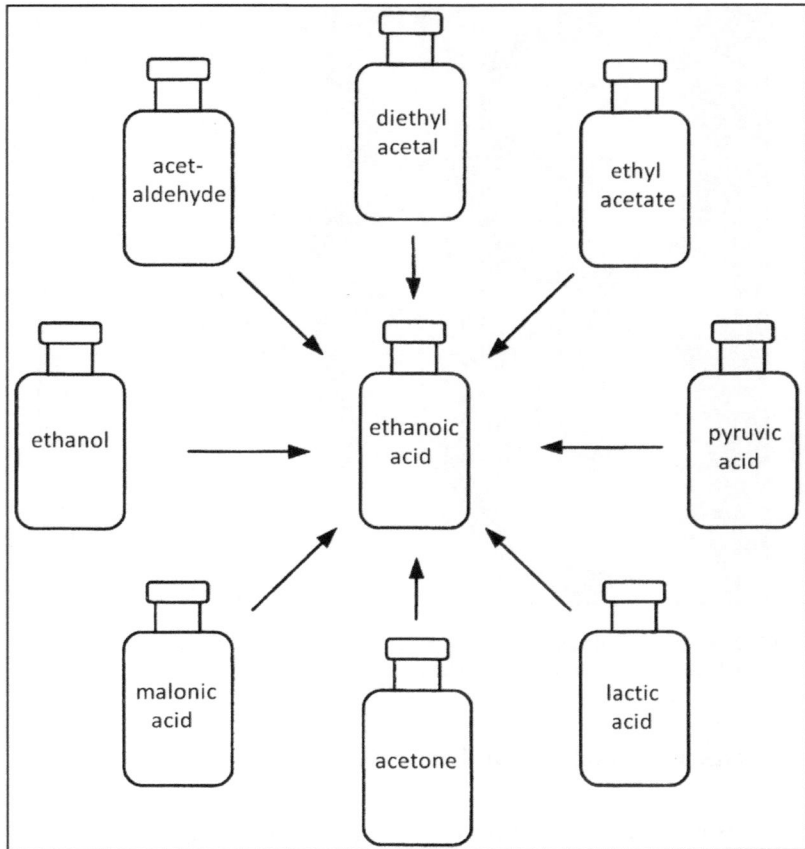

Fig. 9.13 Many synthetic pathways end with acetic acid, starting with ethanol, acetaldehyde, acetal, ethyl acetate, pyruvic acid, lactic acid, acetone, and malonic acid. These reactions can be accomplished by means of the dichromate reagent as an oxidizer [6]. Note that for the oxidation of acetone, more rigorous conditions are needed. For the decarboxylation of malonic acid, no oxidizer, but only heating is needed

the biochemical metabolic processes. Many reactions in the organism lead to the stable acetic acid molecule. Oxidation reactions like ethanol → acetaldehyde → acetic acid, as well as oxidative decarboxylation reactions such as lactic acid → pyruvic acid → acetic acid + CO_2 also play an important role here. In living organisms, the oxidizer is NAD+, corresponding to the dichromate oxidizer in the test tube experiments. So chemistry lessons can support biology lessons by providing an exemplary basis of biochemically relevant substances through in vitro syntheses.

   Not only the above named substances, but others like citric acid, ascorbic acid, amino acids, proteins, fats, and carbohydrates as well as aromatic compounds and dyes can be integrated into the PIN-Concept thus forming a growing system [1].

9.9   Other Associable Experiments and Concepts

The PIN-Concept is based on many interrelated experiments ([1], Chap. 27). Only some of them can be conducted in any given teaching and learning situation. This is not disadvantageous, because the modular design of the PIN-Concept allows various combinations without losing the overall context. Furthermore, the experimental possibilities are not exhausted at all through this pool of experiments. Since the PIN-Concept is build on system genesis and system growth it offers many opportunities for the integration or association of further experiments. These can be found in chemistry educational literature and add to the PIN-Concept beyond the context of their original intention. Experiments that can be integrated into the net of fundamental organic substances (Figs. 9.5, 9.12 and 9.15) are of particular interest. The following are stated exemplarily:

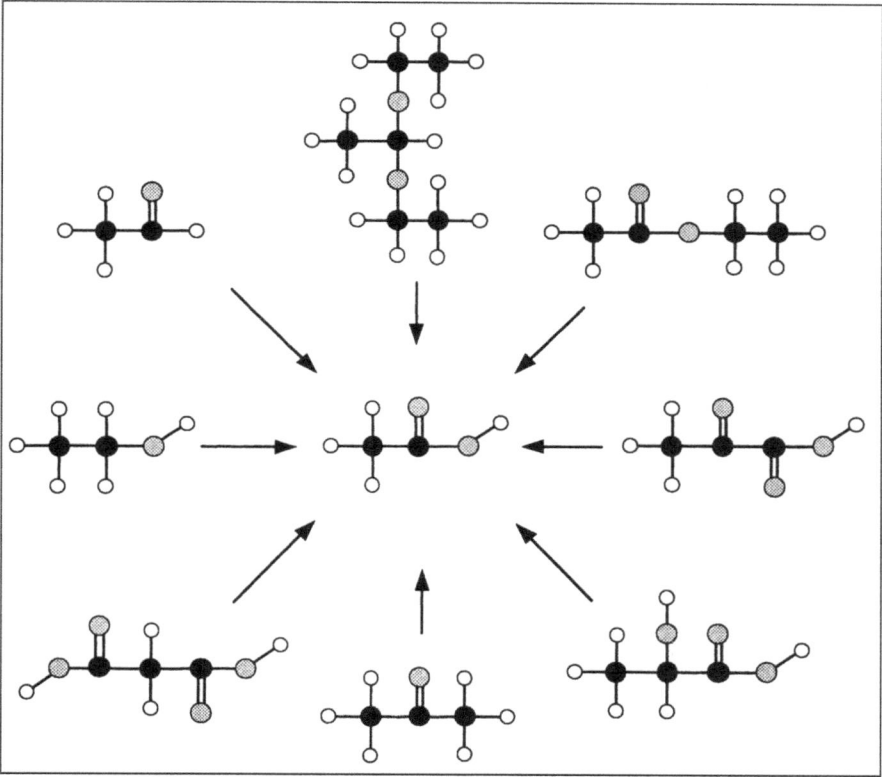

Fig. 9.14 Interpretation of the acetic acid syntheses on a molecular level [6]. Compare with Fig. 9.13. Acetal is first hydrolyzed into acetaldehyde and ethanol, and ethyl acetate into acetic acid and ethanol. The hydrolysis products are then oxidized into acetic acid. Ethanol is first oxidized into acetaldehyde, and lactic acid into pyruvic acid

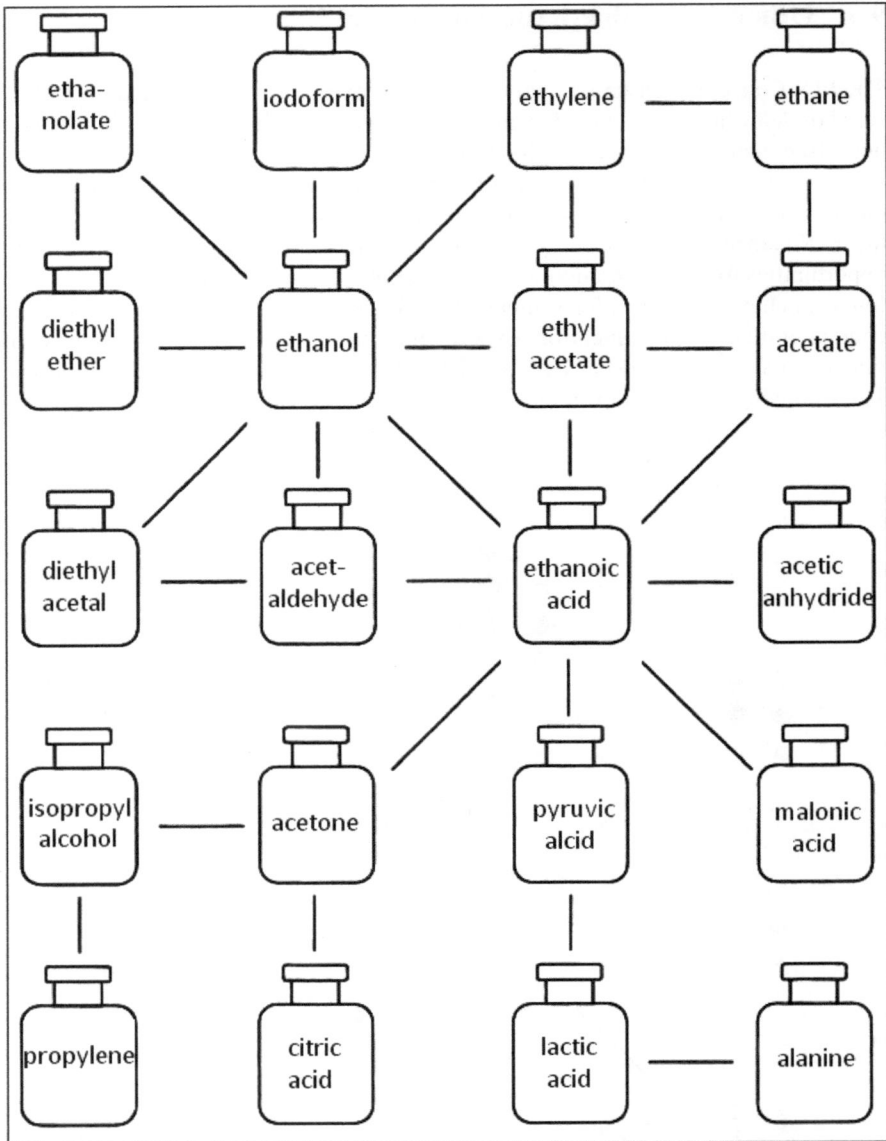

Fig. 9.15 A map of synthetic relations between some basic substances, which are fundamental for the genesis of a growing system in the PIN-Concept [1]. In some cases, the syntheses can be discovered by the students in both directions

- Experiments for the structural analysis of the ethanol molecule [45–48] and for the determination of formulae for other organic liquids [49–52], lesson plans for the concrete operational development of the homologous series of alcohols

[13, 14], and the term "isomerism" [15] with the help of molar mass determination under uniform conditions [16].

- Historical problem-oriented teaching units using the example of ethanol [53] and acetaldehyde [54, 55].
- Experiments on the ether synthesis considering mechanistic aspects [56–58] as well as experiments about ethylene [59–63].
- Experiments on the Kolbe electrolysis of acetates and other salts of carboxylic acids [64–69] as well as on the ester equilibrium [70, 71] and complex carboxylic acids such as citric acid [72–76] (also see [8]), lactic acid [77–79], sorbic acid [80], and ascorbic acid [81–84].

9.10   The PIN-Concept for Teacher Training

The PIN-Concept has proven to be of value in the education of prospective chemistry teachers (middle and high school level) at the universities of Muenster and Leipzig for many years. Prospective teachers like the integrative interconnection of chemistry, experiments, and didactics as complementary aspects of teacher education. It is especially advantageous when the prospective teachers learn to reflect on their own learning process on a metacognitive level and when the experimental instruction is organized in an inquiry-based way (with flexible integrated seminar sessions). In the beginning, it is sometimes difficult for prospective teachers to work with unfamiliar substances. Some of them want to know the results of the experiments a priori. Others have learned a lot about formulae and reaction mechanisms at high school, but little about the use of empirical data that can be observed in the laboratory. These prospective teachers have a hard time to put themselves into the position of their future students at schools, who do not have a consolidated systematic of chemistry, but after a short time even these prospective teachers realize that constructivist-oriented learning processes cannot be replaced by timesaving verbally explained contents, at least not in introduction or elaboration phases. Only what students work on themselves leaves a trace in their minds which can be activated in other situations and contexts.

One prospective teacher expressed her experience with the PIN-Concept at the end of a series of experiments in the laboratory like this:

> I had the feeling to explore something really by myself. I did not think that I could find out so much with so little preknowledge – an important experience, also for my self-confidence.

Furthermore, the PIN-Concept was used successfully for the methodical training of experienced practicing teachers on many occasions (1 or 2 day courses). When these teachers were asked for their impressions, they typically used words like these:

> The PIN-Concept is interesting, motivating, logical, understandable, student oriented, practicable, convincing, but also time consuming and demanding. It fosters cognitive and experimental competences, and the yield of interrelated knowledge and process oriented skills is very high. And most important: It is never dull, it is fun to learn chemical thinking in this way.

9.11   The PIN-Concept for Grammar Schools

Harsch and Heimann [2] conducted an empirical study with grade 11 students at three grammar schools to prove the efficiency of the PIN-Concept in practice. The involved chemistry teachers (one woman, three men) received a complete teacher manual, which covered a teaching unit of about 30 h. The manual consisted of comprehensive lesson plans with all required experimental instructions, spectra, and exercises with answers. The four teachers were acquainted with the philosophy of the PIN-Concept, the experiments, the exercises, and the methodical variation possibilities at the Institute for Chemistry Education at the University of Münster. They were instructed in a 2-day course to teach the whole unit independently at their schools. The length of the lessons was only given roughly so that the teachers could adapt it to the needs at their school. They had the chance to focus on topics that seemed especially important to them and to include other aspects, such as spectroscopic methods for structural analysis. The sequence could be changed if this was appropriate. This specifically applies to the exercises, which they should integrate flexibly into the course flow. The teaching unit was divided into three sections and covered 11 steps, which are displayed in Table 9.5.

On average the four teachers needed about 35 h for this comprehensive program. It should be mentioned that the time need differed significantly from class to class (min. 29 h, max. 40 h).

Table 9.5 Content of the empirical study for the students with grade 11 at grammar schools [2]

Step	Content
Step 1.1	Classification of six unknown substances A–F with the help of test reactions.
Step 1.2	Analysis of household and everyday substances (e.g., vinegar, alcohol, spot remover, glue).
Step 1.3	Discovering synthesis relations between the substances A–F (oxidation with dichromate; ester synthesis and hydrolysis).
Step 1.4	Exercises for using data from test reactions and understanding of their meaning.
Step 2.1	Qualitative elemental analysis of the substances A–F (detection of C/H/O atoms).
Step 2.2	Structural elucidation of A–F with the help of simplified ^{13}C-NMR and mass spectra.
Step 2.3	Understanding structure–properties relations by using formulae and functional groups for the interpretation of experimental data.
Step 2.4	Exercises for understanding synthetic pathways and for predicting possible results.
Step 2.5	Mixed exercises for the integration of different methods and observations.
Step 3.1	Discovering the homologous series of alcohols and understanding its definition at the macroscopic level (substances) and at the submicroscopic level (formulae).
Step 3.2	Structure elucidation of a natural substance (glycolic acid) by using data from different investigative methods.

Altogether, 11 learning steps had to be absolved

9.11.1 Evaluation by the Teachers

One might say that the time needed for this teaching unit is very high, and that there is not enough time for it in normal chemistry classes. But this provokes the question: What do we want to achieve in a chemistry class? The involved teachers at least unanimously had the opinion that the high time need definitely paid off. The quality of the initiated learning processes, the interconnection of contents and methods, as well as the experimental discovery and the profound understanding of the students led to this opinion.

The teachers expressed the following opinions in the concluding discussion, which was recorded on tape [2]:

- "In the beginning there was a lot of work (preparation of solutions, testing of experiments). But once you get into it, it works like clockwork."
- "The motivation was generally good, even in theory phases, this is unusual."
- "High motivation by student experiments and by independent work."
- "The students liked working without prior knowledge and some of them were even enthusiastic to feel successful for the first time in their chemistry class."
- "Good and less good students were well integrated in the groups. Even students without prior knowledge had a chance, and even very quiet students put their hands up."
- "28 out of 36 students continued chemistry class. That are a lot more students than last year. Obviously the concept worked very well."
- "I would like to do this again and I am happy that I got to know another concept. I had a lot of fun."

9.11.2 Evaluation by the Students

How did the 58 students (26 boys and 32 girls) evaluate the PIN-Concept?

The students were asked the following questions about every step after each of the three sections (see Table 9.6). The answers had to be selected on a five-point scale:

- How much did you understand in this step?
- How well could you follow in this step?
- How interesting did you find this step?

The results of this questionnaire will be explained exemplarily using the steps 1.1 and 2.2 and 3.2 (see Table 9.6). In the Tables 9.6–9.8, the percentage of students that ticked the respective category is presented.

Table 9.6 shows that the three steps were evaluated very positively. On average about 70% of students said that they found these steps interesting or even very interesting; they understood much or very much and they could follow well or very well.

To get a general idea of the evaluation of all 11 steps by the students, a number between +2 and −2 was assigned to every category. For example:

very much (+2), much (+1), middle (0), not so much (−1), little (−2).

	Question	Reply
Table 9.6 Evaluation of learning step 1.1 (classification of six unknown substances A–F with the help of test reactions) by the students (percentile)	How much did you understand?	Very much: 42%
		Much: 32 %
		Middle: 21%
		Not so much: 5%
		Nearly nothing: 0%
	How well could you follow?	Very good: 34%
		Good: 41%
		Middle: 18%
		Not so good: 7%
		Not at all: 0%
	How interesting did you find this step?	Very interesting: 27%
		Interesting: 43%
		Medium: 25%
		Not so interesting: 5%
		Uninteresting: 0%

With the help of these numbers, the percentages for every step were translated into an index by averaging. This index could range between +2 (max. approval) and −2 (max. rejection). The results are shown in Figs. 9.16 and 9.17. It is evident that the students voted positively for every step in the cognitive dimension (comprehension) as well as in the affective dimension (interest). It is apparent that the boys consistently assessed their comprehension higher than the girls (Fig. 9.16, below). It will be discussed later, if this subjective assessment reflects the actual test results.

With regard to the interest (Fig. 9.17) the gender-specific differences were not that distinctive. The girls assessed the more phenomenological and everyday life oriented steps more positive. The boys preferred the more theory oriented steps in the middle of the unit (elementary analysis, spectroscopy, structure–property relations). At the end of the unit the interests of both boys and girls assimilated again.

It is remarkable that boys and girls both found working on exercises (steps 2.4 and 2.5) the least interesting. The unpopularity of exercises is a well-known pedagogical problem, which deserves closer attention. Exercises may not be waived in any case. The teachers confirmed this claim, but it is really striking that the girls' comprehension problems were mostly experienced during the phases of exercises and problem solving. This point needs further clarification in future studies.

9.11.3 Pretest Results of the Students

Little prior knowledge is needed for the PIN-Concept. Therefore, the aim of the pretest was not to check the prior knowledge in organic chemistry, but general chemistry knowledge. One of the exercises will be shown below.

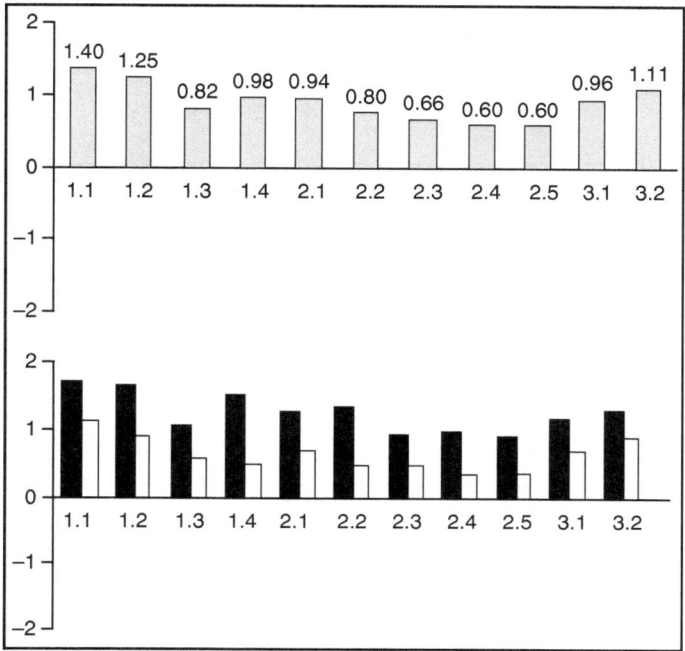

Fig. 9.16 Students' evaluation for the question: How much did you understand in the single steps? Meaning of the step numbers, see Table 9.5. Meaning of the ordinate numbers: +2 = full understanding; −2 = no understanding at all. *Above*: all students, *below*: boys (*black*) and girls (*white*). Data from ([2], p. 147)

In problem 6, the students were asked to classify the following pure substances according to their properties ([2], p. 91):

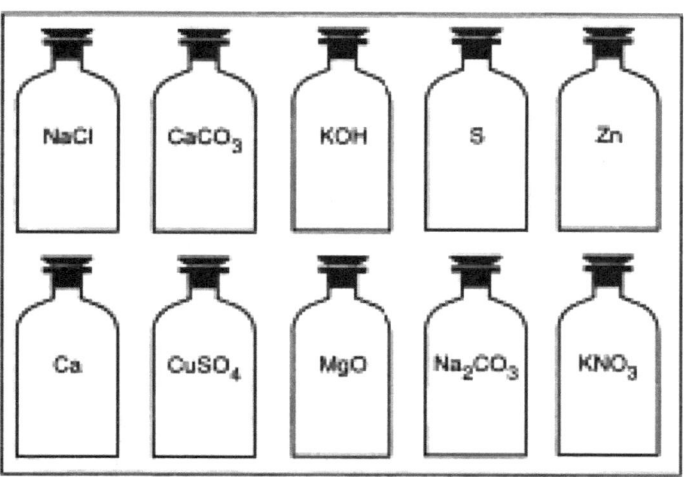

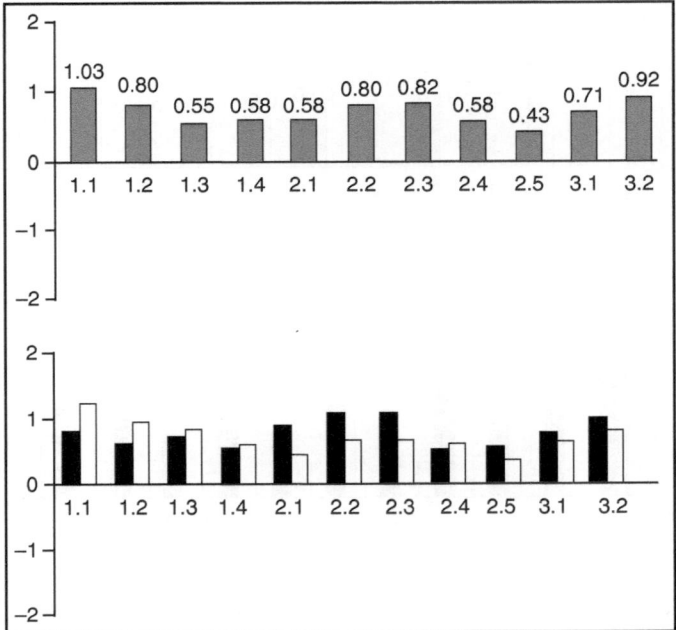

Fig. 9.17 Students' evaluation of the question: How interesting did you find the single question? Meaning of the ordinate numbers: +2 = very interesting; −2 = totally uninteresting. *Above*: all students, *below*: boys (*black*) and girls (*white*). Data from ([2], p. 143)

The following answers were obtained (number of students in percent that gave the correct answer for the item):

– Which of these pure substances are white solids?

NaCl	MgO	KOH	$CaCO_3$	KNO_3	All correct
71%	57%	12%	10%	7%	0%

– Which substances are poorly soluble in water?

Zn	S	MgO	$CaCO_3$	All correct
50%	40%	19%	7%	0%

– Which substances react alkaline in aqueous solution?

KOH	Ca	MgO	Na_2CO_3	$CaCO_3$	All correct
22%	12%	10%	10%	7%	0%

– Which substances react with acids generating a gas?

Ca	Zn	$CaCO_3$	Na_2CO_3	All correct
14%	12%	10%	7%	0%

– Which substances conduct when in aqueous solution?

NaCl	Na$_2$CO$_3$	CuSO$_4$	KNO$_3$	KOH	All correct
57%	16%	12%	9%	7%	0%

– Which substances are built of ions?

NaCl	CuSO$_4$	CaCO$_3$	MgO	KOH	All correct
50%	36%	28%	22%	17%	3%

On average, the general chemistry knowledge of the students leaves a lot to be desired. This holds true for both girls and boys. Overall, the pretest results of the boys were significantly better than those of the girls ($p < 0.001$ according to the U-test of Mann–Whitney).

9.11.4 Posttest Results of the Students

Towards the end of the teaching unit the students were given an unannounced posttest, which they had to fill out anonymously. The aim was to test acquired skills and content knowledge that was spontaneously accessible without prior repetition or grade pressure. Transfer of knowledge was required to solve these problems. Two examples might illustrate what students were able to accomplish and what not.

Problem 1. Six molecular formulae of substances were given in problem 1. The students were not explicitly familiar with these formulae. These formulae represented bifunctional molecules with different orientation of the functional groups. The molecules could be assigned to the following substances: succinic acid, 2-hydroxypropanoic ethyl ester, ethylene glycol, malonic acid monomethyl ester, 2-hydroxypropionic acid, and methyl propanoate. The students were asked to identify all functional groups and to predict the behavior of the corresponding substances in the test reactions. The following results were obtained:

– Eighty percent of students recognized all alcoholic groups, 61% all carboxylic groups, and 40% all ester groups. Altogether, 47% of the students recognized at least 8 of the overall 11 functional groups correctly.
– Seventy-six percent of the students predicted the reaction of the substances in the cerium nitrate test (group test for alcohols) correctly, 74% the reaction in the BTB test (group test for carboxylic acids), and 59% the reaction in the Rojahn test (group test for esters).
– The majority of the students could predict the results of all tests correctly, although a considerable percentage of them had trouble to identify the ester group in the molecule.

Problem 2. In a chemical cabinet you find an unlabeled bottle, which contains an unknown pure substance. The molecules of this pure substance have one of the following ten structural formulae:

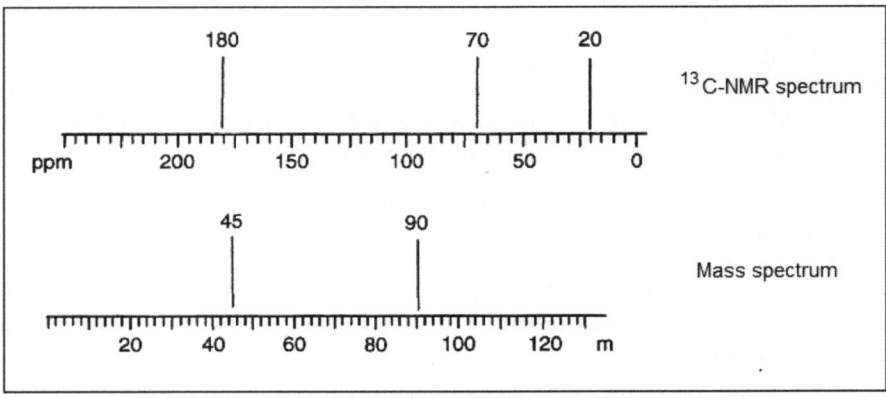

To solve this problem, you have the following experimental data:

1. The unknown pure substance shows a positive BTB test and a positive cerium nitrate test, but a negative Rojahn test.
2. The ^{13}C-NMR spectrum and the mass spectrum of the pure substance are:

This exercise is designed to test whether the students were able to use the different methods, which they had learned during the unit, for problem solving.

Table 9.9 Students' results for problem 2 (percentages rounded to whole numbers)

Points	All students	Boys	Girls
0	16 (%)	12	19
1	10 (%)	4	16
2	16 (%)	12	19
3	3 (%)	0	6
4	5 (%)	8	3
5	50 (%)	65	38

Table 9.10 Overall results in the follow-up test [2]

Points	All students	Boys	Girls	Class 2	Classes 1 + 3 + 4
0–12	9 (%)	0	16	31	0
13–16	22 (%)	19	25	25	21
27–40	41 (%)	31	50	31	45
41–53	17 (%)	31	6	6	21
54–66	10 (%)	19	3	6	12

Percentages rounded to whole numbers

They received one mark for every correctly used method. For the consistent combination of these methods and the correct result (formula number 8 = lactic acid) including appropriate argumentation they received two marks; added together 5 points. The results of this exercise are displayed in Table 9.9.

Half of the students solved this difficult problem completely – a remarkable result. The boys had a significantly better test result than the girls (significance level $P = 0.033$ for the categorized, and $P = 0.037$ for the uncategorized number of points in the U-test according to Mann–Whitney).

The overall test results (Table 9.10) of the boys showed a significantly better performance than the girls' ($P = 0.002$ for the categorized, $P = 0.006$ for the uncategorized number of points in the U-test according to Mann–Whitney). This indicates that the subjective assessment of the boys and the girls concerning the comprehensibility of the unit (Fig. 9.16) was also reflected in the objective test results. Note that this difference could be observed already in the pretest.

The performance of one class (class 2) remained notably behind the performance of the other three classes. This abnormality, which was also reflected in the higher time need of this class (40 h), possibly had an organizational reason. Some of the students were missing repeatedly and some of the lessons had to be canceled. Since problems like this happen in school life, these results remained in the evaluation.

It becomes apparent that the PIN-Concept can be used successfully in fundamental chemistry classes in grade 11, even when the conditions are bad. This positive conclusion holds true for boys and girls, despite the gender-specific differences, which need further investigation.

9.12   Conclusion

The PIN-Concept intends to support process skills and interconnected thinking. The students should not just take over ready-made knowledge, but should actively participate in the construction of knowledge in an inquiry-based learning process. Thereby they should gain insight into scientific working and argumentation. This process-oriented approach is a lot harder and more time consuming than a strictly content-oriented approach, but it is also much more fun and leads to a deeper methodical understanding. The results of this empirical study in fundamental classes of grade 11 show that these aims can be achieved. The PIN-Concept imparts rewarding learning experiences: "I liked to be kind of a scientist..." ([2], p. 157). This is how one student explained her experience with the PIN-Concept. These cognitively and affectively positive overall impressions build the foundation of what will remain from high school chemistry once the students graduate and forget about most of the details. The image of chemistry in our society and the educational value of it are based on the quality of the learning processes, which students experience themselves in chemistry lessons.

The philosophy of the PIN-Concept may be summarized as follows: "Not the systematic of organic chemistry itself, but the construction of this systematic based on experimental experience contributes to general knowledge. The active, constructive acquisition of knowledge organizes the students' mind and leaves deeper impressions than the simple reproduction of knowledge. The systematic knowledge becomes relevant, if this process is successful. Otherwise it will be rejected." [85].

Cary expressed the same idea as follows: "Education is ultimately the flavor left over after all the facts, formulae and diagrams have been forgotten." [86]

Experimental Details

1 Test Reactions as Analytical Tools

1.1 Cerium Nitrate Test

– *Reagent*: 40 g cerium ammonium nitrate are dissolved in 100 ml nitric acid ($c = 2$ mol/l).
– *Procedure*: 1 ml cerium ammonium nitrate reagent is to be diluted with 2 ml water. Add 5 drops of the test substance and shake well. Let rest for 5 min and observe.
– *Result*: A color change from yellow to red (or orange) indicates molecules with

alcohol groups. $- \overset{|}{\underset{|}{C}} - OH$

– *Note*: Alcohol molecules make red colored complexes with hexanitratocerate ions.
– Diethyl acetal is hydrolyzed (acid catalysis) to ethanol and acetaldehyde; the alcohol makes the test positive.

Sometimes, the red color, indicating an alcohol group, disappears immediately or slowly, and the solution becomes colorless. This indicates that cerium ions Ce^{4+} have been reduced to Ce^{3+}, and that the test substance has been oxidized.

For example, ethanol is oxidized to acetaldehyde (after heating), 2-propanol to acetone, lactic acid to acetaldehyde and carbon dioxide, citric acid to acetone and carbon dioxide. These reactions can be discovered by the students when they make the cerium nitrate test with these substances in a larger scale in a synthesis apparatus (Fig. 9.3) and investigate the distillates with the reactions.

1.2 Bromothymol Blue Test (BTB test)

– *Reagent*: 0.02 g bromothymol blue and 0.6 g sodium hydroxide have to be dissolved in 100 ml ethanol.
– *Procedure*: 1-ml test substance is to be added to 1-ml BTB reagent. Mix it well and observe.
– *Result*: A color change from blue to yellow/orange indicates molecules with carboxylic acid groups. $-\overset{\overset{\displaystyle O}{\|}}{C}-OH$
– *Note*: The test is also positive with inorganic acids. Therefore, the detection of carboxylic acid groups is only conclusive in the absence of inorganic acids.
– If the test substance is not soluble (e.g., stearic acid), heating is helpful for detecting the carboxylic acid groups.

1.3 Phenolphthalein Test (Rojahn test)

– *Reagent 1*: 0.1 g phenolphthalein is dissolved in 100 ml ethanol.
– *Reagent 2*: Sodium hydroxide solution ($c = 3$ mol/l).
– *Procedure*: 1 ml of the test substance and 3 drops phenolphthalein solution are added to 1 ml ethanol. Sodium hydroxide solution is then to be added drop wise (constant shaking) until a permanent pink color is observed which even resists vigorous shaking. (It is very important to only add as much sodium hydroxide solution as needed to change the indicator to pink.) The test tube with the solution is then to be put into a water bath (40°C). It needs to be taken out after every minute to shake it. The test can be stopped, when a discoloration is observed. Otherwise, the test tube has to be observed for a maximum of 10 min.
– *Result*: Discoloration of the pink phenolphthalein solution indicates molecules with ester groups. $-\overset{\overset{\displaystyle O}{\|}}{C}-O-\overset{\overset{\displaystyle |}{}}{\underset{|}{C}}-$

– *Note*: Esters are hydrolyzed in the alkaline test solution, and the consumption of hydroxide ions is indicated by phenolphthalein.

For the detection of ester groups in fat molecules, propanol is needed as a solvent instead of ethanol.

Lactic acid gives a positive test result because of intermolecular esterification. This is a nice exception, which can be used as a problem for inquiry learning.

This test was introduced by Carl August Rojahn (1889–1938) who was a Professor for Pharmacy and Food Chemistry at the University of Halle (Germany).

1.4 Dinitrophenylhydrazine Test (DNPH Test)

– *Reagent 1*: A mixture of 186 ml water, 33 ml hydrochloric acid (37%), and 1 g 2,4-dinitrophenyl-hydrazine are to be stirred well for 15 min. Undissolved ingredients are then to be removed by filtration.
– *Procedure*: To 1 ml DNPH reagent in a test tube is to be added one drop of the test substance. A stopper is to be put on the test tube which is to be shaken for 15 s and observed for another 45 s.
– *Result*: A yellow or orange precipitate (sometimes only a milky turbidity) indicates molecules with aldehyde or ketone groups.

$$-\overset{\displaystyle O}{\overset{\|}{C}}-H \qquad -\overset{}{\underset{|}{C}}-\overset{\displaystyle O}{\overset{\|}{C}}-\overset{}{\underset{|}{C}}-$$

– *Note*: the carbonyl groups of aldehydes or ketones react with the amino group of the DNPH reagent and make an insoluble yellow or orange hydrazone.

For the detection of carbonyl groups in sugar molecules, heating with a boiling water bath (5 min) is necessary.

Diethyl acetal is hydrolyzed (acid catalysis) to the ethanol and acetaldehyde; the aldehyde makes the test positive.

1.5 Fehling Test

– *Reagent 1*: 7 g $CuSO_4 \cdot H_2O$ are dissolved in 100 ml water.
– *Reagent 2*: 35 g sodium potassium tartrate and 10 g sodium hydroxide are dissolved in 100 ml water.
– *Procedure*: 1 ml reagent 1 and 1 ml reagent 2 are mixed. The mixture is deep blue or purple and clear. Then 1 ml test substance and a boiling chip are added. After shaking the sample is put into a boiling water bath. The test can be stopped when a precipitate or turbidity is observed. Otherwise it is observed for a maximum of 5 min.
– *Result*: A red or reddish brown precipitate of copper oxide Cu_2O at the bottom of the test tube (often a small amount) indicates molecules with an Acetyl group.

$$H_3C-\overset{\displaystyle O}{\overset{\|}{C}}-$$

– *Note*: Aldehydes are oxidized by copper ions Cu^{2+} in alkaline solution to acid ions; Cu^{2+} ions are reduced to Cu^{1+}, and these ions form insoluble red copper oxide Cu_2O.

Not only glucose but also fructose gives a positive test result because of rapid isomerization in the alkaline solution, in contrast to saccharose.

Diethyl acetal gives a negative test result because acetals are not hydrolized in alkaline solutions. (Note that saccharose is also an acetal!)

This test reaction was discovered by Hermann Fehling (1812–1885) who was a Professor for Chemistry at the University of Stuttgart (Germany).

1.6 Copper Sulfate Test

– *Reagent 1*: Copper sulfate in water ($c = 0.1$ mol/l).
– *Reagent 2*: Sodium hydroxide in water ($c = 3$ mol/l).
– *Procedure*: 1 ml of the liquid test sample (or a small spatula amount of a solid sample) is dissolved in 1 ml copper sulfate solution. The mixture must be clear. Then 5 drops of sodium hydroxide solution are added drop by drop (continuous shaking).
– *Result*: A color change from light blue to deep blue or purple without a permanent precipitate (the solution must be clear after adding 5 drops of sodium hydroxide solution) indicates molecules with alcohol groups on adjacent carbon atoms (diol

$$-\overset{\displaystyle |}{\underset{\displaystyle |}{C}}-\overset{\displaystyle |}{\underset{\displaystyle |}{C}}-\text{groups}).$$
$$\quad\text{OH} \quad \text{OH}$$

– *Note*: In the copper sulfate test, the test substance (diol) has the same function as the tartrate ions (they are diols) in the Fehling test: They form deep blue and stable chelate compounds with Cu^{2+} ions in alkaline solution.

1.7 Bromine Test

– *Reagent*: 10 drops of bromine are dissolved in 250 ml water.
– *Procedure*: 2 ml test substance (or 1 spatula) are dissolved or suspended in 2 ml bromine water. The solution is shaken well. (Close the test tube with a stopper and use gloves.)
– *Result*: Decolorization indicates molecules with double bonds (alkenes).

$$\overset{\diagdown}{\diagup}C = C\overset{\diagup}{\diagdown}$$

– *Note*: Decolorization occurs by addition of bromine molecules to the double bonds of the test molecules. (The addition products are colorless.)

If bromine is only extracted from the water phase into the organic phase without decolorization (e.g., with alkanes or esters as test substances), the test is negative.

1.8 Iodoform Test (Lieben Test)

- *Reagent 1*: 12.7 g iodine and 25.4 g potassium iodide dissolved in 100 ml water.
- *Reagent 2*: Sodium hydroxide in water ($c = 3$ mol/l).
- *Procedure*: To 0.5 ml test substance are added 1 ml iodine solution (brown) and without delay sodium hydroxide solution in 1 ml steps, if the substance is still brown.
  After every 1 ml step, the test tube has to be shaken, until the solution becomes yellow and clear. Then the test tube is observed for 5 min.
- *Result*: A yellow or pale precipitate (often only a slight turbidity) indicates molecules with special functional groups, namely:

Type 1: Acetyl groups of aldehydes and ketones

$$H_3C - \overset{\overset{\displaystyle O}{\|}}{C} -$$

Type 2: Special alcohol groups with the structure

$$H_3C - \overset{\overset{\displaystyle OH}{|}}{\underset{\underset{\displaystyle H}{|}}{C}} -$$

Type 3: Special ester groups with the structure

$$H_3C - \overset{\overset{\displaystyle O - \overset{\overset{\displaystyle O}{\|}}{C} -}{|}}{\underset{\underset{\displaystyle H}{|}}{C}} -$$

- *Note*: Esters of type 3 are hydrolyzed in the alkaline test solution to alcohols of type 2, and these are oxidized by iodine to acetyl groups of type 1, which after iodination of the methyl group are cleaved to iodoform I_3CH and an acid ion in the alkaline solution.

This complex and valuable test reaction has been discovered by Adolf Lieben (1836–1914) who was a Professor for Chemistry at the Universities of Paris (France), Palermo, Turin (Italy), and Vienna (Austria).

1.9 Iron Chloride Test

- *Reagent 1*: 8 g iron chloride $FeCl_3 \cdot 6H_2O$ dissolved in 100 ml water.
- *Reagent 2*: Amyl alcohol (1-pentanol).
- *Reagent 3*: 0.1 g phenolphthalein dissolved in 100 ml ethanol.
- *Reagent 4*: Sodium hydroxide solution ($c = 3$ mol/l).
- *Reagent 5*: Hydrochloric acid ($c = 0.5$ mol/l).

– *Procedure*: To 10 drops of the test substance and 2 drops phenolphthalein, sodium hydroxide solution is to be added drop by drop until the phenolphthalein indicator just changes to pink. Then the solution is discolored without delay by adding one drop of hydrochloric acid (or a few, if necessary). If the solution warmed up, it has to be cooled down to room temperature.
  Now, 2 drops of iron chloride solution and 1.5 ml amyl alcohol are to be added. The stoppered test tube has to be shaken vigorously for 5 s. Wait until the two phases separate.
– *Result*: The test is positive if an orange color can be observed in the upper or lower phase. This indicates carboxylic acid molecules with short chains. With formic acid and acetic acid, the lower phase is orange; with propanoic acid and butanoic acid, the upper phase is orange.
– *Note*: The iron chloride test does not work directly with the carboxylic acids, but with their salts. The purpose of the first step of the procedure is therefore to neutralize the carboxylic acids. (This step is not necessary for the salts, of course.) The acid ions form orange complexes with the iron ions. The complexes with propanoate and butanoate prefer the unpolar amyl alcohol (upper phase), whereas the more polar complexes with formiate and acetate prefer the lower water phase.
  The iron chloride test is negative with mineral acids, in contrast to the BTB test. However, it is limited to short-chain carboxylic acids.

1.10 Dichromate Test

– *Reagent*: 2.35 g potassium dichromate are to be dissolved in 150 ml water. Then 8 ml concentrated sulfuric acid have to be added. After stirring, fill up with water to 200 ml.
  This reagent is poisonous. Use gloves. Pay special attention to solid dichromate, because it is carcinogen. Do not inhale dichromate dust.
– *Procedure*: To 2 ml dichromate reagent are added 4 drops test substance and a boiling chip. The sample is to be put into a boiling water bath for 5 min.
– *Result*: A color change from orange to green or brown indicates molecules which are oxidizable (e.g., primary and secondary alcohols, aldehydes, esters, and other compounds) under the test conditions. Alkanes, ketones, and mono carboxylic acids (exception: formic acid) are not oxidizable.
– *Note*: The dichromate test should be used only by experienced students, who are aware of the necessary safety conditions. Another possibility is to restrict this test to teachers' demonstrations, or even omit it because it is not strictly necessary as an analytical tool. (Note that for synthesis the dichromate reagent can be substituted by potassium permanganate.)

2. Syntheses

2.1 Esterification

With the purified product (ca. 15 ml), only the Rojahn test is positive. Additionally, the iodoform test should be made. It is positive as it should be for ethyl acetate, but only a slight turbidity is observable.

– *Note*: In the same way, propyl propionate can be synthesized from propanol and propanoic acid. This time, the iodoform test is negative as it should be. The yield is ca. 20 ml.
– *Procedure*: In an Erlenmeyer flask (100 ml), 20 ml ethanol, 20 ml ethanoic acid, and 5 ml sulfuric acid (conc.) are mixed. Shake the mixture well. It warms up spontaneously. Let it stand for 5 min and pour it then into another Erlenmeyer flask (100 ml) containing 75 ml water. Immediately, an organic phase separates in the neck of the flask. It smells like an ester.
Test the crude product by means of analytical tests (Rojahn test, cerium nitrate test, BTB test, iron chloride test).
In order to purify the crude product, transfer it with a pipe into a flask (100 ml), add the double volume of sodium carbonate solution ($c = 1$ mol/l) and shake the unstoppered flask well, until the gas evolution (CO_2) is finished. Then stopper the flask and shake well. Let the phases separate.
Investigate the upper phase (purified product) with the tests.
– *Result*: With the crude product, the four tests are all positive. From this it can be concluded that an ester has been produced which is still contaminated with alcohol and acids.

2.2 Ester Hydrolysis

– *Procedure*: In a flask (250 ml), put 60 ml sodium hydroxide solution ($c = 3$ mol/l) and 15 ml ethyl acetate. Stopper the flask and shake it vigorously for 2 min. Then pour the reaction mixture by means of a funnel into a distillation apparatus (Fig. 9.3) and distil with a boiling water bath for 15 min. (Cover the upper part of the apparatus with an aluminum foil.) Investigate both the distillate and the residue with the analytical tests. Note that the residue must be acidified with sulfuric acid ($c = 1.5$ mol/l) before making the tests. (Mix 5 ml residue with 4.5 ml sulfuric acid.)
– *Result*: Ethyl acetate is hydrolyzed in alkaline solution into ethanol (distillate) and acetate (residue). In the distillate, only the cerium nitrate test, the iodoform test, and the dichromate test are positive. This is the pattern of ethanol.
In the residue, the iron chloride test is positive (lower phase orange) indicating the presence of ethanoic acid (or acetate). The BTB test is also positive, but it is not conclusive for carboxylic acids because the residue was acidified with sulfuric acid.

– *Note*: Propyl propanoate is more stable than ethyl acetate. It is hydrolyzed only very slowly. In this case, it is better to let the mixture stand for a week and then shake it again vigorously before distilling it. But even then, the ester is not completely hydrolyzed.

2.3 Oxidation

– *Procedure*: In a synthesis apparatus (Fig. 9.3), 12 drops of the oxidizable substance (ethanol, 1-propanol, 2-propanol, ethyl acetate, diethyl acetal) are mixed with 14 ml dichromate synthesis reagent (82.0 g $K_2Cr_2O_7$ dissolved in 12.5 ml water and 1.5 ml conc. H_2SO_4). Use a funnel to put the reagent into the apparatus and a magnetic stirrer. Heat the oil bath to ca. 120°C and distil at this temperature for 10 min. Cool the distillation receiver with ice water. A charcoal absorber is not necessary in this case. Investigate the colorless distillate with the tests.
– *Result*: In all cases, the reaction mixture changes its color immediately from orange to dark green or brown. This indicates that the substances are oxidized. With ethanol, ethyl acetate, and diethyl acetal as educts, the distillate gives a positive test only with the BTB test and with the iron chloride test (lower phase orange). This indicates that the educts have been completely converted into ethanoic acid.
With 1-propanol, propanoic acid is found in the distillate. (In the iron chloride test, the upper phase is orange.)
With 2-propanol as an educt, the distillate makes the DNPH test and the iodoform test positive; all other tests are negative. It can be concluded that 2-propanol has been oxidized to acetone.
– *Note*: The dichromate synthesis reagent has not the same concentration as the reagent for the analytical dichromate test. Pay attention to the safety recommendations (see 1.10).

References

1. Harsch G, Heimann R (1998) Didaktik der Organischen Chemie nach dem PIN-Konzept. Vom Ordnen der Phänomene zum vernetzten Denken. Springer, Heidelberg
2. Harsch G, Heimann R (2001) "Selber etwas Wissenschaftler spielen, das fand ich gut ..." Ein evaluiertes Unterrichtskonzept zur Organischen Chemie in der gymnasialen Oberstufe. Schüling, Münster
3. Harsch G, Heimann R (1995) Organische Chemie im Vorfeld der Formelsprache. Chemkon 2:151–157
4. Heimann R, Harsch G (1997) Die Behandlung der Carbonylverbindungen nach dem PIN-Konzept. Chemkon 4:71–76
5. Harsch G, Heimann R (1996) Wenn das Ganze mehr ist als die Summe seiner Teile: Polyfunktionelle Verbindungen im Chemieunterricht. MNU 49:219–227
6. Heimann R, Harsch G (1998) Viele Wege führen zur Essigsäure. PdN-Chemie 47:26–29

7. Harsch G, Heimann R (1994) Der Estercyclus – ein experimentelles Projekt zur Schulung des vernetzten Denkens und Handelns. Chemie in der Schule 41:7–18

8. Heimann R (1999) Einem unbekannten Stoff auf der Spur – Isolierung. Analyse und Strukturaufklärung eines Naturstoffes am Beispiel der Citronensäure. PdN-Chemie 48:26–31

9. Harsch G, Heimann R (1995) Schulung des analytischen Denkens am Beispiel von Kohlenhydratnachweisen in Lebensmitteln. PdN-Chemie 44:19–23

10. Heimann R, Harsch G (1998) Der experimentelle Weg vom Olivenöl zum Traubenzucker. Die Chemie der Fette und Kohlenhydrate nach dem Phänomenologisch-Integrativen Netzwerkkonzept. Teil 1: Vom Fett zutn Glycerin. Tcil 2: Vom Glycerin zum Traubenzucker. MNU 51:32–38, 95–99

11. Heimann R, Harsch G (1999) Untersuchungen von Kohlenhydraten in Pflanzen. Ein Weg zur Förderung naturwissenschaftlicher Denk- und Handlungskompetenz. MNU 52:226–232

12. Harsch G, Heimann R (1996) Mischungsexperimente nach Plan. Naturwissenschaftliche Kompetenzschulung am Beispiel der Polarität der Alkohole. Chemie in der Schule 43:142–146

13. Heimann R (2000) Fliissigkeiten in abgestufter Bewegung: Mischungsexperimente fur die Overheadprojektion. PdN-Chemie 49:32–33

14. Heimann R (2000) Der experimentelle Weg zum Begriff der homologen Reihe am Beispiel der Alkohole. Chemie in der Schule 47:14–16

15. Heimann R (2000) Experimentelle Wege zur Isomene. MNU 53:103–108

16. Heimann R, Harsch G (2000) Die Ermittlung der molaren Massen organischer Flüssigkeiten unter einheitlichen Bedingungen. Chemkon 5:73–76

17. Heimann R, Harsch G (1997) NMR-Spektroskopie und Massenspektroskopie im Unterricht – Möglichkeiten zur Schulung naturwissenschaftlicher Denk- und Handlungskompetenz. PdN-Chemie 46:8–14

18. Harsch G, Heimann R, Jansen E (1992) Die Sprache der Phänomene. Eine Überblicksmatrix zur qualitativen organischen Analytik im Unterricht. Chemie in der Schule 39:358–363

19. Harsch G, Heimann R (1995) Konkretheit und Verknüpfung in aktuellen Chemieschulbuechern am Beispiel der Organischen Chemie. Chimica didactica 21:149–167

20. Harsch G, Heimann R (1996) Das PIN-Konzept: Ein Phänomenologisch-lntegratives Netzwerkkonzept zum Aufbau einer erfahrungsgesteuerten Wissensstruktur im Bereich des organisch-chemischen Grundlagenwissens. In: Gräber W, Bolte E (eds) Fachwissenschaft und Lebenswelt. Chemiedidaktische Forschung und Unterricht, vol 153. IPN Monographie, Kiel, pp 73–108

21. Harsch G, Heimann R (1997) Organic chemistry as precursor for formula language. In: Gräber W, Bolte C (eds) Scientific literacy. An international symposium. IPN Monographie, Kiel, pp 415–38

22. Schlösser K (1975) Muß die Organische Chemie immer so spät im Chemieunterricht einsetzen? NiU 436–440

23. Wenck H (1980) Organische Chemie in der Sekundarstufe I – Hypothesen und Erfahrungen. In: Härtel H (Hrsg) Zur Didaktik der Physik und Chemie. Schroedel, Hannover, pp 166–169

24. Wenck H, Kruska G (1989) Wird der Chemieunterricht durch frühzeitige Behandlung der Organischen Chemie attraktiver? NiU PC 37:4–9

25. Christen HR (1987) Warum nicht mit der Organischen Chemie beginnen? Vortrag Sommersymposium der Chemiedidaktiker in NRW, Bielefeld

26. Sumfleth E (1988) Lehr- und Lernprozesse im Chemieunterricht. Lang, Frankfurt

27. Sumfleth E (1985) Systematisierungshilfen – von Gedächtnisstrukturen abgeleitete Lernhilfen. Chimica didactica 10:141–153

28. Sumfleth E (1985) Lernhilfen für den problemlösenden Unterricht. Chimica didactica 11:63–88

29. Aebli H (1980/81) Denken: Das Ordnen des Tuns. Klett-Cotta, Stuttgart, 2 Bde

30. Miller GA (1956) The magical number seven, plus or minus two – some limits on our capacity for processing information. Psychol Rev 63:81–97

31. Johnstone AH, Letton KM (1982) Recognising functional groups. Educ Chem 19:16–19

32. Johnstone HA, Wham AJB (1982) The demands of practical work. Educ Chem 19:71–73
33. Johnstone HA (1984) New stars for the teacher to steer by. J Chem Educ 62:847–849
34. Muckenfuß H (1995) Lernen im sinnstiftenden Kontext. Entwurf einer zeitgemäßen Didaktik des Physikunterrichts. Cornelsen, Berlin
35. Huntemann H, Paschmann A, Parchmann I, Ralle B (1999) Chemie im Kontext – ein neues Konzept für den Chemieunterricht? Darstellung einer kontextorientierlen Konzeption für den 11. Jahrgang. Chemkon 6:191–196
36. Parchmann I, Ralle B, Demuth R (2000) Chemie im Kontext – Eine Konzeption zum Aufbau und zur Aktivierung fachsystematischer Strukturen in lebensweltorientierten Fragestellungen. MNU 53:132–137
37. Gräber W, Stork H (1984) Die Entwicklungspsychologie Jean Piagets als Mahnerin und Helferin des Lehrers im naturwissenschaftlichen Unterricht. MNU 37:193–201, 257–269
38. Stork H (1979) Zum Verhältnis von Theorie und Empirie in der Chemie. Der Chemieunterricht CU 10:45–61
39. Reiners C (1992) Naturwissenschaftliche Erklaerungen. Rezepte oder Konzepte für die Chemiedidaktik? PdN-Chemie 1/41:41–44, 2/41:43–46
40. Liebig J (1844) Chemische Briefe. Akad. Verlangshandlung von C.F. Winter, Heidelberg
41. Herron JD (1975) Piaget for chemists. Explaining what good students cannot understand. J Chem Educ 52:146–150
42. Shayer M, Adey P (1989) Towards a science of science teaching. Cognitive development and curriculum demand. Heinemann, Oxford
43. Lawson AE (1985) A review of research on formal reasoning and science teaching. J Res Sci Teach 22:569–617
44. Häußler P, Bünder W, Duit R, Gräber W, Mayer J (1998) Naturwissenschaftsdidaktische Forschung. Perspektiven für die Unterrichtspraxis. IPN, Kiel, v.a. Abschnitt 6.2
45. Flint A, Jansen W (1990) Ethanol – Probleme der Aufklärung der Konstitutionsformel und des S_N2-Reaktionsmechanismus im Chemieimterricht der gymnasialen Oberstufe. PdN-Chemie 39:35–40
46. Geske D, Sandner A, Pauly C, Anft M, Kaminski B, Matuschek G, Flint A, Jansen W (1991) Neuer Weg zum Beweis der Konstitutionsformel des Ethanolmoleküls. Chemie in der Schule 2:86–91
47. Hallstein H (1991) Die experimentelle Ermittlung der Konstitutionsformel des Ethanols. Ein Beitrag zur, "Rettung" eines grundlegenden Schulversuchs. MNU 44:371–376
48. Thiemann F, Flint A, Jansen W (1994) Zur Ermittlung der Konstitutionsformel des Ethanolmoleküls. MNU 8:478–482
49. Fickenfrerichs H, Jansen W, Kenn M, Peper R, Ralle B (1981) Die Ermittlung der Summenfortneln leicht verdampfender organischer Flüssigkeiten. PdN-Chemie 30:362–367
50. Wegner G (1993) Ermittlung der Molekülformeln organischer Verbindungen. Chemie in der Schule 40:268–272
51. Wegner G (1994) Ermittlung der Molekül- und Konstitutionsformeln flüssiger organischer Stoffe. Chemie in der Schule 41:49–55
52. Wegner G (1994) Bestimmung der molaren Masse von Alkoholen (Ethanol und Methanol) und anderen leicht verdampfbaren Flüssigkeiten. Chemkon 1:134–137
53. Flint A, Jansen W, Peper R, Fickenfrerichs H (1987) Die Strukturaufklärung des Ethanols – eine an der geschichtlichen Entwicklung orientierte Unterrichtseinheit. NiU-PC 35:28
54. Matuschek C, Jansen W, Peper-Bienzeisler R, Fickenfrerichs H (1985) Aldehyde – eine an der Entdeckungsgeschichte orientierte Unterrichtskonzeption. PdN-Chemie 34:7–19
55. Hermanns R (1985) Erfahrungsbericht zur Unterrichtskonzeption, Aldehyde. PdN-Chemie 34:19–23
56. Kaminski B, Flint A, Ralle B, Jansen W (1992) Der Reaktionsmechanismus der Ether bildung aus Ethanol und Schwefelsäure im Chemieunterricht. MNU 45:490–498
57. Kaminski B, Flint A, Jansen W (1993) Vereinfachter Versuch der Ethersynthese. NiU-Chemie 17:32–33

58. Wiederholt E, Meinhardt E, Fahrney V (1993) Reaktionsprodukte von Ethanol mit Schwefelsäure. Gaschromatographische Analyse. PdN-Chemie 3:14–17

59. Schmidt HJ, Küppershaus E (1978) Der experimentelle Einstieg in die Organische Chemie über das Ethylen. PdN-Chemie 12:309–316

60. Armbrust R, Jansen W (1976) Darstellung von Aethen und Propen durch Cracken von n-Hexan im Schulversuch. PdN-Chemie 12:321–329

61. Jansen W, Pöpping J, Ralle B, Peper R (1981) Vom Ethen und Propen zu Aldehyden. Ketonen und Säuren. Technische. Synthesen im Schulversuch. NiU-P/C 29:98–102

62. Ralle B, Bode U (1991) Die Hydrierung einfacher Kohlenwasserstoffe bei Raumtemperatur. PdN-Chemie 40:18–23

63. Ralle B, Bode U (1993) Hydrierung von Ethin und Ethen. Bestimmung der Reaklionsenthalpie. PdN-Chemie 2:29–33

64. Kolbe H (1849) Untersuchungen über die Elektrolyse organischer Verbindungen. Annalen der Chemie und Pharmacie 69:257–294

65. Becker HJ (1977) Zum Nachweis des bei der Elektrolyse einer Natriumethanatlösung gebildeten Ethans. PdN-Chemie 7:179–183

66. Becker HJ (1979) Zur Darstellung von Ethan durch Elektrolyse einer Natriumethanatlösung. PdN-Chemie 12:321–323

67. Becker HJ (1999) Ethandarstellung im Schulversuch. Chemkon 6:26

68. Oetken M, Hogen K (1997) Die Kolbe-Synthese. Chemkon 4:83–84

69. Menig J, Bader HJ, Flintjer B (1998) Unerwartete Reaktionswege bei der Kolbe-Elek- trolyse. Organische Elektrochemie im Chemieunterricht. Chemkon 5:174–180

70. Schlösser K, Schmidt H (1979) Probleme bei der Planung des chemischen Gleichgewichts – Unterrichtseinheit zum Estergleichgewicht. NiU-P/C 1:13–29

71. Steiner D, Härdtlein M, Gehring M (1997) Das Estergleichgewicht. Möglichkeiten und Grenzen eines Schulversuchs. Chemkon 4:110–116

72. Sumfleth E, Ruhmann H (1984) Ein Vorschlag zur Strukturierung der Inhalte des Chemieunterrichts in der Sekundarstufe II am Beispiel der Citronensäure. – Nachweise und Isolierung. MNU 37:224–227

73. Sumfleth E, Gramm A, Dannat P (1986) Analytik der Citronensäure – eine Unterrichts reihe für die Jahrgangsstufe 13. MNU 39:415–426

74. Sumfleth E, Crispien K-D (1987) Ein Vorschlag zur Erarbeitung der organisch-chemi schen Reaktionsmechanismen ausgehend vom Beispiel der Citronensäure. MNU 40:229–231

75. Blume R, Wiechoczek D (1996) Die Gewinnung von Citronensäure mit einem Kationenaustauscher. MNU 5:289–291

76. Oswald B, Hildebrand A, Wenck H (2000) Citronensäurecyclus – Für den Chemieunterricht zu schwierig? PdN-ChiS 49:24–29

77. Dietrich V (1999) Zur Behandlung der Milchsäure im Chemieunterricht. PdN-Chemie 7/ 48:6–11

78. Huntemann H, Parchmann I (2000) Biologisch abbaubare Kunststoffe. Einordnung in ein neues Konzept für den Chemieunterricht. Chemkon 7:15–21

79. Menzel P (1993) Ethyllactat – ein umweltschonendes Lösungsmittel. PdN-Chemie 3:20–21

80. Stübs R, Wegner G, Lifson K (1996) Der Konservierungsstoff Sorbinsäure im chemischen Schulexperiment. Chemkon 3:129–133

81. Blume R, Bader HJ, Plauschinat M (1982) Neue Aspekte der Ascorbinsäure – Chemie. PdN-Chemie 10:289–298

82. Haselhoff H-P, Mauch J (1989) Das Vitamin-C-Projekt. Eine Unterrichtseinheit für das naturwissenschaftliche Praktikum. Diesterweg u. Sauerlaender, Frankfurt u Aarau

83. de Rijke PJ, van der Veer W (1992) Ascorbinsäure – quantitative Untersuchungen von Vitamin C einschließlich qualitativer Schulversuche. PdN-Chemie 41:21–31

84. Deifel A (1993) Die Chemie der L-Ascorbinsäure in Lebensmitteln. ChiuZ 27:198–207

85. Harsch G, Heimann R, Heinrich S (2002) Wie erzieht man Schueler zum komplexen Denken? Ein Unterrichtsbaustein für die gymnasiale Oberstufe am Beispiel der Dicarbonsaeuren. Chemkon 9(Heft 1):6–12

86. Cary WR (1984) State of the art in the high school curriculum. J Chem Educ 61:856–857

Chapter 10
Structure-Oriented Approach in Chemical Education

Chemists consider the chemical structure to be very important and essential for the understanding of substances and their properties; for them, the structure of matter is of highest significance. Modern chemistry lessons should therefore integrate exemplary chemical structures and the use of corresponding structural models. Following these preliminary considerations we will develop and describe a conception, which is called "structure-oriented chemistry."

10.1 Chemical Structures in Modern Chemistry

Today's chemists are mostly interested in the chemical structure for explaining the properties of produced substances or for the development of new substances. The newest edition of the German "GDCh-Nachrichten aus der Chemie" [1] shows an example (see Fig. 10.1): "Lithium-organyl compounds are mostly regarded as monomers – but the real structure is more sophisticated. Taking the example of methyl lithium, one observes that it is insoluble in nonpolar solvents like pentane or hexane. An explanation may be the structure of tetramers like $(MeLi)_4$; they consist of Li-tetrahedrons whose faces are occupied by methyl groups. The Li–C forces knot those tetrahedrons together and create a three-dimensional network. That is the reason for insolubility of lithium organyl compounds in nonpolar solvents." [1]

The reader should know that in publications of inorganic or organic chemistry every structure of a new crystalline substance or molecule is generally illustrated by model drawings and discussed (see Fig. 10.1). If students are supposed to get an insight into procedures of today's chemists, they should have the chance to reflect on structures of specially chosen substances in their chemistry lessons. The understanding of the structure of matter and the regrouping of atoms, ions, or molecules in reactions can be developed on the base of Daltons atomic model. Chemical symbols for the composition of substances and equations can be derived from chemical structures. These approaches would represent modern chemistry lessons because they connect to fundamental ideas of today's scientists. Later in their

H.-D. Barke et al., *Essentials of Chemical Education*,
DOI 10.1007/978-3-642-21756-2_10, © Springer-Verlag Berlin Heidelberg 2012

Fig. 10.1 Structure of
polymeric methyl lithium
with knotted $(MeLi)_4$ units [1]

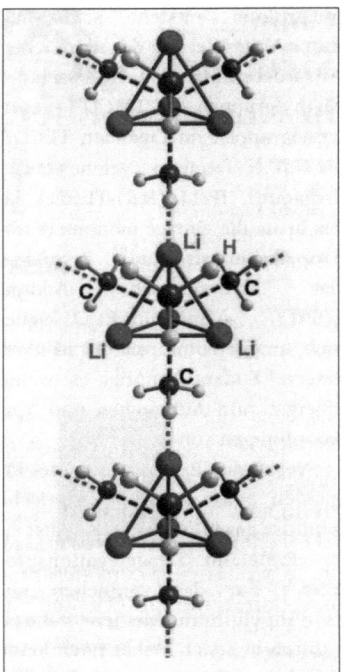

lessons, students will grab the idea of the nucleus-shell model of one atom or ion and will understand changes from atoms to ions and vice versa.

10.2 Chemical Structures as the Basis for the Interpretation of Reactions

How can chemical reactions be interpreted in chemistry lessons in an adequate way? Grosser [2] suggests one possible way, when he writes: "No matter in which way students are introduced to chemistry – they will always see experiments with metals or other solids. Often new solids are produced in these experiments. However, these experiments are rarely described with models of solids in the textbooks. 'MgO' or 'FeS' are produced (sometimes also 'molecules') or 'HgO' is being decomposed. Structures of these substances remain unclear and are not used to support an easier understanding of chemical reactions."

"Every observed reaction is usually described on the blackboard like this: Fe + S→FeS. In this kind of notation, it is not being mentioned that these and other reactions are mostly regroupings of "atoms" and/or "groups of atoms" in space. In these equations, it is also not expressed that it is not a single Fe atom that reacts with a single S atom. In fact, a tremendous amount of particles is involved in the reaction. Moreover, it is not mentioned that – in the case of iron sulfide again – two crystal structures are being destroyed during the reaction and a new crystal structure is being produced." [2]

The fact is taken as a basis that a visible solid crystal consists of a giant arrangement of atoms, ions, or molecules: a chemical structure. The chemical reaction or the formation of new substances can be described as a restructuring of involved smallest particles, the phenomena in chemical reactions can be explained by the connection of two possible interpretations – chemical structure and chemical bonding (see Fig. 10.2):

1. Structures of the reactants are being destroyed and new structures form new substances
2. Bonds are broken in the reactants using up energy, new bonds are formed in the new substances with energy release

Structure and bonding in substances are linked inseparably. Dalton's atomic model is sufficient for the illustration of chemical structures in elementary lessons: the occurrence of new substances is being explained with the formation of new structures, neglecting all questions of bonding. For mental models regarding chemical bonding, a differentiated nucleus-shell atomic model is necessary – usually these models are introduced in advanced lessons and can be a useful addition to the structure-oriented interpretation.

If concrete models are possible, learners will develop their mental model for the structure of matter with more success than without those models (see Chap. 7). Johnstone [3] proposed to differentiate three levels of interpretation (see Fig. 10.3); after observing the substance or chemical reaction (macro level), the learner should grab the infinite structure of an ionic lattice or the finite molecular structure of a molecule (submicro level). Through shortening the models to structural formulae and common formulae, they will understand formulae or chemical equations better than through other ways (representational level).

Simple structures of metal crystals can be illustrated with the help of the particle model of matter in beginning lectures of chemistry; silver particles are arranged in a

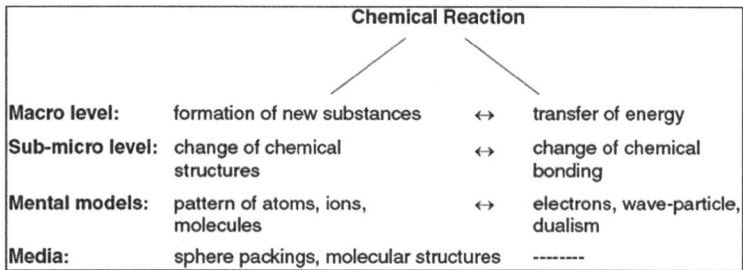

Fig. 10.2 Interpretation of chemical reactions on the base of chemical structure and bonding

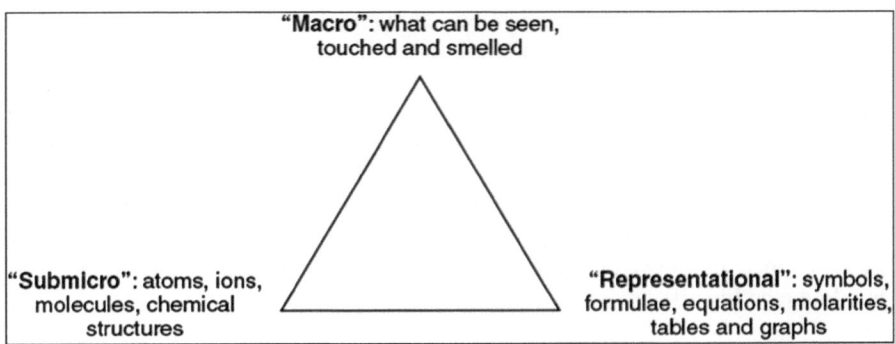

Fig. 10.3 Johnstone's triangle for chemical education [3]

silver crystal like balls in a close-packing of spheres. This model provides a simple and appropriate image of the silver crystal structure, but also an idea of an infinite crystal structure and of a giant chemical structure. Since solids play an important role in elementary lessons, certain structures can be illustrated easily this way. The silver structure can be described with close-packed Ag atoms on the level of Dalton's atomic model. The description of metal bonding can be added later with the help of the differentiated atomic model of nucleus and shell.

Illustrations of finite structures of molecules are also possible on the level of Dalton's atomic model: different balls represent models for different kinds of atoms. Molecular structures can be imagined and build on this basis; for example, a molecular model for an ethanol molecule is built by three kinds of spheres that represent C atoms, O atoms, and H atoms. Besides these educational reasons, there are also aspects of educational psychology that suggest the interpretation of chemical reactions with the help of chemical structures. According to Piaget, students at the age of 13–16 are in the concrete operational stage: bonding theories are too abstract and not comprehensible for them, but illustrations with packing of spheres or molecular models are possible at this stage of development.

From the viewpoint of educational psychology, the arguments in the direction of chemical bonds have to be set on a higher formal operational stage. If protons, neutrons, and electrons are being introduced with the differentiated atomic model, it is extremely difficult to make an arrangement of electrons in shells or energy levels clear and understandable; electron density distributions, probabilities of finding electrons, or the wave-particle duality would have to be introduced. Scientifically, it seems impossible to illustrate electrons as small particles or with the help of balls for concrete models; today physicists, chemists, or theorists calculate energies or energy distributions to determine the charge distribution in lattices or molecules.

10.3 Ions on the Level of Dalton's Atomic Model

The big class of salts cannot be integrated easily into structure-oriented chemistry lessons, because ions – the smallest particles of salts – do not exist in Dalton's atomic model. The introduction of the term "ion" calls for the differentiated atomic model and we shall illustrate this concern with one example.

During the electrolysis of salt solutions or molten salts, deposition of substances at the electrodes can be observed. The decomposition is explained with the transformation of ions to atoms transferring specific numbers of electrons. The charges of ions can be derived by comparing the number of protons in the nucleus and the number of electrons in the shell. Differences of properties between salts and involved elements are described by differences in the atomic and ionic structure.

The downside of this approach is the fact that the term "ion" cannot be used in chemistry lessons before the introduction of the differentiated atomic model. Until then, the smallest particles of salts are not called "ions," but "imaginary molecules, chemical units, units of compounds, compound particles, or even atoms." [3] This circumscription of the term "ion" leads to vague concepts, in most cases to the well-known molecule concepts: empirical studies show that even high school students in grade 11 or 12 after several years of chemistry lessons still think, that salts and salt solutions are built of molecules [4].

It is possible to introduce ions earlier than usual, e.g., with the freezing point depression [5]. Initially, it has to be noted that the freezing point depression rises proportionally with the increase of the concentration of dissolved particles. The freezing point depression of, e.g., 1-M ethanol solution or 1-M glucose solution is $-1.9°C$.

However, the freezing point depression of a 1-mol sodium chloride solution is $-3.8°C$, and of 1-mol calcium chloride solution it is $-5.7°C$ [5]. It can be derived from these measurements that 1 l of a sodium chloride solution contains 2 mol dissolved particles, 1 l of calcium chloride solution contains 3 mol dissolved particles. If the fact is known that salt solutions conduct electric current and explained with dissolved particles that carry electric charges, the name "ion" can be introduced and combined with the following conclusions from freezing point depressions of 1-M solutions [5]:

$$1\,mol\,NaCl \rightarrow 1\,mol\,Na^+(aq)\,ions + 1\,mol\,Cl^-(aq)\,ions = 2\,mol\,ions\,per\,litre$$

$$1\,mol\,CaCl_2 \rightarrow 1\,mol\,Ca^{2+}(aq)\,ions + 2\,mol\,Cl^-(aq)\,ions \ldots = 3\,mol\,ions\,per\,litre$$

If the ions are thought of as smallest particles carrying electric charges, they can be identified by ionic symbols and like atoms introduced as basic particles of matter according to Dalton's model . They can be visualized by spheres with specific diameters and integrated into the Periodic Table of the Elements (PSE) as smallest particles of salts, as Christen suggests [6]. Sauermann and Barke [7] adopted this kind of periodic table (see Fig. 10.4, and www.thinking-chemistry.org).

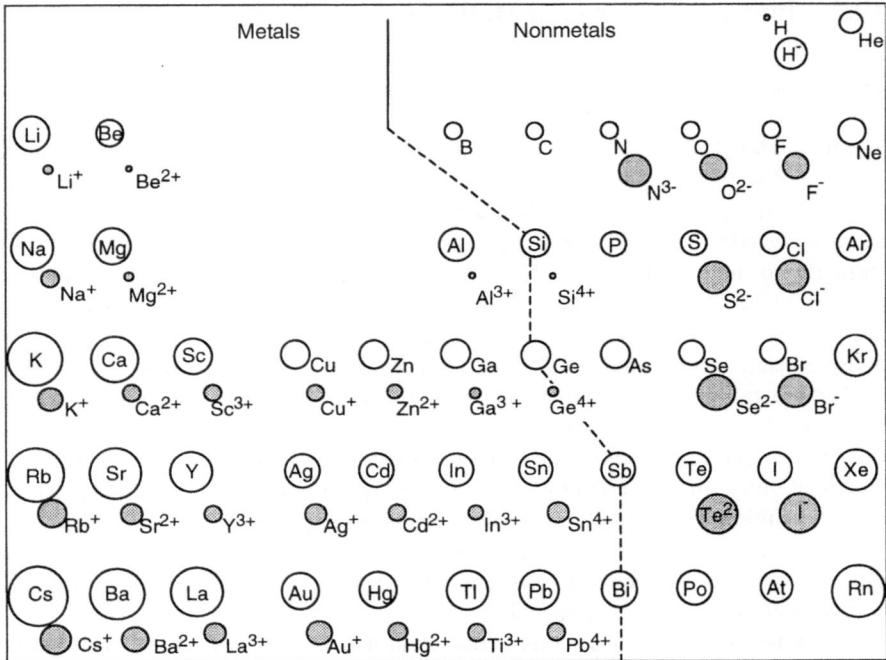

Fig. 10.4 Atoms and ions as basic particles of matter [7]

10.4   Atoms and Ions as Basic Particles of Matter

The periodic system in its shortened form (see Fig. 10.4) shows some of those fundamental particles; listing all subgroups of elements would lead to a complete collection of all fundamental particles of matter. The ancient idea of the Greek natural philosophers has been finished by having a modular system of smallest particles to build up all substances in the universe by combination of these particles. This modular system was foreshadowed when Dalton made the first step by connecting the term "element" with the term "atom" and claiming as many kinds of atoms as there are elements. In 1808, he made the following statements about atoms as smallest basic particles of matter in his book "A New System of Chemical Philosophy":

1. There are as many kinds of atoms as there are elements. They can be distinguished from one another by their different masses
2. All atoms of one element have the same mass and behave the same way in the universe
3. Atoms cannot be created or destroyed nor get lost by chemical reactions
4. The bonding of atoms is reproducible and leads to certain portions of matter

5. Substances can be distinguished by the kind of atoms and by their spatial arrangement

Since Arrhenius postulated in 1884 that ions are the other big group of fundamental particles of matter, Dalton's statements have to be applied to ions as well. It can be claimed that all matter of the universe is built of atoms or ions. Two substances are identical, when they correspond in the kind of atoms or ions and in their spatial arrangement, in their chemical structure.

Accordingly today's chemists have to:

− Detect the kind of atoms or ions bound in a substance.
− Analyze the spatial arrangement of the atoms or ions with different methods of instrumental analysis.

On the other hand, one has to know, which kind of atoms or ions have to be visualized and how they are connected spatially for a substance to have mental models or to build concrete models of the structure. When the model of a sodium crystal is being built mentally, Na atoms have to be connected in a way that they form a cubic body-centered lattice. When the model for a sodium chloride crystal is being built, spheres as models for Na^+ ions and bigger spheres as models for Cl^- ions have to be connected to a cubic face-centered structure.

In reality, fundamental particles can only be connected when they show the same quality of bonding forces. Experiments and laboratory experiences from the last centuries led to three classes of basic particles with different types of attractive forces:

1. Metal atoms
2. Nonmetal atoms
3. Ions

Metal atoms with the same type of attractive forces are placed on the left side of the PSE and nonmetal atoms with another type of attraction are placed on the right side (see Fig. 10.4) – therefore, the H atom and H^- ion is placed on the right side of PSE close to the 7th group. Ions with the third kind of attractive forces are placed on both sides; cations on the left side together with their metal atoms, anions on the right side together with their kind of nonmetal atoms. Special rules of combination show that atoms with the same type of attractive forces can be combined (see Table 10.1).

Nondirectional bonding forces are forces, which a particle reveals spherically symmetrical in all directions of space: such particles form crystal lattices. Any number of metal atoms, for example, can be connected in a metal structure.

Table 10.1 Rules of combination for atoms and ions as fundamental particles of matter [7]

Position in PSE	Kind of articles	Bonding	Structure
"Left and left"	Metal atoms	Spatially undirected	Metal structure
"Left and right"	Ions	Spatially undirected	Ionic structure
"Right and right"	Nonmetal atoms	Spatially directed	Molecule or atomic lattice

The close packing of spheres might serve as a concrete model: one sphere is surrounded by as many spheres as space permits, which means one sphere is surrounded by exactly 12 other spheres.

Directional bonding forces are forces in which a particle only reveals into certain directions. Such particles, for example, nonmetal atoms, connect to form molecules. Only a limited, but specific number of atoms are bonded by directional forces. One C atom and four H atoms are connected in one methane molecule (CH_4). The C atom is therefore called tetravalent; the H atom is called monovalent. Colored spheres with a special number of push buttons or connecting sticks are used in common molecular modeling kits that serve for modeling small molecules, for example, to teach molecular structures in Organic Chemistry.

Combinations and arrangements of metal atoms, ions, and nonmetal atoms will be described and a few examples of the resulting fundamental structures will be presented below. Other examples and in-depth explanations can be found in publications of Sauermann and Barke: the formation of alloys by metal atoms can be found in volume 2 [8], the bonding of nonmetal atoms in volume 3 [9], and the bonding of ions in volume 4 [10]. Many special chemical structures are shown in other publications (e.g., Wells [11]).

10.5   Combinations of Atoms and Ions in Giant Structures and Molecules

On the basis of the periodic table (see Fig. 10.4), atoms and ions should be linked theoretically to form chemical structures according to the above-mentioned rules of combination (see Table 10.1). Metals, alloys, and salts form infinite structures that can be illustrated with close-packing of spheres and crystal structures.

10.5.1   Combination of Metal Atoms ("Left and Left in Periodic Table")

The nondirectional bonding forces of metal atoms lead to metal lattices. The arrangement of one kind of metal atoms is generally possible in three different ways, leading to three lattice structures, which can be illustrated easily with packing of spheres (note: a few metals have structures different to the ones covered here). Figure 10.5 lists names and symbols for these structures, coordination numbers, structure models, and examples.

Two close-packed arrangements of spheres have the coordination number 12. The base is a triangular pattern with one sphere being surrounded by six other spheres in one layer, three spheres touch this sphere in the layers above and below: altogether, each sphere has 12 neighbors (see Fig. 10.6). But there are two different

Crystal structure		Coordination number	Structural model	Examples
hexagonal	⬡•	12	hexagonal closest packing of spheres	Magnesium, zinc Mg type
cubic face-centred	⊠	12	cubic closest packing of spheres	Copper, silver, gold, Cu type
cubic body-centred	◻•	8	body-centred packing of spheres	Alkaline metals, tungsten W type

Fig. 10.5 Main crystal structure types of metals

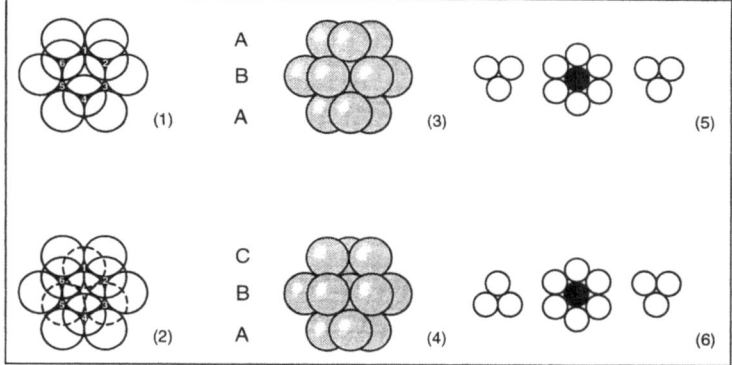

Fig. 10.6 Coordination polyhedrons for both close-packed arrangements of spheres: on top for hexagonal closest packing, below for cubic closest packing

possibilities to stack three spheres on a layer of seven spheres: either the interstices 1, 3, and 5 or the interstices 2, 4, and 6 are filled. The result is two different sphere packings, which can be characterized with sequences ABAB... and ABCABC.... Only the layers with triangular patterns are to be counted with ABA or ABCA.

While Fig. 10.6 shows both polyhedra with the coordination number 12, Fig. 10.7 highlights the packings that are based upon a triangular pattern and have the shape of triangular pyramids (see 1 and 2 in Fig. 10.7). Both elementary units of the pyramids are shown: on one hand packing of spheres (see 1a and 2a), on the other hand crystal lattices (see 1b and 2b). Additionally one can see that the ABCA packing has a square pattern of spheres, when one line of spheres is taken off the edge of the triangular pyramid (see 3 in Fig. 10.7).

If spheres are being stacked starting from a square pattern, the result is a square pyramid, which also has the coordination number 12 (see 1 and 2 in Fig. 10.8). The faces of the square pyramid have the triangular pattern, which can be continued by

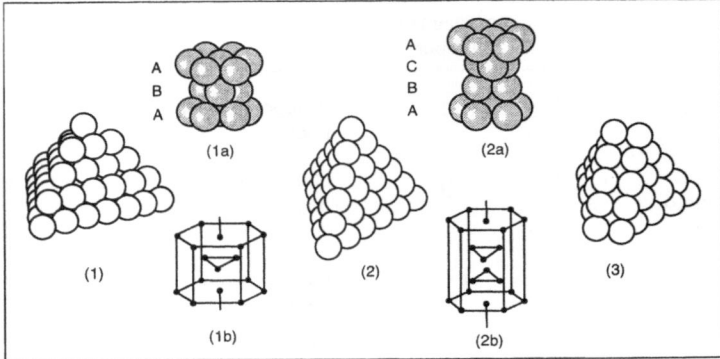

Fig. 10.7 Both forms of close-packed arrangements of spheres, starting with a triangular pattern

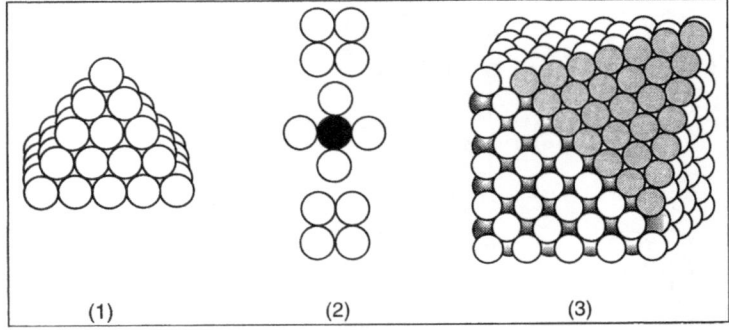

Fig. 10.8 ABCA-close-packing of spheres beginning with a square pattern

in the sequence ABCA (see 1 as well as 3 in Fig. 10.8). The ABCA-packing of spheres can be built starting with the triangular pattern as well as with the square pattern: both types have the identical structure.

In the close packing of spheres, starting from a square pattern (see Fig. 10.8), one can see a cubic packing of spheres, which apart from spheres on the corners also holds one sphere in the middle of each face (see Fig. 10.9). This elementary unit is therefore called face-centered cube, and the corresponding packing of spheres ABCA is called cubic face-centered or cubic close-packed. The elementary unit can be described in two ways: either starting from the square pattern with $5 + 4 + 5 = 14$ spheres, or starting from the triangular pattern with $1 + 6 + 6 + 1 = 14$ spheres (see Fig. 10.9). It is difficult to imagine the elementary cube being a part of the ABCA cubic close-packing, especially when it is shown as the triangular pyramid (see (2) in Fig. 10.7): the cube stands on an apex with the body diagonal perpendicular to the plane. This can be made clearer by constructing an elementary cube and building it into the packing of spheres (see also Chap. 6, M6.5).

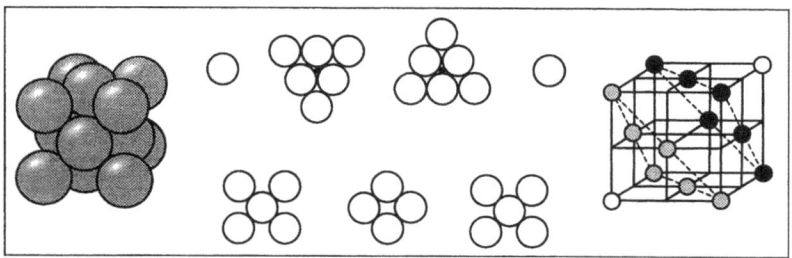

Fig. 10.9   Cubic face-centered cube as the elementary unit of the ABCA-packing

There are two kinds of close-packing of spheres (see Fig. 10.7): the cubic close-packing of spheres and the hexagonal close-packing of spheres. It derives its name from the special hexagonal elementary unit of $7 + 3 + 7 = 17$ spheres (see 1a and 1b in Fig. 10.7). The hexagonal packing cannot be described with another elementary unit of a higher symmetry.

A third metal structure exists with the coordination number 8 that does not describe a close packing of spheres: it is the cubic body-centered structure. The model for this structure is the cubic body-centered packing of spheres. The elementary unit of this packing is a cube, which consists of nine spheres: the cube center is occupied by one sphere that touches the eight spheres on the corners, these spheres do not touch (see Fig. 10.10).

The elementary units that show the symmetry of these sphere packings, were found in search of the smallest segment of each packing of spheres: Fig. 10.10 displays these three elementary units. Another special segment, which forms the whole structure by shifting it in all three directions in space, is called the unit cell. It is obtained by vertical and horizontal cuts through the centers of the spheres in the elementary units. The number of full spheres in each unit cell can be received by adding together all parts of the unit cell (see Fig. 10.10).

The unit cells of the cubic structures make it easy to understand that the overall structure can be built by putting together many unit cells in all three dimensions in space. It should also be pointed out that there are other unit cells for other structures. This issue can be compared to patterned wallpapers: there are also different possibilities to find a segment that builds the whole pattern of the wallpaper by a shift in two directions in plane.

It has to be pointed out that the pictured unit cell of the hexagonal close packing (see Fig. 10.10) cannot be shifted into all directions in space to build the lattice without gaps. Only one third of the displayed cell is used to build the whole lattice (see dashed lines in Fig. 10.10). This one third is the true unit cell: it contains one third of six spheres (see Fig. 10.10), namely two full spheres when all parts are added together.

Metal crystals can have different structures at different temperatures. It has been known for thousands of years that iron can be deformed if it is heated to glowing red and hardens when it cools down. Today we can explain that iron crystals undergo a

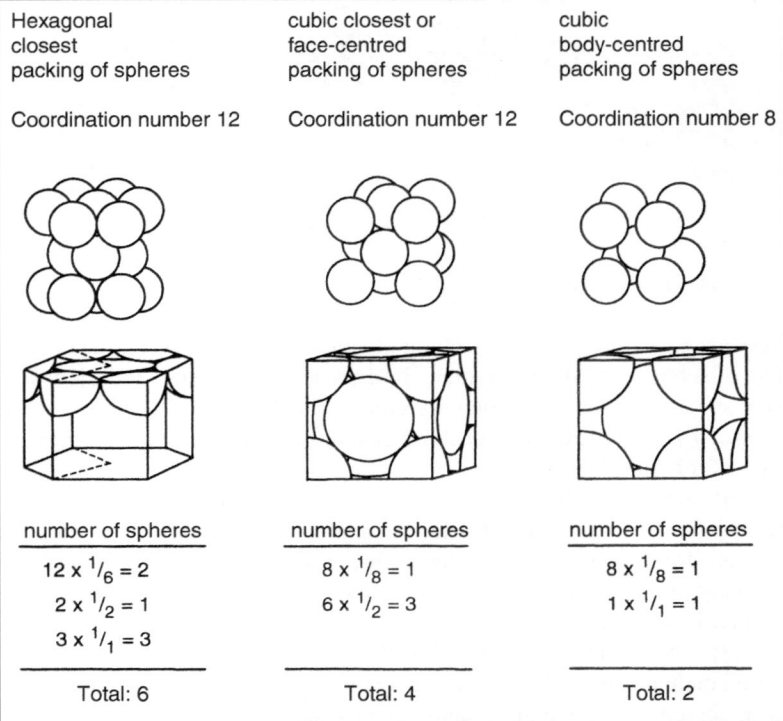

Fig. 10.10 Elementary units and unit cells of the three metal structures

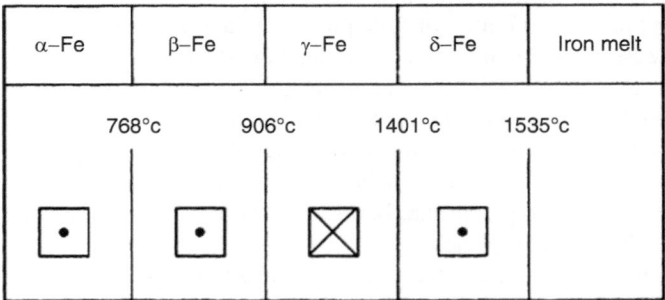

Fig. 10.11 Different iron structures depending on temperature

structural change at 906°C: the cubic body-centered structure changes to a cubic face-centered structure [11]. Figure 10.11 shows different modifications of iron and corresponding structures: β-iron only differs from isostructural α-iron by the disappearance of magnetism at 768°C.

The ductility of γ-iron in the given temperature range can be explained with the high symmetry of the cubic face-centered structure: close-packed layers of atoms in a triangular pattern can be found in the direction of every of the four body diagonals. These layers can be shifted easily in four directions when forces have an impact on the metal crystals. The hexagonal packing only has one such shift direction; therefore, metals with a hexagonal structure like magnesium and zinc are not very ductile. Metals with a cubic face-centered structure like gold, silver, copper, or γ-iron can be rolled out to thin wire or very thin sheets ("gold leaf").

The shape memory effect of memory metals like "Nitinol" (NiTi), a special alloy of nickel and titanium, is also based on a diffusionless and therefore lightning-fast change of structure at certain temperatures. But the memory effect cannot be explained without the formation and translation of twin crystals [8, 12].

$$3 \ [12 \, h] \quad 3 \quad [8c]$$
$$\text{Mg} \qquad \text{Fe}$$
$$\infty \qquad\quad \infty$$

$$\{\text{Mg } 12/12\}\text{G}$$
$$\{\text{Fe } 8/8\}\text{G}$$

It is possible (but not usual) to give information about the different structures of metal crystals with the help of structural symbols. The written symbols state that a three-dimensional infinite packing of metal particles is given: Mg atoms form a hexagonal packing with the coordination number 12, while Fe atoms form a cubic packing with the coordination number 8. These symbols are called Parthé symbols. Other symbols show the coordination number in curly brackets and indicate that a three-dimensional lattice is given: Niggli-symbols.

10.5.2   Metal Reactions: Regrouping of Metal Atoms

Sodium and mercury react vigorously when a small piece of sodium is crushed in a drop of mercury [13]. This reaction can be understood as a regrouping of Na atoms and Hg atoms into a new metal lattice with both kinds of atoms distributed statistically (see Fig. 10.12).

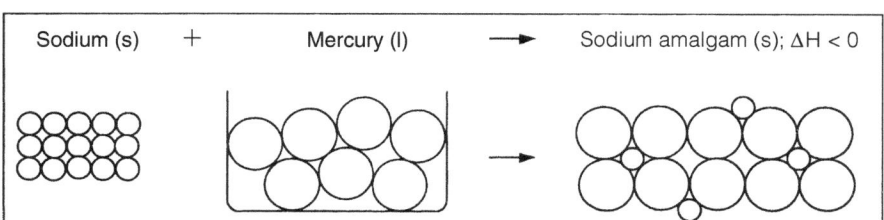

Fig. 10.12 Model drawing for the regrouping of metal atoms by forming sodium amalgam [13]

Alloys of mercury are called amalgams. Dentists use silver–tin amalgams for dental fillings: they fill the tooth with the freshly prepared ductile mixture, where it hardens within 1 h and forms crystals, which enclose the Hg atoms in the special crystal structure. Gold amalgam plays an important role for gold diggers, mercury is used for the separation of gold dust and sand: a generous amount of mercury has to be added to the suspension of gold dust and sand in water and then shaken; gold dissolves in mercury. The gold amalgam has to be washed out and vaporized later in the camp fire: gold crystals remain.

Alloys are compounds with overall metallic character. Smelts of pure metals have to be mixed and crystallized by cooling to produce mostly solid alloy crystals. Some important alloys are listed with their compositions and densities; steel is an alloy that contains up to 1.7% of carbon (see Table 10.2).

If the atoms have more or less the same size, they form substitutional mixed crystals. If they have different sizes, the small atoms occupy sites between the bigger atoms: interstitial mixed crystals (see Fig. 10.13). Two kinds of metals with the same structure and more or less the same atomic radii cause the metals to form a continuous solid solution. For example, gold and copper, both metals with a cubic face-centered structure and nearly same atomic radii, can alloy in every possible

Table 10.2 Familiar alloys, their compositions and densities

Alloys	Components (%)	Density (g cm^{-1})
Steel	Fe, C (bis 1.7%)	7.8
Chromium–nickel steel	55 Fe, 25 Cr, 20 Ni, 0,5 Si	7.9
Inox steel	71 Fe, 20 Cr, 8 Ni, 0.2 Si/Co/Mn	7.3...7.4
Brass, white	50 Cu, 50 Zn	8.2
Brass, yellow	70 Cu, 30 Zn	8.4
Brass, red	90 Cu, 10 Zn	8.8
Bell bronze	80 Cu, 20 Sn	8.7
Duralumin	92 Al, 5 Cu, 2 Mg, 1 Mn	2.8
Electron	95,7 Mg, 4 Al, 0.3 Mn	1.8
Constantan	60 Cu, 40 Ni	8.9
Nickel–silver	64 Cu, 30 Zn, 6 Ni	8.3
Platinum–iridium	5 Pt, 95 Ir	22.4

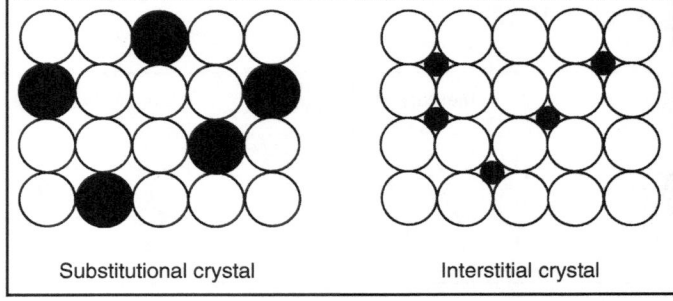

Fig. 10.13 Models of substitutional (*left*) or interstitial (*right*) metal alloy structures

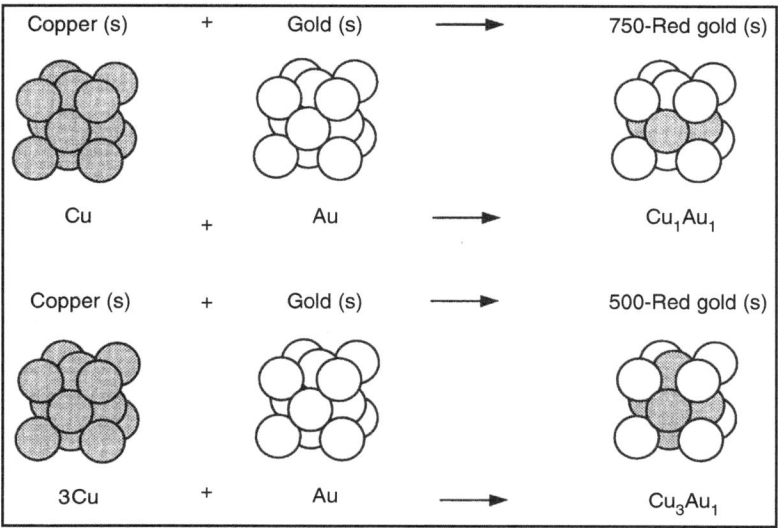

Fig. 10.14 CuAu and Cu₃Au super structures of red-gold alloys [14]

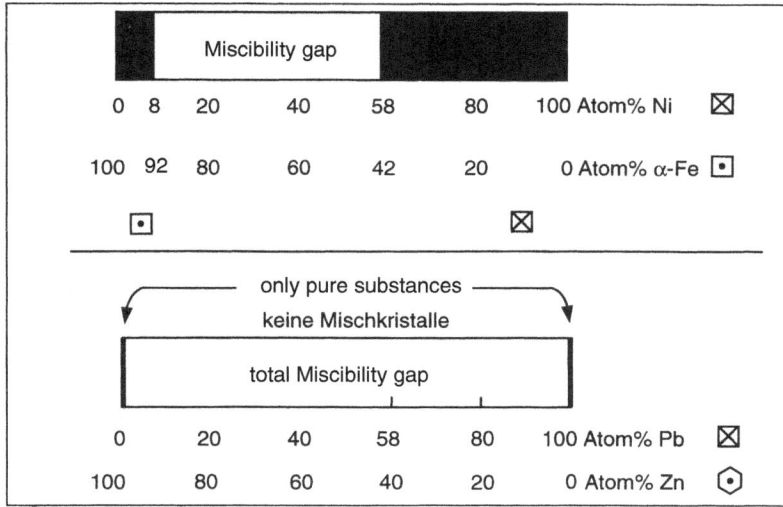

Fig. 10.15 Examples for composition bands and miscibility gaps for mixtures of metals

ratio. For ratios of Cu atoms and Au atoms of 1:1 or 3:1 exactly, the atoms do not form a random distribution in the mixed crystals, but rather order periodically [14]: they form line phases with super structures (see Fig. 10.14).

If metals with different lattice structures alloy, substitutional mixed crystals of both structures as well as miscibility gaps may occur: composition bands for the examples nickel–iron and lead–tin describes this behavior (see Fig. 10.15). A 100%

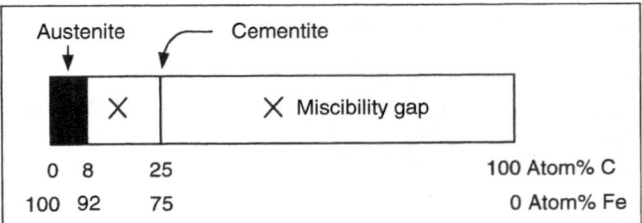

Fig. 10.16 Austenite and cementite in the composition band of iron and carbon (730°C)

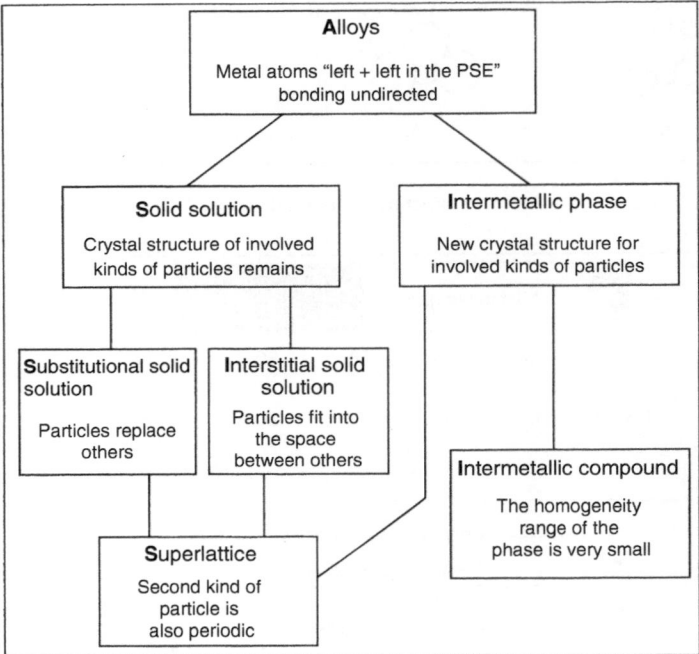

Fig. 10.17 Summary of pathways to the formation of alloys and intermetallic compounds

miscibility gap occurs for the example lead–tin: only the pure metals crystallize from the melt, a heterogeneous mixture of lead crystals and tin crystals appears. One last example shows that, apart from a miscibility gap, other compounds may be formed: cementite, Fe_3C (see Fig. 10.16).

A final diagram will schematically summarize the multitude of possible pathways to the formation of alloys and intermetallic compounds (see Fig. 10.17).

10.5.3 Combination of Ions ("Left and Right in Periodic Table")

The nondirectional bonding forces of ions lead to the formation of ionic lattices, when ions of two or three different kinds are bonding. The structure of ionic lattices basically depends on two factors (1) ionic charge, (2) radius ratio of the ions.

Ionic charge: According to attraction and repulsion forces an ionic structure is in equilibrium: equal numbers of negative charges are grouping around every positively charged ion (cation), and equal numbers of positive charges are around every negatively charged ion (anion). If the charge on cations and anions is equal, an ionic lattice with the ratio 1:1 results (see 1 in Fig. 10.18). If the cations are charged 2+ and the anions are charged 1−, the ion ratio has to be 1:2 (see 2 in Fig. 10.18). If three kinds of ions form a lattice, the ion ratios have to be such that the ionic compound is electrically neutral (see Table 10.3).

Size ratio of ions: Usually ionic lattices can be described with close packing of the big negatively charged ions (anions), where the small, positively charged ions (cations) occupy the interstitial sites between the anions. According to the radius ratio rule different interstitial sites are occupied, depending on the size difference (see Fig. 10.19). If the cation is relatively big, the anions form a simple cubic packing in which the cations occupy the *hexahedral sites*: cesium chloride is one example for this kind of ionic structure.

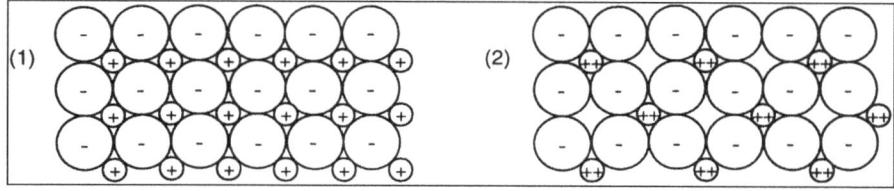

Fig. 10.18 Mental models for charge balanced arrangements of ions (ratio 1:1 and 1:2)

Table 10.3 Examples for ion ratios in ionic structures and corresponding formulae

Charge of		Ratio of numbers	
Cations	Anions	Cations:anions	Examples for ion lattices
1+	1−	1:1	$NaCl$, $CsCl$, $NaNO_3$, NaH_2PO_4
2+	2−	1:1	CaO, ZnS, FeS, $BaSO_4$
2+	1−	1:2	CaF_2, $MgCl_2$, $Fe(OH)_2$
4+	2−	1:2	TiO_2, PbO_2
1+	2−	2:1	Li_2O, Na_2S, Na_2SO_4
3+	2−	2:3	Al_2O_3, Fe_2O_3, $Fe_2(SO4)_3$
1+, 2+	1−	1:1:3	$KMgCl_3$
2+, 2+	2−	1:1:2	$CuFeS_2$
2+, 3+	2−	1:2:4	$MgAl_2O_4$
2+, 4+	2−	1:1:3	$CaTiO_3$

Coordination number	Size ratio	Packing of spheres	Coordination polyhedron	Chemical structure	Examples
12	$r_1 : r_2 = 1$ (Cubo-octahedron gap)		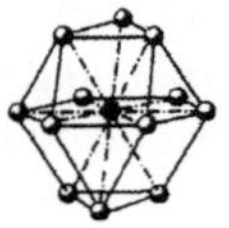	cubic face-centred (cubic closed)	Cu, Ag, Au, Pt, Ca, Sr, Al, Pb
8	$r_1 : r_2 \geq 0{,}732$ (Hexahedron gap)	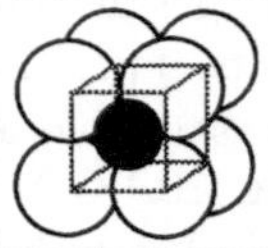	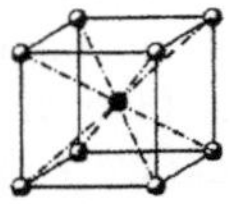	cubic body-centered CsCl-Structure	CsCl, CsBr, CsI, NH$_4$Cl, NH$_4$Br
6	$r_1 : r_2 \geq 0{,}414$ (Octahedron gap)	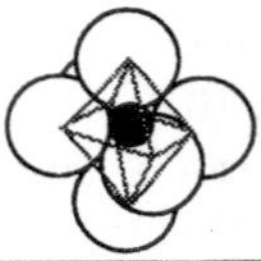	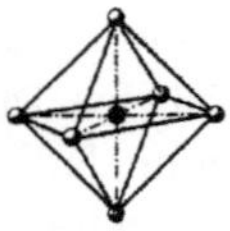	cubic face-centered NaCl-Structure	NaCl,Mgo, CaO, Li-,Na-,K-,Rb-halides, CsF, Alkali metal hydrides
4	$r_1 : r_2 \geq 0{,}225$ (Tetrahedron gap)	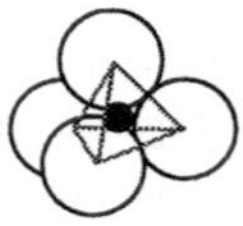	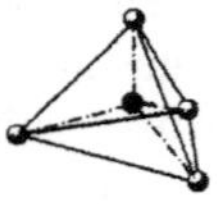	ZnS-Structure Li$_2$O-Structure	ZnS, CuCl, AgI, Li$_2$O, CaF$_2$

Fig. 10.19 Examples for the formation of ionic structures with cubic unit cells depending on radius ratios of the involved ions

If the cations only have about half the size of the anions they occupy the *octahedral sites*, which can be found in both close-packed arrangements of spheres. Sodium chloride is one example: chloride ions form the cubic close packing and sodium ions occupy all the octahedral sites, the coordination number is 6.

Smaller radius ratios lead to an occupation of *tetrahedral sites*: in zinc sulfide, the sulfide ions with the charge 2– form the cubic, close packing, and half of the tetrahedral sites are occupied by the same number of zinc ions with the charge 2+, the coordination number is 4. In lithium oxide, all tetrahedral gaps are filled because the lithium ion is charged 1+ (see Fig. 10.19).

The gaps in an ionic lattice may be all occupied. The sodium chloride lattice can be described as a cubic close packing of Cl$^-$ ions with the octahedral sites completely occupied by Na$^+$ ions. If the charge numbers of the ions are different, only a fraction of the interstitial sites is occupied. The Cl$^-$ ions form a cubic close packing in the cadmium chloride structure, for example, but only half of the octahedral sites are occupied by Cd^{2+} ions (see Table 10.4).

Since there are twice as many tetrahedral sites as there are octahedral sites in cubic close packing, all tetrahedral sites are occupied in the ionic structure of lithium oxide: Li$^+$ ions and O^{2-} ions exist in the ratio of 2:1 (Li$_2$O). Only half of the tetrahedral sites are occupied in the ionic lattice of zinc sulfide: the formula is ZnS (see Table 10.4).

Table 10.4 Description of salt structures based on the filling of interstitial gaps

Packing of the anions	Occupation by cations	Examples
Cubic closed	all OG	NaCl, KCl, AgCl, MgO, CaO, FeO, PbS
Cubic closed	1/2 OG	$MgCl_2$, $CaCl_2$, $ZnCl_2$, $FeCl_2$, $CdCl_2$, CaH_2
Cubic closed	All TG	Li_2O, Na_2O, K_2O
Cubic closed	1/2 TG	ZnS (zinc sulfide), CuCl, AgI, HgS
Cubic closed	1/2 OG, 1/8 TG	$MgAl_2O_4$ (spinelle), Fe_3O_4
Hexagonal closed	All OG	NiAs, FeS
Hexagonal closed	2/3 OG	Al_2O_3 (corunde), Fe_2O_3
Hexagonal closed	1/2 OG	CdI_2, PbI_2, $Mg(OH)_2$, $Ca(OH)_2$
Hexagonal closed	1/2 OG	ZnS (wurtzite), ZnO, CdS
Simple cubic	All HG	CsCl, CsBr, CsI, NH_4Cl
Simple cubic	1/2 HG	CaF_2, PbF_2, BaF_2
Cubic closed by S_2^{2-} dumb-bell	All OG	FeS_2 (pyrite)

OG octahedral gap, *TG* tetrahedral gap, *HG* hexahedral gap

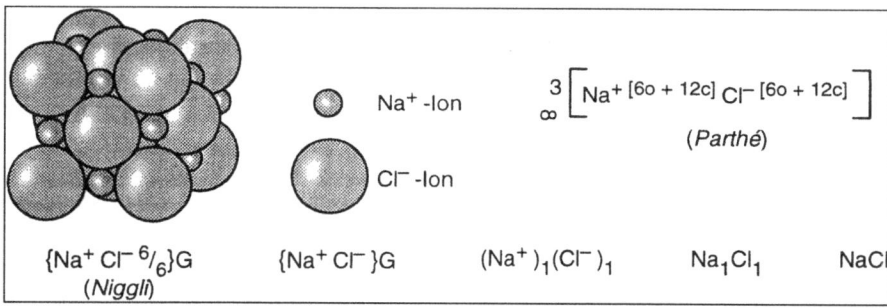

Fig. 10.20 Chemical symbols for the NaCl-structure with decreasing information content

Ionic structures: The mentioned structures of ionic lattices are known from instrumental analysis, especially from X-ray structure analysis. In the simplest case, a Laue diagram can be obtained by bombarding an aligned salt crystal with a collimated beam of X-rays for some time [15]: the pattern of the diffracted beams on a photo plate reflects the symmetry of the structure. Experts are even able to produce a three-dimensional structural model of the crystal. Other methods of X-ray structure analysis can be found in the specific literature.

Symbols for ionic structures: The structural symbol and the stoichiometric symbol of a salt can be derived from the known ionic structure by shortening the structural model. An example is shown for the common structure of sodium chloride (see Fig. 10.20). Derived from the 3D sphere packing, the Parthé symbol not only outlines the infinite (∞) and three-dimensional (3) structure but also includes the coordination of other ions (6o: six Cl^- ions coordinate one Na^+ ion in octahedral symmetry), and in addition the coordination of the same kind of ions (Na^+ ions) is given (12c: twelve Na^+ ions coordinate one in the cubic face-centered geometry).

The Niggli symbol is limited to the coordination number of the surrounding oppositely charged ions (6/6) and the type of infinite structure (G for German "Gitter" = lattice). Another possibility for shortening the structure symbol is the indication of the types of ions in curly brackets (see Fig. 10.20): the brackets symbolize the solid salt, the "G" (German "Gitter") the three-dimensional structure.

The result of rigorous shortening is the symbol with the least information: the stoichiometric symbol NaCl. Since teachers mostly work with stoichiometric symbols, and since these even suggest an abbreviation of the substance name, the learner cannot build up any mental model of the structure at all. The consequences are often misconceptions concerning salt structures [4]: NaCl molecules, NaCl formula units, etc.

The ratio of two different types of ions can be determined in three different ways:

1. A closest sphere packing of spheres shows two types of gaps: the octahedron gaps (OG) and the smaller tetrahedron gaps (TG). The ratio of their numbers is: spheres:OG:TG = 1:1:2. The fraction of occupied octahedral or tetrahedral gaps defines the ion ratio (see Table 10.4).
2. The ratio of ions can be counted directly on an appropriate segment of the packing of spheres, but not on the usual elementary NaCl cube: it has a ratio of 14:13. If another layer is added to the cube, the ratio is 18:18. Only experts are able to determine those segments.
3. The unit cell is the smallest unit of an infinite crystal structure that forms the whole lattice by translation in all three directions in space. It does not only show all lattice symmetries but also the ion ratio. The NaCl unit cell shows the ratio 4:4 or shortened to 1:1 (see Fig. 10.21).

The examples NaCl and ZnS show that different structures can occur with the same ion ratio (see Fig. 10.21) – so it is necessary to not only take the formulae

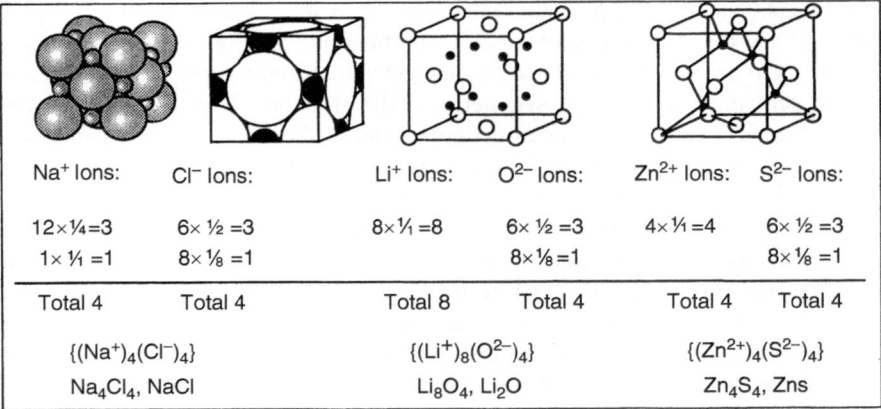

Na$^+$ Ions:	Cl$^-$ Ions:	Li$^+$ Ions:	O^{2-} Ions:	Zn^{2+} Ions:	S^{2-} Ions:
12×¼=3	6× ½ =3	8× ¹⁄₁ =8	6× ½ =3	4× ¹⁄₁ =4	6× ½ =3
1× ¹⁄₁ =1	8× ⅛ =1		8× ⅛ =1		8× ⅛ =1
Total 4	Total 4	Total 8	Total 4	Total 4	Total 4
{(Na$^+$)$_4$(Cl$^-$)$_4$}		{(Li$^+$)$_8$(O^{2-})$_4$}		{(Zn^{2+})$_4$(S^{2-})$_4$}	
Na$_4$Cl$_4$, NaCl		Li$_8$O$_4$, Li$_2$O		Zn$_4$S$_4$, Zns	

Fig. 10.21 Derivation of chemical symbols for ionic structures from unit cells

but also look better to the chemical structures. If one wants to emphasize that chemical symbols always symbolize smallest structural units of substances (see Fig. 10.21), unit cells and their symbols like Na_4Cl_4, Li_8O_4, or Zn_4S_4 should be used and studied [14].

In other compounds the molecules are the smallest structural units, for example, the molecular symbol C_6H_6 is chosen for the benzene molecule. This unit is obvious for the smallest particle of benzene and nobody would recommend shortening this symbol to the stoichiometric symbol C_1H_1 or CH. It would be consequent to discuss according to salts those formulae of the unit cells.

10.5.4 Reactions of Salts: Regrouping Ions

The dissolution of salts in water is usually accompanied by energy transfer – therefore, it is a chemical reaction. It can be interpreted as a separation of ions from the salt's ionic lattice and simultaneous hydration of the ions by water molecules:

$$\text{ammonium nitrate (s)} \rightarrow \text{aq} \rightarrow \text{ammonium nitrate (aq);}\quad \text{endothermic}$$

$$\left\{ (NH_4^+)_1 (NO_3^-)_1 \right\} \rightarrow \text{aq} \rightarrow NH_4^+(\text{aq}) + NO_3^-(\text{aq});\quad \Delta H > 0$$

$$\text{sodium hydroxide (s)} \rightarrow \text{aq} \rightarrow \text{sodium hydroxide (aq);}\quad \text{exothermic}$$

$$\left\{ (Na^+)_1 (OH^-)_1 \right\} \rightarrow \text{aq} \rightarrow Na^+(\text{aq}) + OH^-(\text{aq});\ \Delta H < 0$$

For the dissolution of salts in water, lattice energy is used for separating the ions of an ionic structure. Parallel to the separation of ions, a shell of H_2O molecules forms around the ions: this process releases the hydration energy. The first shown reaction above is endothermic – therefore, the lattice energy has to be higher than the released hydration energy. For exothermic dissolution processes, this is the other way around. One example shows the dissolution of sodium hydroxide in water: the released hydration energy is higher than the lattice energy required.

Precipitation reactions: If solutions that react to insoluble salts are mixed, the corresponding solids precipitate, while the ions of easily soluble substances remain dissolved. Such a precipitation process is a regrouping of ions (see Fig. 10.22). The formation of characteristically looking precipitates confirms the existence of certain kinds of ions: these reactions can therefore be employed as analytical tests for ions (qualitative analysis). Drying and weighing the obtained precipitates allows calculating the amounts of precipitates (quantitative analysis, gravimetry). Some examples for precipitation reactions are listed below:

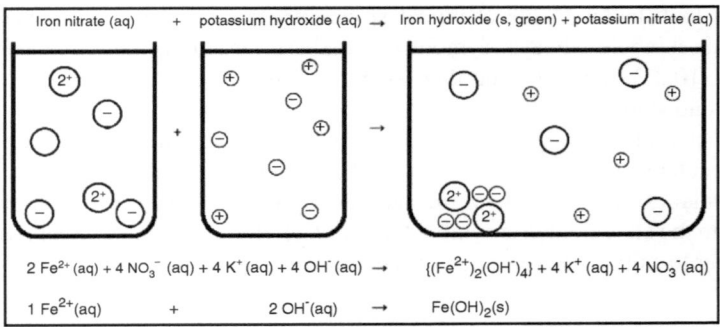

Fig. 10.22 Beaker model for the regrouping of ions in a precipitation reaction

$$Ag^+(aq) + Cl^-(aq) \rightarrow AgCl\,(s, white)\quad Cd^{2+}(aq) + S^{2-}(aq) \rightarrow CdS\,(s, yellow)$$

$$Ba^{2+}(aq) + SO_4^{2-}(aq) \rightarrow BaSO_4(s, white)\quad Ca^{2+}(aq) + CO_3^{2-}(aq) \rightarrow CaCO_3(s, white)$$

$$Hg^{2+}(aq) + 2I^-(aq) \rightarrow HgI_2(s, red)\quad Fe^{2+}(aq) + 2OH^-(aq) \rightarrow Fe(OH)_2(s, green)$$

$$Cu^{2+}(aq) + S^{2-}(aq) \rightarrow CuS\,(s, black)\quad Fe^{3+}(aq) + 3OH^-(aq) \rightarrow Fe(OH)_3\,(s, brown)$$

10.5.5   Combination of Nonmetal Atoms ("Right and Right in Periodic Table")

The combinations of nonmetal atoms lead to familiar molecules like CH_4, NH_3, H_2O, or HF and on the other side to giant structures like in crystals of diamond, graphite, or silicon. In these cases directed bonding forces exist between the nonmetal atoms, the concept of bonding valence sorts the bonding capacity of atoms based on their position in the periodic table (see Fig. 10.23): the C atom bonds with four H atoms in the CH_4 molecule. Accordingly the C atom is tetravalent, the N atom is trivalent, the O atom is bivalent, H or F atoms are monovalent. In later conceptions of the nucleus-shell model of an atom, the bonding valence corresponds to the number of bonding electron pairs or electron clouds.

The usual molecular model kits display the bonding valence: models for C atoms have four push buttons, which point at the tetrahedral corners with an angle of 109°. Models for N atoms have three push buttons, for O atoms they have two push buttons etc. Since these molecular model kits are familiar, they will not be explained any further. The question remains as to how experts determine the structure of molecules and thereby bond length and bond angle. On the one hand there are the methods of X-ray analysis, on the other hand there are spectroscopic

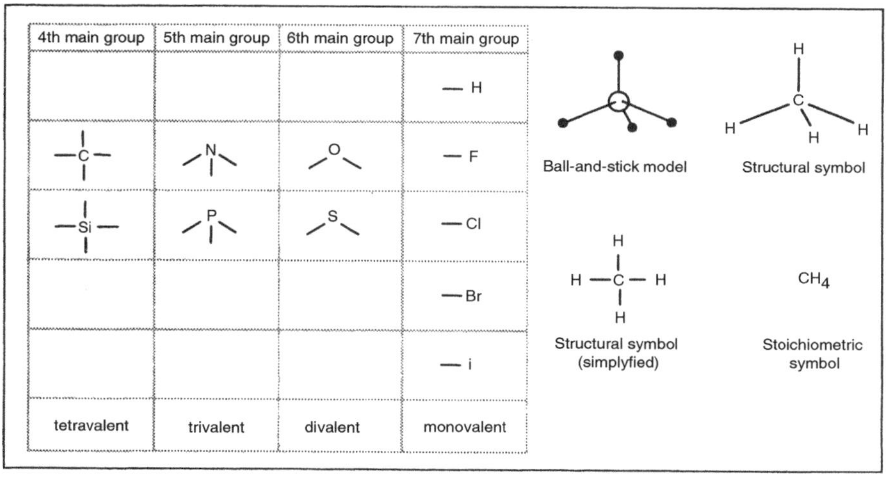

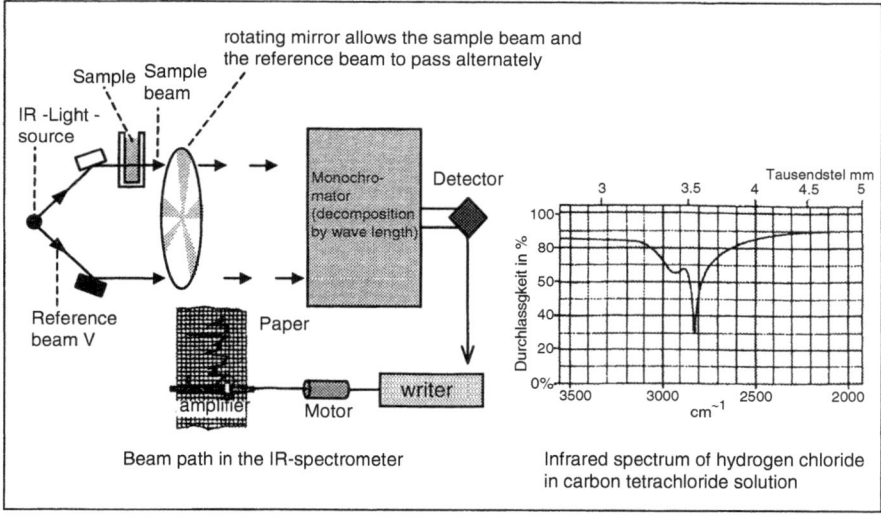

Fig. 10.23 Bonding valences of nonmetal atoms in molecules

Fig. 10.24 Set-up for infrared spectroscopy, with a solution of HCl in CCl₄ as an example

methods. They make use of the energetic interaction that exists between electro-magnetic radiation and the molecules of a specific substance. Infrared spectroscopy, for example, works with long-wavelength infrared radiation (see Fig. 10.24): this beam of radiation excites the molecules in a solution to vibrate in different ways: stretching vibration causes a change of bond length, bending vibration causes a change of bond angles. Specific frequencies are being absorbed.

Absorbed frequencies can be displayed by comparing and calculating the signals of the unchanged reference beam, characteristic absorption peaks form in the IR spectrum. HCl molecules in carbon tetrachloride solution, for example, have a peak at 2,800 cm^{-1} (2,800 vibrations per cm). The corresponding frequency (vibrations per second) of the IR radiation has been absorbed by the molecules. Certain groups of atoms independently absorb IR radiation in the same wave length range, the functional groups show characteristic absorption bands: the molecular structure of a new substance can be derived from those absorption bands in the IR spectrum.

Computers are today able to print three-dimensional pictures of molecular structures, which serves as a template for building molecular models with the help of molecular kits – a molecular model can be build by these and additional information.

Chemical symbols: The combination of nonmetal atoms can lead to molecules, which exist in a linear or bent structures, nonmetal atoms build rings, tetrahedral, octahedral, or any other shape. Combinations of nonmetal atoms can also form infinite structures like giant or honeycomb structures (see Fig. 10.25).

Molecular and giant structures can be described with structural symbols as well as with stoichiometric symbols [9]. Since the structural symbol is often very complicated, one can switch to the stoichiometric symbol, if there is only one possible structure. But as soon as there are isomeric structures, the structural symbol must be used.

The molecular symbol always states the composition of the whole molecule and the arrangement of all atoms in the molecule, like C_6H_6 for the benzene molecule – shortening this symbol to C_1H_1 or CH is not usual. In most cases, half-structural symbols are sufficient: the structure of the ethanol molecule is then stated with the symbol C_2H_5OH or CH_3CH_2OH, the structure of the acetic acid molecule with CH_3COOH (see Fig. 10.26).

10.5.6   Reactions of Nonmetals: Regrouping of Atoms in Molecules

White phosphorus reacts with oxygen to form white crystals of solid phosphorus pentoxide. All three substances are built of molecules, the solid substances phosphorus and phosphorus pentoxide have a molecule lattice. The reaction symbol for the regrouping of molecules can be derived from the corresponding models of the molecular structures (see Fig. 10.27).

The gases hydrogen and oxygen react to form water vapor that condenses at room temperature to liquid water. If a hydrogen–oxygen mixture is being ignited in a eudiometer tube at room temperature, it can be observed that the mixture completely reacts to water, when the volume ratio is 2:1. If this experiment is run in a heated eudiometer tube at temperatures $t > 100°C$, two volume parts of hydrogen react with one volume part of oxygen to produce two volume parts of water vapor. This phenomenon can be explained with the regrouping of H_2 and O_2 molecules to H_2O molecules and illustrated with a model drawing (see Fig. 10.28). The basis of

Molecules	Structural symbol	Stoichiometric symbol
Hydrogen dumb-bell	H – H	H_2
Chlorine dumb-bell	Cl – Cl	Cl_2
Oxygen dumb-bell	O = O	O_2
Nitrogen dumb-bell	N ≡ N	N_2
Eight-membered sulfur ring	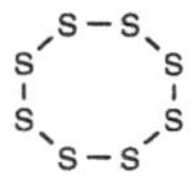	S_8
Selenium chain	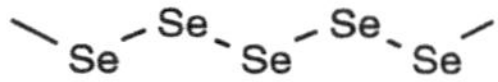	Se_n
Phosphorus tetrahedron (in white phosphorus)		P_4
Phosphorus honeycomb structure (in red phosphorus)		$(P)_n$
Arsenic honeycomb structure	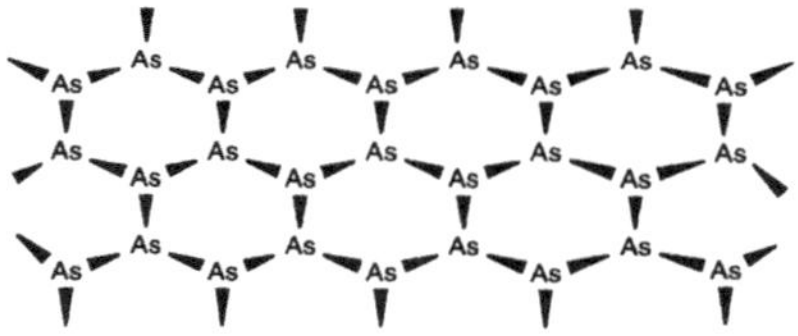	$(As)_n$
Graphite lattice (one sheet)	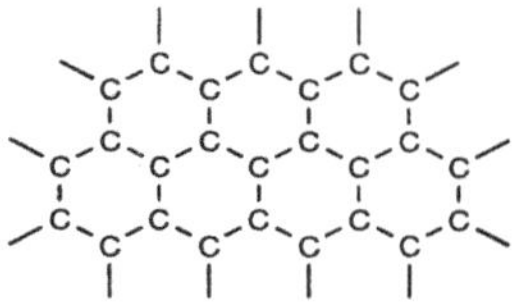	$(C)_n$ Graphite
Diamond lattice	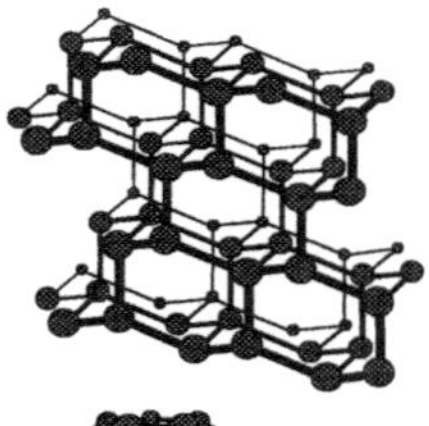	$(C)_n$ Diamond
C_{60} fullerene molecule (without double bonds)	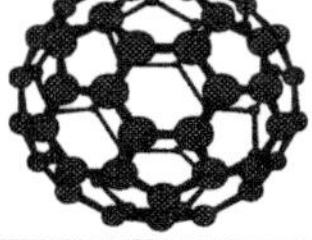	C_{60}

Fig. 10.25 Chemical symbols for structures built of nonmetal atoms [9]

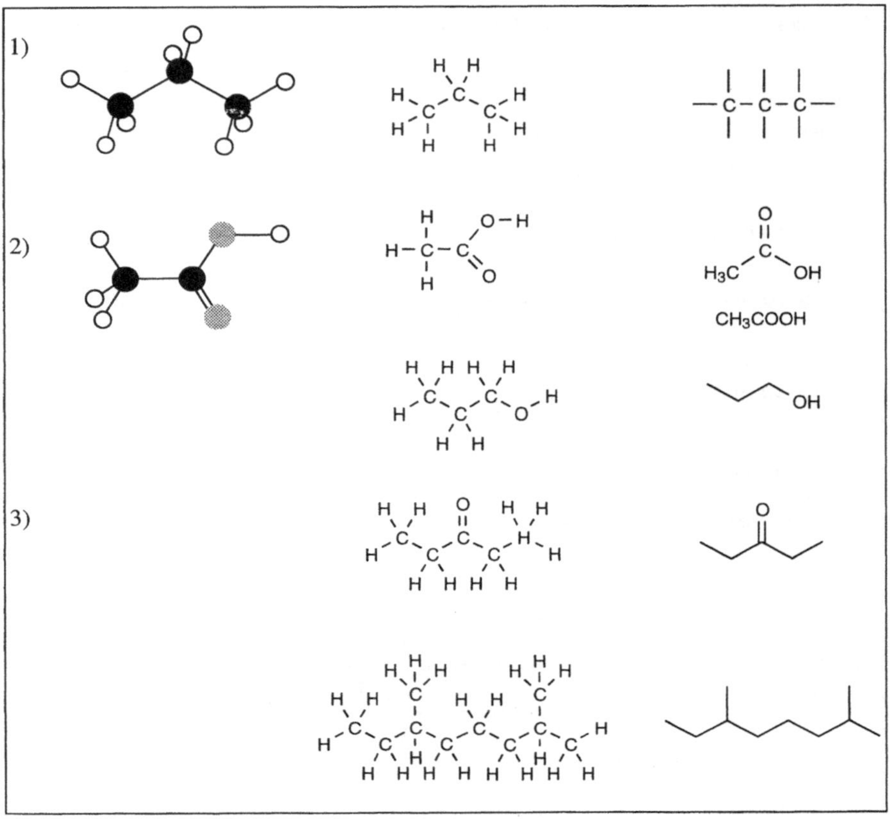

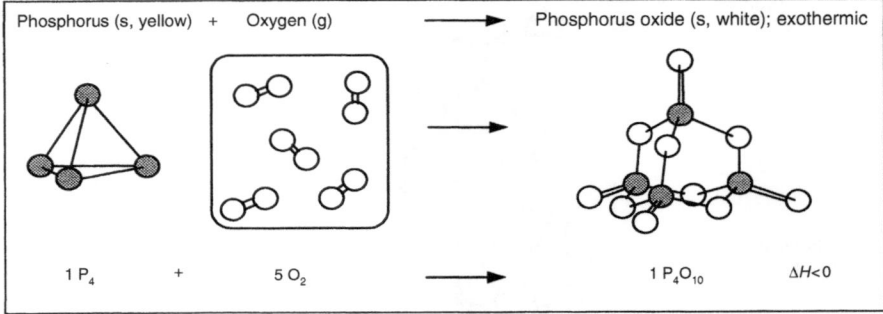

Fig. 10.26 Molecular models, structural and half structural symbols (some examples)

Phosphorus (s, yellow) + Oxygen (g)	⟶	Phosphorus oxide (s, white); exothermic

1 P$_4$ +   5 O$_2$ ⟶   1 P$_4$O$_{10}$ $\Delta H < 0$

Fig. 10.27 Model for regrouping of molecules in the phosphorus–oxygen reaction

this scheme is Avogadro's law: equal-sized boxes in the model drawing depict equal volumes of gas, the same number of molecule symbols in the boxes depicts the same number of molecules in the volume parts of the involved gases.

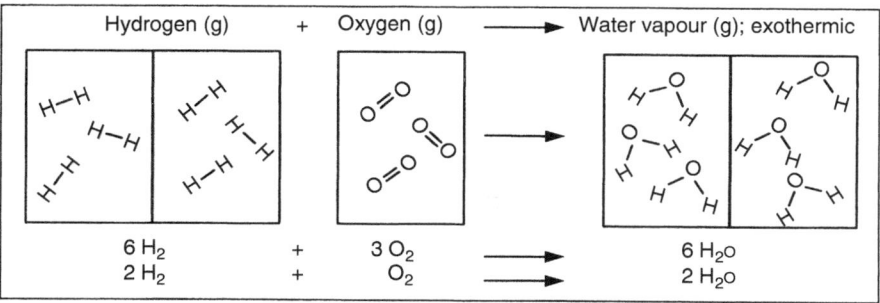

Fig. 10.28 Model for regrouping of molecules in the hydrogen–oxygen reaction ($t > 100°C$)

10.6 Summary of the Structure-oriented Approach

With the intention of formulating important goals of science education, 150 participants from the United States and other countries met; professors of universities, teachers of colleges and high schools, journalists of well-known publishers, and managers of big companies. They wanted "benchmarks" of what students should know or be able to do in science, mathematics, and technology by the end of grades 2, 5, 8, and 12. These grades are suggested as reasonable checkpoints for estimating students progress toward goals of science literacy [16].

In Chap. 4 of the "AAAS-Benchmarks for Science Literacy," they highlighted the importance of understanding the structure of matter and Dalton's atomic model: "The scientific understanding of atoms and molecules requires combining two closely related ideas: all substances are composed of invisible particles, and all substances are made up of a limited number of basic ingredients, or 'elements.' These two merge into the idea that combining the particles of the basic ingredients leads to millions of materials with different properties. *By the end of grade 8*, students should know that all matter is made up of atoms, which are far too small to see directly though a microscope. The atoms of any element are alike, but are different from atoms of other elements. Atoms may stick together in well-defined molecules or may be packed together in large arrays: different arrangements of atoms into groups compose all substances." [16]

The structure-oriented approach of chemistry is doing just that; teaching the atoms and different arrangements of atoms for students in beginner classes of chemistry, which may be grade 7 in some countries or grade 8 in other countries. The novel idea is that of using "atoms *and ions* as basic particles of matter" in the sense of Dalton's atomic model; the ions are given with their positive or negative charge and presented in the periodic table as spheres with different diameters (see Fig. 10.3). From atoms and ions, all substances can be created mentally; metal atoms "left and left in the periodic table" can arrange in crystals of pure metals or alloys, nonmetal atoms "right and right in the periodic table" in molecules or atomic lattices, ions "left and right in the periodic table" will arrange in ionic lattices and form salts (see Table 10.4). With the addition of ions in the periodic table, salts can be described scientifically correctly – without ions, the big group of salts cannot be included correctly into the "millions of materials with different properties." [16]

These mental models can be developed better if concrete models are built; because many metals are composed of metal atoms like spheres in a close packing of spheres, it is easy to build these models using spheres and glue. Many salt crystals like table salt can be described with the bigger anions in a close sphere packing and the smaller cations filling into their gaps. These sphere packings are easy to build and to understand for students through "learning by doing."

For developing mental models for the structure of molecules, molecular models are to be built with standard molecular kits; especially when understanding different isomers of organic molecules they are required. Students will generally be able to understand the idea of "well-defined molecules or large arrays." [16] They may get an overview of the most important arrangements of atoms and ions (see Fig. 10.29): atoms may be free in noble gases or in metal vapor, atoms may be arranged in molecules (oxygen, nitrogen, water, etc.) or giant structures (graphite, diamond, silicon, etc.); ions are combined in ionic structures (table salt and other salts) or may be moving free in molten salts or salt solutions.

Based on Dalton's atomic model, students learn that substances are made up of atoms or ions and understand first reactions without a change in the type of particle. They do not only work with molecules as in traditional chemistry lessons but also on the concept of crystal lattices and their models, e.g., packing of spheres, spatial lattices, and unit cells. If they derive formulae from the unit cell, they realize the importance of smallest structural units and their symbols as reduction of the chemical structure – they form conceptions that differentiate molecules and giant structures. Beyond the knowledge of formulae, they develop mental models of the structure of matter: "Imagination is more important than knowledge," as Albert Einstein once said.

Further instruction: After working on these fundamental structures, one can start to look inside atoms and ions, to see how atoms and ions are made up of nucleus and shell. With the help of the differentiated atomic model and models considering

Basic particles		Chemical structures		Examples
atoms		free atoms		noble gases, metal vapour
		molecules	molecule lattice	ice, iodine, sulfur
			atom lattice	diamond, silicon
		lattice	metal lattice	metals, graphite
ions			ion lattice	salts
		free ions		salt melts, salt solutions

Fig. 10.29 Combinations of basic particles of matter – summary

different kinds of chemical bonding, students are able to understand first reactions including the change of particles from atoms to ions, or ions to corresponding atoms. Important topics can be taught this way:

- Redox reactions and electron transfer
- Acid–base reactions and proton transfer
- Complex reactions and ligand exchange

The development of these concepts for the interpretation of many important chemical reactions continues in grade 10–12 – additionally, considering Organic Chemistry and chemistry in everyday life. Students should not only work with isolated atoms, ions, or molecules but also with the structure of corresponding substances before and after the reaction. With the principle of structure-orientation continuously applied, students will gain a modern understanding of chemistry and thereby the conception of today's chemists.

References

1. Gessner VH, Strohmann C (2010) Zwischen Aesthetik und Verstaendnis. Nachrichten aus der Chemie 7/8, Society of German chemists GDCh
2. Grosser CG (1975) Strukturorientierter Chemieunterricht von Anfang an. Praxis Chemie 24:261
3. Johnstone AH (2000) Teaching of chemistry – logical or psychological? Cerapie 1:9
4. Barke H-D, Hasari A, Yitbarek S (2006) Misconceptions in chemistry. Springer, Heidelberg
5. Barke H-D (1992) Einführung des Ionenbegriffs durch Experimente zur Gefrierpunktser- niedrigung. NiU-Chemie 3:93
6. Christen HR, Baars G (1997) Chemie. Diesterweg, Frankfurt
7. Sauermann D, Barke H-D (1997) Chemie für Quereinsteiger. Band 1: Strukturchemie und Teilchensystematik. Schüling, Münster
8. Sauermann D, Barke H-D (1998) Chemie für Quereinsteiger. Band 2: Struktur der Metalle und Legierungen. Schüling, Münster
9. Sauermann D, Barke H-D (1999) Chemie für Quereinsteiger. Band 3: Moleküle und Molekülstrukturen. Schüling, Münster
10. Sauermann D, Barke H-D (2000) Chemie für Quereinsteiger. Band 4: Ionenkristalle mit einfachen Grundbausteinen. Schüling, Münster
11. Wells AF (1987) Structural inorganic chemistry. Clarendon, Oxford
12. Barke H-D, Sauermann D (1998) Memorymetalle – sie besitzen ein Formgedächtnis. Praxis Chemie 47(Heft 3):7
13. Barke H-D (1991) Die Bildung von Natriumamalgam. Praxis Chemie 40(Heft 6):29
14. Barke H-D, Wirbs H (2000) Chemische Symbole für kleinste Struktureinheiten. Praxis Chemie 49(Heft 2):2
15. Barke H-D, Rölleke R (1999) Max von Laue: ein einziger Gedanke – zwei große Theorien. Praxis Chemie 48(Heft 4):19
16. American Association for the Advancement of Science "AAAS" (1993) Benchmarks for ScienceLiteracy. Project 2061. Oxford University Press, New York

Index

H.-D. Barke et al., *Essentials of Chemical Education*,
DOI 10.1007/978-3-642-21756-2, © Springer-Verlag Berlin Heidelberg 2012